International Horticultural Exposition 2011 Xi'an China
PLANNING

2011西安世界园艺博览会
规划精粹

中国建筑工业出版社
CHINA ARCHITECTURE & BUILDING PRESS

凤凰出版传媒集团
江苏人民出版社

凤凰空间
IFENGSPACE

总　策　划：王　军

编委会主任：杨六齐

编委会副主任：杨民生　王公理　邢忠民　徐军前　门轩
　　　　　　　丁学俊　姚景芳　张余文　成斌　杨希军

编委会成员：陈默　牛天山　张志杰　李新春　霍炳男
　　　　　　李宝塔　赵西兵　杜岩岫　王海若　王天兵

序 作 者：孟兆祯
主　　编：周俭
执行主编：李毅
总 编 审：冯利芳
策划编辑：石莹　林佳艺
执行编辑：曾健　张强军　李智　王磊　庄凌云
美术编辑：杨翠微
文字整理：艾侠　李品一　高媛婧
英文编审：Grant Donald　要欢
英文翻译：International New Landscape 国际新景观

Preface 序

一所
尽可能完美的
现代文人自然山水园

赞2011中国西安世界园艺博览会
修建性详细规划

自1999年昆明首办世界园艺博览会以来，逐步形成了具有中国特色的园博会，不以单纯进行园艺生产交流与交易为主要内容，而以室内外结合，融园艺于园林设计及庭院设计，力求通过展园充分展示本地自然与人文交融的特色。这就依仗于总体规划科学，艺术地创造宜于布置展园的全园环境。上海同济城市规划设计研究院作为项目详细规划设计单位，非常规范地，尽可能完美地完成了这项既光荣而又更艰巨的任务，谱写了现代文人写意自然山水园，满足了现代社会休闲游览生活的需要，并"知行合一"地从理论方面总结了实践的成果。这是可喜可贺的兴造成果，我在此表示诚挚地感谢，感谢所有致力于此的人。

走现代中国传统的创新之路

传统是相传成统，要靠世代接力"承前启后，与时俱进"的持续发展。中国造园名著《园冶》认为传统是适应时代而进的，"时宜得致，古式何裁"。我们不仅有古代的传统，还要在这条文脉上创造现代的传统，这样传统才有新生命发展下去。我们欢迎外国园林设计师参与中国项目设计，但要具有中国特色和地方风格，符合"赏心悦目"的游览综合要求。传统结合现代生活必然就有所创新。

"天人合一"的宇宙观

人有自然性和社会性，这是人类观察万物的根本哲理。"以人为本"是基于治理社会而提出的社会观，不宜扩大到涵自然的宇宙观。我们不能对自然说："以人为本"，说了也为不了本。自然者自其然也。人是自然的成员之一，自然性大于社会性，君不见中国造字，一人为大，但第一大是天，一大为天，自然是君，人为臣。中国人尊重自然，以自然为师，园林艺术"有真为假，做假成真"，真是自然，假是人工，在顺从自然的前提下强调"人杰地灵"和"景物因人成盛概"，所以中国园林追求的境界和评价标准是"虽由人作，宛自天开"。

园主题"天人长安，创意自然"是体现"天人合一"哲理的共性并强调了长安特性。唯人与天调，天人共荣才得长安。朴素的自然是没有意识的，但人可以创意寄托于自然，把社会美融入自然美从而创造园林艺术美。所以管子说："人与天调而后天下之美生"。中国园林艺术的基础是文学，"天人合一"之于文学就是"物我交融"。中国美学家李泽厚先生从美学概括中国园林是"人的自然化和自然的人化"，前半句指科学，后半句是艺术。这和钱学森先生说："中国园林是特殊的艺术部门"完全一致。"宛自天开"和"有真为假"的真谛在于创意于自然或人造自然，区别于朴素的自然。文人写意自然山水园来自山水诗画，以诗情画意创造园林空间。

巧于因借　精在体宜

借景非借贷之借，是凭藉之藉，借藉同义。借景是园林第一要法，凡明旨、相地、立意、布局，理微都是以借景为中心的。事物都有因果关系，觅因成果便是巧于因借的涵义。借景理法主要有"借景随机"、"借景无由"、"触情俱是"和最高境界"臆绝灵奇"。"精在体宜"指精湛之处在于体现用地之异宜。

相中浐灞流域之广运潭地段作为园博会园址是借前人之相地成果予以"古木新花"之发挥。此地是以水为主，山为辅的水景园地，水是创作的灵魂。循理水要领"疏水之去由，察水之来历"，在大量收集和调研用地自然资源和人文资源变迁的基础上进行了山水间架的竖向设计。根据长安作为十三代帝王都城结合以大雁塔为城市地标之宜来确立主景突出式布局类型无疑是借景有据，是很明智的。又借南涵秦岭终南山余脉，用地最高处定为主景长安塔的塔址，依塔山为轴，星位辟广运门短轴，长安花谷"以素药艳"地左呼右拥，夹道掩映。止于水岸之天人长安塔，隔水再压轴线终端。起长安塔如同大块文章言简意赅地树立了令人一见钟情，今生难忘的自然山水地标，统率全园。

从现状测绘图可看出是一块经人为开发干预的次生的土地资源。低丘山与低湿地自然交叉，居中的南北两块是主体。山水格局在原地形基础上作了小调整，不仅挖填土方量基本平衡，更塑造了水与岛的自然形象，从中体现"聚则辽阔，散则潆洄"的理水理法和发挥山之"三远"的山水空间美。胸有丘壑、岬、湾、水口明秀，堤岛星点而结为一体，初具"远观有势，近看有质"之妙。成功山水环境的塑造为布置景点建筑创造了地形基础。景可因境而出了。

先立宾主之位　次定远近之形

山水布局既利用原地形为主，局部改造为辅地树立了主景突出式布局的山水间架，园林建筑布局则应境而生。中国山水画在布局方面有"先立宾主之位，后定远近之形"的理法。园林建筑布局亦然，与绘画形式虽殊而理致则一。

作为全园主景建筑，长安塔在相地立意，因境选型方面是很成功的。安塔之山间名"小终南"，令人心晓循历史上长安以终南山为所依方向轴的传统。塔的选型无疑是传承了大雁塔作为地标的传统。长安的向往和追求更是"久治长安"的古意和"安定团结"的融合。无论人与自然相处或人与人相处，和者为贵，和则长安。初见此塔，为之一震，似曾相见而又未曾相见，充分展现了中国现代传统建筑的灵气和魅力。塔并不拘泥于正对轴线的"倒座"而独取坐北朝南。平面取方正，平正又端庄，尺度相宜，比例恰好，即使从园外的大河景物看也能起到引领空间的作用。对园内是当之无愧的永久地标。材料是新的，模样依然浮图，收分相对较小，更显得端庄，雄浑而孤峙无依。体量大而又晶莹剔透，虚实交呈而无过于敦实之弊。总体规划给塔创造了小山左右交拥，湖水周环间抱的好境界。隔水"天人长安塔"低伏相朌，东邻自然馆附岩相依。其他建筑也都一呼百应，展园都从得景的角度借景，如广州园以框景借塔，云南园以白石桥对塔等。张锦秋院士一如既往地完成了长安塔新作，为园博会主景建筑画上了圆满的句号。登塔俯望：周环山水景物奔来眼底，小坐啜茗，秀色堪餐，成景、得景而皆宜。

规划者是揣着园博会的现代社会实用功能来画山水的。山水间架既立，园内景区划分也就落盘了。各园区自得其山水胜景而园区间又有章法的衔接，彼此取得合宜的关系。清代画家笪重光说："文章是桌面上的山水；山水是地面上的文章"。文章起、承、转、合的章法在此园总体规划中有明显的表露。东北主入口为一"起"。"长安花谷"是"承"，各园区和各展园都是"转"。最后上长安塔举目四望，回忆历程，恰是一"合"，章法不谬也。

建筑景合落座，园路就通达有据了。贵在因山就水，极尽自然。因地广而分内外环，兼得"道莫便于捷"和"妙于迂"之妙。园内都有微地形起伏，"文同观山不喜平"。在地形的基础上广植树木花草，绿地率接近50%，加以展园绿化，葱茏一片，为花展创造了"万绿丛中一点红"的背景基础。广大的江湖地还要呈现出"深柳疏芦"的植物种植主调，这也是发挥广运潭柳堤之美的新意。

求教于众　集思广益

园林艺术虽然追求美轮美奂，但只可以是尽可能的美而不是绝对的美。作为总体设计这是一张优秀的考卷，即使有所瑕疵也是"瑕不掩瑜"。我只是想从科学发展观的角度作一些讨论，这对今后可能是有益的。例如在权衡朝向利弊时未必一定要取倒座，而且主风向为东北风，长风直入主入口对"长安花谷"的空气湿度会有影响而不利于种花。倒座主要借南边自然的小山，不无道理。但用地最宝贵的一方地是长安园所在的岛，不特面积广大而且"坐北朝南，负阴抱阳"，不足之处是有山两边而不集中，高程也不及南山高。南山建塔初高程约为407.3，北山也有398.3的峰值。填高北山完全可以统领全园，况且北山麓有大片平地，造山易于变化，但得坐北朝南则好处无穷。最后就是全园的微地形起伏"有丘无壑"，都是坡连坡，没有坡谷相间的虚实对比，这也是当前人造自然地形的通病，值得探讨。

规划设计的效果还要求教于众，游人批准才算数。我很敬佩项目主持人"知行合一"的学风和严于交待原地形图纸以便了解规划设计与原地形之差。我只是在我学习的学科范围内从赞赏的角度作序表达挚情真意，不当之处诚望业内外人士指正。

孟兆祯

2011.3.9

Foreword 前言

实践 & 思考

2011西安世界园艺博览会即将在我国西安市浐灞生态区举行。为期178天的世园会对浐灞生态区和西安市的发展都将带来巨大的机遇。作为浐灞生态区管委会的成员，我受主编之邀为这本以世园会景观规划为内容的图书撰写前言。不论是对于整个生态区，还是对于我个人而言，2011西安世界园艺博览会的举办都将成为一个值得铭记的历史时刻，而长达五六年的筹办和建设过程，则在我们的人生中留下了不可或缺的印记。思忖许久，既然是写在前面的话，我就以文字的形式记录些许浐灞生态区和世园会之间的渊源。

浐灞生态区是西安市于2004年9月成立的新区，是西安市重点发展的"四区一港两基地"之一。位于西安市主城区的东部，因浐河、灞河自南向北穿过而得名，规划总面积129平方公里，这里自然环境优美，人文荟萃，史上更有"浐灞之间，三辅胜地"之说，有着众多千古绝唱和文化遗存。然而随着历史的变迁和人类活动的加剧，浐灞却变成了生态重灾区，面临着河道污染、垃圾围城、非法挖沙三大生态灾害。经过多方的努力，浐灞生态区的建设目标得到了确立，虽是生态恶化和城市增容双重压力的一种无奈选择，却也是规划者和建设目标者对于城市发展模式主动求变的创新方式。

生态区建立之初，按照西安城市总体规划和西安"十一五"发展规划的要求，浐灞生态区需实现三大任务：一是生态重建，修复浐灞河的生态环境，使其成为西安市的生态补偿区；二是满足疏散老城人口和城市扩容的需要，建设一座宜居新城；三是发展现代服务业，完善西安的城市功能，提高西安的综合承载力，建设一个宜于创业和工作的新城。将构建经济、社会、环境协调发展的宜居、宜创业的生态新城作为目标，并将新型生态城市的建设作为长久的发展方向。从以往国内其他先行城市的发展经验看，大项目和大事件对一个新区的发展往往起着决定性的作用。建区之初，我们更是将大项目的建设提升至新的高度，总部经济区和金融商务区的落户，使区域内的产业结构框架初具规模。欧亚经济论坛的永久落址，2007F1摩托艇世界锦标赛的举办及世园会的申办，都对区域的建设和经济的推动起到了积极的作用。如今的浐灞生态区，不再是生态重灾区，她展现的是生态补给区的重要分工和国际金融区的功能延展。

在过去六年的规划和建设过程中，既是经营者又是执行者的我们，经历了浐灞生态区的处处艰辛，也与市民和建设者一起感受生态区所带来的积极改变。2006年10月19日，西安市政府召开第19次常务工作会议，决定申办世界园艺博览会，成立世界园艺博览会申办委员会，开始西安世园会的申办工作，并将会址定在西安浐灞生态区广运潭。2007年9月4日，第59届国际园艺生产者协会年会在英国布莱顿召开。经过评议，西安获得了2011年世界园艺博览会举办权。2007年12月10日，西安世界园艺博览会筹备委员会办公室成立，负责世园会的相关筹备工作。2008年10月26日，在阿联酋迪拜召开的第60届国际园艺生产者协会年会上，《西安世园会总体策划方案》和《西安世园会概念性规划方案》在会议上得到了审议通过。2008年11月10日，西安世园会建设启动仪式在西安浐灞生态区广运潭景区隆重举行。

从决心申办到世园会建设项目正式启动，西安市政府、陕西省政府、国家林业局，乃至外交部、商务部、花卉协会和贸促会为此付出了诸多努力，为浐灞生态区的破茧重生创造了契机。在此，我作为浐灞生态区管委会的一员，对各级领导与社会各界给予的帮助，深表感谢！

如今本届世界园艺博览会大幕即将拉开，我们已做好一切准备。作为建设者和实现者，我们也将这数年的积累以开园迎宾的方式接受检阅。用世园会的盛况展现我们对于建设新型中国生态大区的成功实践。

西安浐灞生态区管委会主任
2011.3.16

目录

编委会	002
序	004
前言	006
目录	008

项目背景 010
- 概况综述 012
- 世界园艺博览会知识 014

主题演绎 016
- 区位 018
- 现状分析 020
- 主题阐述 022

场址布局 024
- 规划理念 026
- 水系与水景规划 028

场址空间 030
- 空间结构 032
- 入口区 034
- 核心区 052 长安塔
  - 060 创意馆
  - 064 自然馆
- 滨水区 068 椰风水岸
  - 076 欧陆风情
  - 084 灞上人家
- 服务区 090 灞柳驿

生态园艺 098
- 园艺规划 100
- 五大主题园艺景观设计 102
- 景观广场 106
- 景观亭 126
- 展园设计 132
- 国际园 134
- 世界庭院 139
- 大师园 142 山之迷径花园
  - 146 植物学家花园
  - 148 通道园
  - 152 大挖掘园
  - 156 山水·中国地图园
  - 158 迷宫园
  - 162 万桥园
  - 166 黄土园
  - 172 四盒园
- 创意园 178 风的诗歌
  - 182 芬香花园
  - 190 城市景观干预
  - 192 天空之城
  - 198 流园
  - 204 潘帕斯印象
  - 206 编织自然
  - 208 强化的河畔
  - 212 生态平台
  - 217 生态–节气
- 港澳台北园 218
- 内地园 219
- 专类园 239
- 企业园 252
- 植物种植规划 260

支撑体系 266
- 交通规划 268
- 服务设施规划 292
- 基础设施规划 314
- 安全保障与消防规划 318
- 控制指挥中心系统规划 322
- 景观照明规划 326

建设中的西安世界园艺博览会	334
鸣谢	338
后记	340
版权页	342

Contents

Editorial Committee	**002**
Preface	**004**
Foreword	**006**
Contents	**008**

Background **010**
- Summarize **012**
- Get to Know about International Horticultural Exposition **014**

Theme **016**
- Location **018**
- Site Analysis **020**
- Theme Statement **022**

Planning **024**
- Planning Concept **026**
- Water System and Waterscape **028**

Architecture **030**
- Space **032**
- Entrances Zone **034**
- Core Zone **052**
  - **052** Chang'an Tower
  - **060** Theme Pavilion
  - **064** Greenhouse
- Waterfront Zone **068**
  - **068** Southeast Asian Street
  - **076** European Avenue
  - **084** Romance by the Ba River
- Service Zone **090**
  - **090** Ba Liu Hotel

Ecological Gardens **098**
- Functional Partition **100**
- Landscape Design for Five Theme Gardens **102**
- Landscape Plazas **106**
- Landscape Pavilions **126**
- Garden Planning **132**
- International Gardens **134**
- World Gardens **139**
- Master Gardens **142**
  - **142** Garden with a Labyrinth in the Mountain
  - **146** Botanist Garden
  - **148** Passageway Garden
  - **152** Big Dig Garden
  - **156** Landscape Garden of Map of China
  - **158** Garden of Labyrinth
  - **162** Garden of Bridges
  - **166** Loess Garden
  - **172** Quadrangle Garden
- Creativity Gardens
  - **178** Poem of Song
  - **182** Aromatic Garden
  - **190** City Landscape Intervention
  - **192** City of Sky
  - **198** Floating Garden
  - **204** Pampas Impression
  - **206** Weaving Nature
  - **208** Reinforced Riverside
  - **212** Ecological Platform
- Ecology-Solar Term **217**
- Hong Kong, Macau & Taipei Gardens **218**
- Domestic Gardens **219**
- Feature Gardens **239**
- Enterprises Gardens **252**
- Planning for Plants **260**

Supporting **266**
- Traffic Planning **268**
- Planning of Service Facilities **292**
- Infrastructure Facilities **314**
- Safety Insurance & Fire Control **318**
- Command Center Planning **322**
- Lighting System **326**

Expo Site under Construction **334**
Acknowledgement **338**
Postscript **340**
Copyright **342**

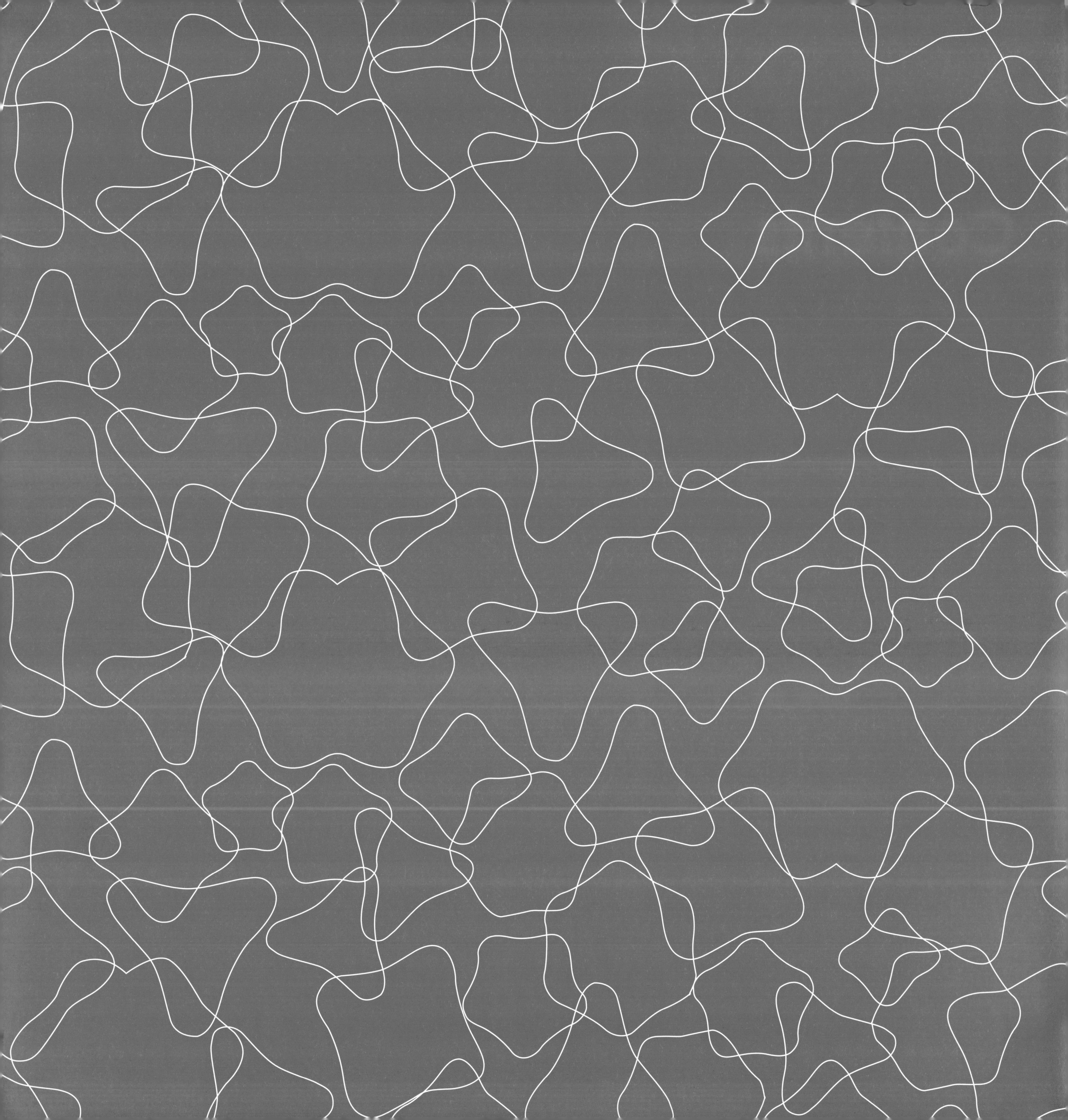

BACKGROUND
项目背景

作为历史悠久的国际大型会展活动，世界园艺博览会是最高级别的专业性国际博览会之一，也是最具国际影响力的世界级盛事之一。

As a grand international historic exposition, International Horticultural Exposition is regarded as the highest level in the profession with great influences across the world.

Summarize
概况综述

城市

西安古称长安,位于北半球亚欧大陆东南部,中国的几何中心附近,是享誉世界的历史文化名城,世界四大古都之一。历史上曾有13个王朝在此建都,书写了中华古代文明史中最为辉煌灿烂的篇章。西安现为陕西省省会,是中国西北地区金融、商贸、科技、教育中心。

"秦中自古帝王州"——西安地区之所以被13个王朝选做建都之地,其优越的地理条件是个重要因素。西安位于陕西省关中平原中部的渭河两岸,南依著名的中国南北分界线——秦岭山脉,北靠渭北荆山黄土台塬,东西两面是沃野千里的膏腴之地——关中平原,号称"八百里秦川"。

从全国范围来看,西安位于我国大陆腹地,中国大地原点即位于西安市北直线距离45公里的泾阳县永乐镇。历史上,它位于中国内地与边疆的交界地域,作为国都,兼具对内统治与对外抗衡机能。其本身又是"左殽函,右陇蜀,南有巴蜀之饶,北有胡苑之利,阻三面而守,独以一面东制诸侯"的"天府之国"。今天,陕西东连山西、河南、湖北,西靠甘肃、宁夏,北连内蒙古,南有重庆、四川,是我国东西部的连接点,西部大开发的前沿基地。

西安有3,100年的建城史,1,100年的建都史。唐朝以后,西安虽然不再成为国都,但仍作为历代政府的西北重镇及政治、经济、文化中心,发挥着重要作用。20世纪后期,随着中国经济的发展,古老的西安重新焕发了青春的光彩,经济、文化蓬勃发展,人民生活水平得到了空前的提升,一个国际化大都市形象呼之欲出。

2011西安世界园艺博览会申办过程

成立于2004年的浐灞生态区,从建设伊始,就秉承生态化理念,以建设生态化、人文化、第三代新城为目标,从治理环境、恢复生态入手,书写了古城西安城市建设的新篇章。浐河和灞河是浐灞生态区内两条最为重要的河流,多年来饱受环境污染之苦——两岸垃圾遍地,蚊虫孳生,由于两岸排污,河水中一度鱼虾绝迹,加上多年来河道中挖沙,造成河道严重下切,河床被严重破坏。浐灞生态区首先在浐河、灞河展开了环境治理工程,治污、截污,修建了"内硬外软"的灞河生态大堤,形成了万亩水面,河道水质逐渐好转,水面上鸥鹭云集,两岸柳绿花红。随着基础设施建设的逐步完善,浐灞生态区已具备了举办大型活动的基本条件。

2006年,西安市正式向国际园艺生产者协会提出申办2011年世界园艺博览会的申请,并于2007年9月,在英国布莱顿第59届国际园艺生产者协会(AIPH)年会上,正式获得了2011年世界园艺博览会的举办资格,会址确定于西安市浐灞生态区广运潭景区内。

The City

Xi'an, named Chang'an in ancient China, lies on the geographical center of China, the southeast Eurasia of the Northern Hemisphere. Xi'an is one of the Four Great Ancient Capitals, a historic and cultural city worldwide, having held that capital position under a long dozen of dynasties in Chinese history. As the capital of Shaanxi province, Xi'an is the center of finance, commerce, technology and education, as well as one of the largest cities in northwest China.

Xi'an was chosen as the capital in ancient China for its favorable geographical location. The city borders the northern foot of the Qinling Mountains to the south, and the banks of the Wei River to the north. It lies on the Guanzhong Plain in central China, on a flood plain created by the eight surrounding rivers and streams. Not far to the north is the Loess Plateau.

Xi'an is an internal city on the central region of the whole China, which is only 45 kilometers from Yongle Town, Jingyang County- the geographic center of China. In history, Xi'an is an area between the border region and the mainland. As the capital, it played a crucial role in governing the whole country and defending against external aggression. Xi'an is "located behind Xiao Pass and Hangu Pass, connects Long Plain and Shu Plain. Land of thousands of miles rich in harvest can be found here, as if this place belongs to the nation of the heaven". Near Shanxi, Henan, Hubei provinces to the east, Gansu, Ningxia provinces to the west, Inner Mongolia to the north and Chongqing, Sichuan to the south, Xi'an is in the middle of the east and west China, chosen as the leading area of Development of the Western Region.

Xi'an has a history of 3,100 years in urban development and 1,100 years as a capital. After Tang Dynasty, Xi'an continued playing an important role in politics, economy, and culture as the most important city in northwest China. As China developed rapidly in late 20th century, the historic Xi'an showed a new image, with expanding urban area, emerging hi-rise buildings and the fast development of economy and culture. Xi'an is developing into an international metropolis.

The Bidding of International Horticultural Exposition

Chan-Ba Ecological District was established in 2004, which incorporates ecological control into urban construction and cultural development for this new town. The two most important rivers of the district, Chan River and Ba River, suffered a lot from the environment pollution and pollution discharge. The ecological environment was badly destroyed by the pollution and excessive sand dredging. The ecological environment and the city are improved by comprehensively managing the river and applying the rules of ecological reconstruction to the valley to increase the overall urban bearing capacity. On the other hand, development consistent with the reality of the region is encouraged by building infrastructure to form a situation of varied development with other regions, enriching the urban culture, raising the quality and value of the city.

As the infrastructure construction of Chan-Ba Ecological District proceeds, the area becomes a desired place for great events.

In 2006, the city government of Xi'an proposed a bid to host International Horticultural Exposition 2011 and got the approval from International Association of Horticultural Producers (AIPH) at the 59th annual meeting of International Association of Horticultural Producers (AIPH) in Brighton in England. The site of the International Horticultural Exposition 2011 Xi'an was designed on Guangyun Lake, Chan-Ba Ecological District, Xi'an.

Get to Know about International Horticultural Exposition

世界园艺博览会知识

世界园艺博览会是最高级别的专业性国际博览会之一，也被称为世界园艺节，是一项具有悠久历史和较大影响的国际性活动。

International Horticultural Exposition is the highest level exposition in the profession worldwide, also called International Horticultural Festival.

世界园艺博览会的类型

世界园艺博览会是由国际园艺花卉行业组织——国际园艺生产者协会(AIPH)批准举办的国际性园艺展会，分为A1、A2、B1、B2四大类。

A1类——大型国际园艺展览会。这类展览会举办每年不超过一个。A1类展览会时间最短3个月，最长6个月。在展览会开幕日期前6-12年提出申请，至少有10个不同国家参展。此类展览会必须包含园艺业的所有领域。

A2类——国际园艺展览会。这类展览会每年最多举办两个，当两个展会在同一个洲内举办时，它们的开幕日期至少要相隔3个月，展期最少8天，最多20天，至少有6个不同的国家参展。

B1类——长期国际性园艺展览会。这类展会每年度只能举办一届，展期最少3个月，最多6个月。

B2类——国内专业展示会。

2011西安世园会类型为"A2+B1"，会期为2011年4月28日至10月22日，历时178天。

Types of International Horticultural Exposition

The International Horticultural Exposition is approved to host by International Association of Horticultural Producers (AIPH), falling into four categories: A1, A2, B1 and B2.

A1 refers to Large International Horticultural Exposition. An exhibition at the A1 level is allowed to be organized only once per year. Its duration may be no less than three months and no more than six months. The application must be handed in 6-12 years before the opening date of the Exposition and at least 10 countries must agree to participate. Lastly, the Exposition must showcase all aspects of horticulture.

A2 stands for International Horticultural Exposition. These exhibitions can not be organized more than twice a year. Also, there must be an interval of at least three months between the opening date of two exhibitions held on the same continent. Their duration may be no less than 8 days and no more than 20 days, and at least six different countries must participate.

B1 stands for Permanent International Horticultural Expositions. These exhibitions are allowed to be organized only once per year. Their duration may be no less than three months and no more than six months.

B2 stands for National Specialized Expositions.

International Horticultural Exposition 2011 Xi'an from April 28th to October 22nd, is the A2+B1 type exposition, lasting for 178 days.

世界园艺博览会部分举办国和举办城市一览

年代	举办国	举办城市	名称
1960	荷兰	鹿特丹	鹿特丹国际园艺博览会
1963	德国	汉堡	汉堡国际园艺博览会
1964	奥地利	维也纳	奥地利世界园艺博览会
1969	法国	巴黎	巴黎国际花草博览会
1972	荷兰	阿姆斯特丹	芙萝莉雅蝶园艺博览会
1973	德国	汉堡	汉堡国际园艺博览会
1974	奥地利	维也纳	维也纳国际园艺博览会
1976	加拿大	魁北克	魁北克国际园艺博览会
1980	加拿大	蒙特利尔	蒙特利尔园艺博览会
1982	荷兰	阿姆斯特丹	阿姆斯特丹国际园艺博览会
1983	德国	慕尼黑	慕尼黑国际园艺博览会
1984	英国	利物浦	利物浦国际园林节
1990	日本	大阪	大阪万国花卉博览会
1992	荷兰	路特米尔	海牙国际园艺博览会
1993	德国	斯图加特	斯图加特园艺博览会
1994	法国	圣·丹尼斯	圣·丹尼斯国际园艺博览会
1995	德国	哥特布斯	哥特布斯国际园艺博览会
1996	意大利	热亚那	热亚那国际园艺博览会
1997	比利时	利戈	利戈国际园艺博览会
1997	加拿大	魁北克	魁北克97国际花卉博览会
1999	中国	昆明	昆明世界园艺博览会
2000	日本	淡路	淡路花卉博览会
2002	荷兰	阿姆斯特丹	芙萝莉雅蝶园艺博览会
2003	德国	罗斯托克	罗斯托克国际园艺博览会
2004	日本	静冈	滨名湖国际园艺博览会
2005	德国	慕尼黑	慕尼黑联邦园艺展
2006	泰国	清迈	清迈世界园艺博览会
2006	中国	沈阳	沈阳世界园艺博览会
2010	中国	台北	台北国际花卉博览会
2011	中国	西安	西安世界园艺博览会

Previous International Horticultural Expositions

1960 Floriade, Rotterdam, Netherlands

1963 IGA, Hamburg, Germany

1964 International Horticultural Exposition, Vienna, Austria

1969 International Horticultural Exposition, Paris, France

1972 Floriade, Amsterdam, Netherlands

1973 IGA, Hamburg, Germany

1974 International Horticultural Exposition, Vienna, Austria

1976 International Horticultural Exposition, Quebec, Canada

1980 International Horticultural Exposition, Montreal, Canada

1982 Floriade, Amsterdam, Netherlands

1983 IGA, Munich, Germany

1984 International Garden Festival, Liverpool, United Kingdom

1990 International Gardens and Greenery Exposition, Osaka, Japan

1992 Floriade, Zoetermeer, Netherlands

1993 IGA, Stuttgart, Germany

1994 International Horticultural Exposition, Saint Denis, France

1995 Federal Gardening Exhibition, Cottbus, Germany

1996 International Horticultural Exposition, Genoa, Italy

1997 International Horticultural Exposition, Liege, Belgium

1997 International Horticultural Exposition, Quebec, Canada

1999 International Garden Festival, Kunming, China

2000 Florart, Awaji, Japan

2002 Floriade, Haarlemmermeer, Amsterdam, Netherlands

2003 IGA, Rostock, Germany

2004 Pacific Flora, Hamamatsu, Japan

2005 BUGA, Munich, Germany

2006 Royal Flora Ratchaphruek, Chiang Mai, Thailand

2006 International Horticultural Exposition, Shenyang, China

2010 Taipei International Flora Exposition, Taipei, China

2011 International Horticultural Exposition, Xi'an, China

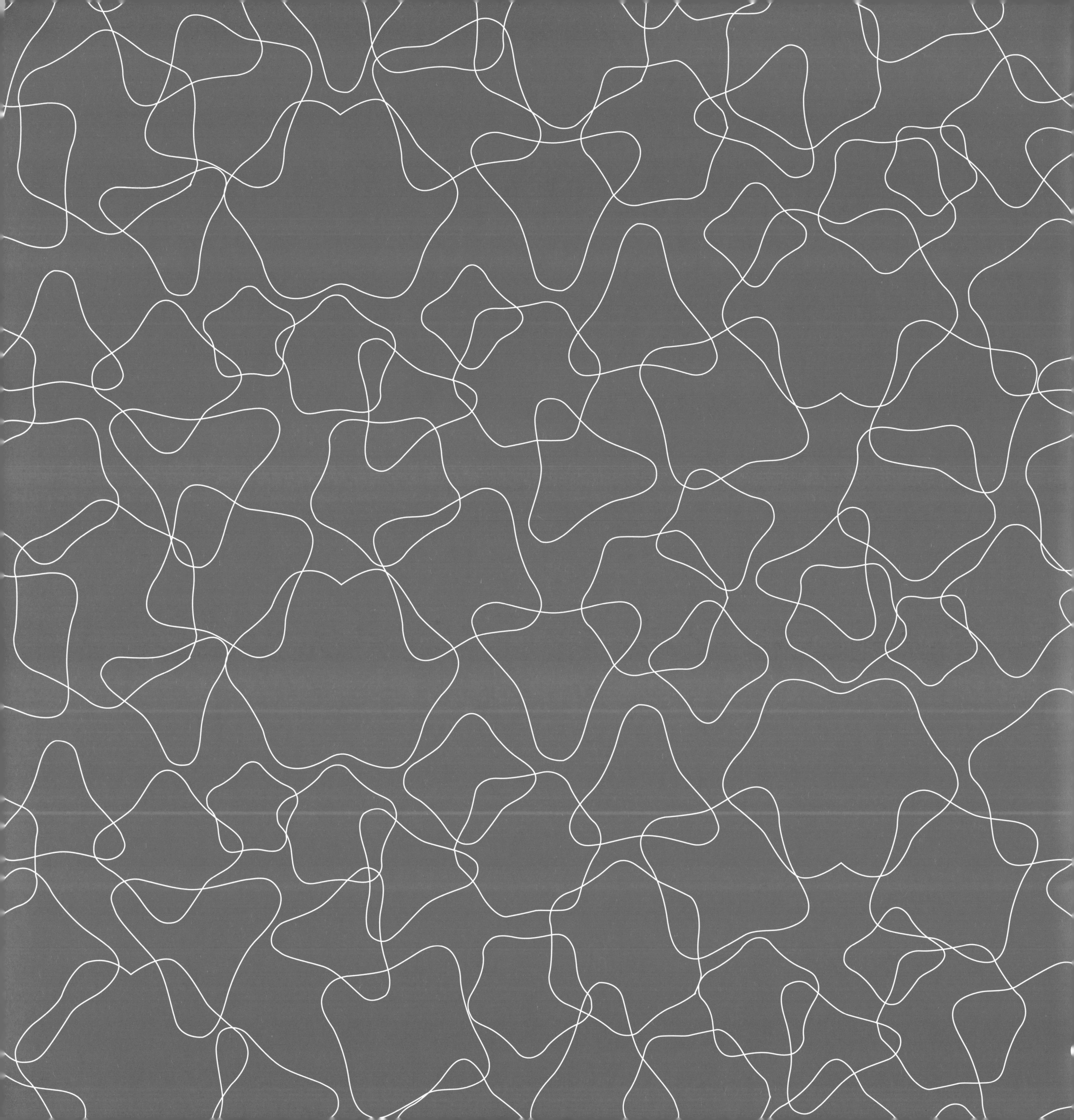

THEME
主题演绎

2011西安世界园艺博览会以"天人长安·创意自然——城市与自然和谐共生"为主题内涵,意在提倡城市与自然和谐共生,在尊重自然的前提下使其为人类服务,延续着"人与自然"的主题。西安世园会占地总面积4,180,000 m^2,共设置室外展园109个。举办日期为2011年4月28日至10月22日,会期178天。

With the theme of "Eternal Peace & Harmony between Nature & Mankind, Nurturing the Future Earth - a City for Nature, Co-existing in Peace", Xi'an International Horticultural Exposition aims to improve the mutual development of urban and natural environment, showing the respect for the human and nature. This chapter introduces the history and spirit of International Horticultural Exposition and makes analyses on the master plan of the Exposition Site. The Site in Xi'an covers an area of 4,180,000 m^2 and sets 109 outdoor exhibition gardens. The 178 days' Exposition lasts from April 28th to October 22nd.

018 | 019

Location
区位

会址面积约4,180,000 m²。西临灞河,南接陇海铁路,东至东三环路,北接世博大道,交通便利,自然条件非常优越。

The Expo Site is 4,180,000 m². Along the Ba River, the site is adjacent to the Longhai Railway in the south, Dongsanhuan in the east and Expo Avenue in the north. The natural conditions and transport are very favorable here.

Site Analysis
现状分析

自然条件

气候
西安市介于北纬33°39′～34°45′与东经107°40′～109°49′之间，属北半球暖温带半湿润季风气候。气候特点是温暖湿润，雨量适中，四季分明。1月份最冷，平均气温−0.5°C～−1.3°C，平均最低温度−3.8°C；7月份最热，平均气温26.3°C～27°C，平均最高气温32.2°C；年平均气温13.6°C。"自古长安多秋雨"，降水主要集中在7、8、9三个月，年平均降水量300mm～500mm，年平均蒸发量为904.7毫米。

地形地貌
西安市南北地形高差悬殊，地形地貌十分特别。南部是著名的中国南北分界线——秦岭山脉，平均海拔2,000米左右，北部为黄土高原，海拔400-600米。地势南高北低，世园会会址处于灞河低阶地，海拔介于376-393米之间。

河流
西安地区河流众多，其中最著名的河流有八条，所以自古就有"八水绕长安"的俗谚。这八条河流分别是：北面的泾河、渭河，南面的潏河、涝河，西部的沣河、滈河，东部的浐河、灞河。

本区域有浐河、灞河、渭河三条比较大的河流，其中灞河与本届世界园艺博览会关系非常密切。

历史人文

广运潭的修建与历史上第一次博览会
盛唐天宝年间，国力强盛，经济发达，社会稳定，人口增加。据历史记载，当时的长安城人口已达百万，关中一地的物产已经不能满足京畿地区的需要。

为了疏通漕运，改善全国到长安的水运通道，唐玄宗于天宝三年，委任水陆转运使建陕郡太守韦坚，在浐灞河旁开辟湖面码头，为停船之所。两年后工程完成，唐玄宗亲自视察了完工的浩大水面。经韦坚的精心策划，在此开展了历史上第一次博览会，南方各地的漕船纷纷贡上各地特产。玄宗观后龙心大悦，钦赐潭名为"广运"。

广运潭建成后，成为桅杆如林、万商云集的码头，江南的粟米、丝帛等物产，源源不断地运往长安，年运粮达400万石，大大提升了漕渠的运输能力，有力保障了长安的社会经济发展。

灞柳、灞桥与送别文化
本届世界园艺博览会会址被称作"灞上"地区，其历史渊源一直与"送别"有关。中国古代史上，该地区扼长安东门户，战略位置突出。《雍录》上指出："此地最为长安要冲，凡自西东两方而入出山尧、潼两关者，路必由之。"秦代，王翦伐荆，始皇送至灞上。秦末，刘邦驻军灞上，秦王子婴在此地降于轵道旁，宣告了秦王朝的结束。

隋开皇二年，在隋文帝建设大兴城的同时，在灞河上修建了我国已知时代最早、跨度最大的一座多孔石拱桥。桥上雕梁画栋，岸边绿柳成荫。阳春三月，游人如织。长亭折柳，桥上舟车往来，河上游船穿梭，灞桥兴盛一时。

唐朝时，在灞河两岸修筑堤岸，植柳万株，游人纷至，蔚为壮观。那时，灞桥旁设有驿站，由长安东去之人，多在此送别。唐人杨巨源有诗云"杨柳含烟灞岸春，年年攀折为行人。好风倚借低枝便，莫遣青丝扫路尘。""折柳伤别，灞柳风雪"成为关中八景之一。

Natural Conditions

Climate

Xi'an ranges in latitude from 33°39'N to 34°45'N and in longitude from 107°40'E to 109°49'E. The city has a warm, semi-humid monsoon climate that is characterized by warmness, humidity, moderate rainfall and four distinct seasons. January in Xi'an is the coldest month with an average temperature of -0.5°C to -1.3°C and the extremely low temperature is about -3.8°C. By contrast, July is the hottest with an average temperature of 26.3°C to 27°C and the average highest temperature is about 32.2°C and the annual average temperature is 13.6°C. The rainfall mainly concentrates in July, August and September. The annual average amount of precipitation and evaporation are separately 300 mm to 500 mm and 904.7 mm.

Topography

There is a sharp altitude difference between the south and north of Xi'an, leading to the special landform of the city. The south lies Qinling Mountains with the average altitude of 2,000 meters; it is the famous south/north dividing line in China. The north of Xi'an is the Loess Plateau, of which the altitude is 400-600 meters. Obviously, the terrain declines from south to north. The Expo Site is in the low terrace of the Ba River, where the altitude is about 376-393 meters.

Rivers

There are a great many rivers in Xi'an and eight of them are the most famous. Hence, there is the old saying that "eight rivers running around the city of Chang'an". The eight rivers include Jing River and Wei River in the north, Yu River and Lao River in the south, Feng River and Hao River in the west, and Chan River and Ba River in the east.
Chan River, Ba River and Wei River are three major rivers around the Expo Site, and Ba River is tightly related to the Expo.

History & Culture

Guangyuntan: The First Expo in Human History

During the Tianbao period of the Tang dynasty (742-756AD), the country developed prosperously and the population of Chang'an reached a great number of several millions. The products from Guanzhong area couldn't meet the need of Chang'an area. Thus Emperor Xuanzong held a large-scale water transport exposition and trade fair here to encourage commerce and trade and display region's capability in smooth water transport. This event in ancient times marked the beginning of the world expo. Special local products from southern areas were delivered to the capital by water, Xuanzong gave the name Guangyuntan to the lake to improve the water transport. Guangyuntan became a booming port after constructed and improved the water transport a lot, so the social and economic development of Chang'an was stimulated with prosperous trade between southern and northern parts.

Ba Willow, Ba Bridge and Farewell Culture

The actual site of the International Horticultural Exposition 2011 Xi'an is known as "Bashang", which is always in deep relation with "farewell". During Sui Dynasty, the Emperor made people construct a multi-arched stone bridge, engraved with delicate pictures and planted with green willows on both sides. Tourists were always crowded with many poets chanting verse to show pathetic parting mood.

Theme Statement
主题阐述

2011西安世界园艺博览会主题

天人长安·创意自然——城市与自然和谐共生。

本届世界园艺博览会的主题解析

从"天人长安,创意自然"的主题释义,"天"指自然,"人"指城市,"长安"是千年古都西安的历史名称,也是国家繁荣与安泰的象征。"天人长安"意味着城市自然和谐共生;"创意自然"是在尊重自然和不破坏自然的前提下,利用自然,修复自然,使自然为人类服务。

本届世界园艺博览会的主题表达

2011西安世园会园区总面积4,180,000 m², 其中水域面积1,880,000 m²; 标志性建筑有广运门、长安塔、创意馆和自然馆;主题园艺景点分别为长安花谷、五彩终南、丝路花雨、海外大观和灞上彩虹;并设有灞上人家、椰风水岸和欧陆风情三处特色服务区;同时将设置展示来自国内外的精美艺术品、雕塑以及珍禽、珍稀动物等,将让人们充分领略园林、园艺、建筑、艺术之美。

Theme

The theme of the Expo is "Eternal Peace & Harmony between Nature & Mankind, Nurturing the Future Earth - a City for Nature, Co-existing in Peace".

Theme and Interpretation

The theme of the Expo is "Eternal Peace & Harmony between Nature & Mankind, Nurturing the Future Earth - a City for Nature, Co-existing in Peace". (Chinese version: Nature and People in One in Chang'an, Nature Creativity- a City for Nature, Co-existing in Peace). "People" represents the city, and "Chang'an" is the ancient name of Xi'an, an ancient capital with a long history and a symbol for national prosperity and security. Hence, "Nature and People in One in Chang'an" embodies the harmonious co-existence between the city and nature. "Nature Creativity", that is "nurturing the future earth", refers to: on the basis of respecting and protecting nature, allowing nature to serve humankind by utilizing and restoring it.

Theme Explanation

The Expo Site covers a total area of 4,180,000 m², of which 1,880,000 m² is water area. The landmarks include Guangyun Entrance, Chang'an Tower, Greenhouse and Theme Pavilion. The major horticultural scenic spots are the Chang'an Flower Valley, Colorful Plants from Qinling Mountains, Flowers along the Silk Road, Overseas Collections and Flower Rainbow over the Ba River. The three characteristic zones settled on the site refer to Romance by the Ba River, Southeast Asian Street and European Avenue. At the same time artworks, sculptures and rare birds, rare animals home and abroad will be on show to make people fully enjoy the beauty of gardening, architecture, and art.

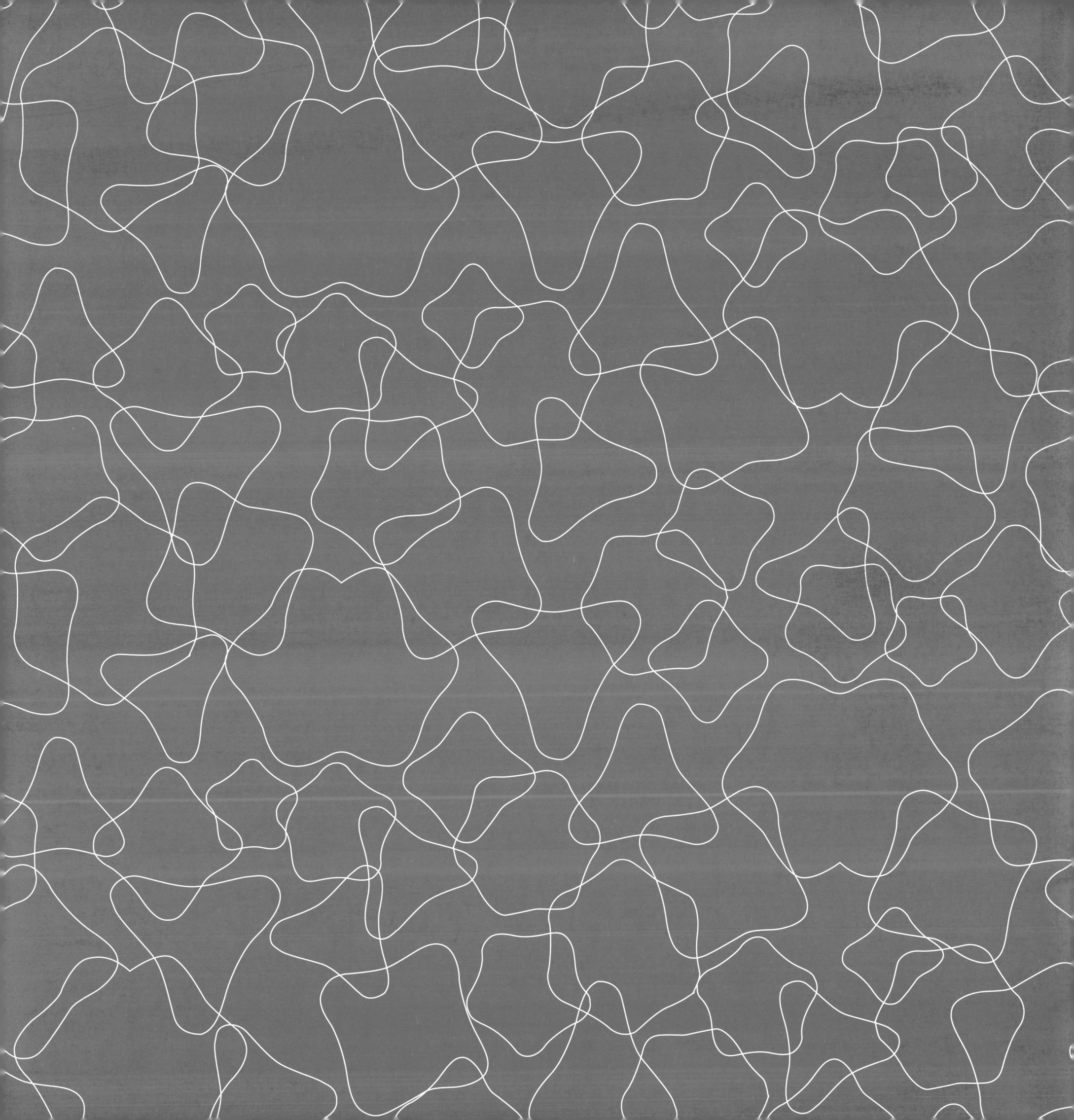

PLANNING
场址布局

2011西安世界园艺博览会通过规划理念、指导思想、布局结构、水系与水景规划等多方面缜密的场址布局，从而全面阐释"天人长安，创意自然"的主题。

The site of the International Horticultural Exposition 2011 Xi'an is planned meticulously from various aspects, such as planning concept, guiding ideology, layout structure and water & waterscape system planning, to fully interpret the theme of the Expo "Eternal Peace & Harmony between Nature & Mankind, Nurturing the Future Earth - a City for Nature, Co-existing in Peace".

Planning Concept

规划理念

规划设计原则

功能完善，满足展会期间功能运作；主题突出，完美阐释展会理念；特色鲜明，具有原创性、独特性和地域性；经济性和可持续利用，节约投资。

规划设计理念

天人长安　创意自然
传承历史　体现创新
绿色和谐　健康共享
生态浐灞　科技世园

指导思想

以科学发展观和生态文明理念为指导，体现天人合一、城市与自然和谐共生思想；以园艺植物、花卉和园艺新技术、新产品和新趋势的展示为主体，营造浓郁节日氛围。

规划目标

通过2011西安世界园艺博览会，从多个方面展示古都西安的文化、历史、科技、人文精神，园区彰显中华传统文化魅力，荟萃世界园艺精品，向全世界展示一个文脉深厚、与时俱进、时尚和谐的新西安，从而全面阐释"天人长安、创意自然"的主题。

• Principle of Planning and Design

Satisfy the functional operation during the Expo with complete functions; present perfect interpretation of the Expo's philosophy through a prominent theme; show striking originality, uniqueness and regionalism; emphasize economical efficiency and sustainable use; save the cost.

• Philosophy of Planning and Design

Nature and People in Harmony, Nature Creativity
Inherit History, Reflect Innovation
Green Environment, Healthful Sharing
Ecological Chan Ba, Technological Expo

• Guiding Ideology

Take scientific thought of development and the concept of ecological civilization as the guiding ideology; embody the harmonious co-existence of nature and mankind in city; create a full-bodied festive atmosphere with the exhibition of horticultural plants, flowers, new horticultural technologies, new products and trends as the main part.

• Planning Objective

The International Horticultural Exposition 2011 Xi'an aims to display the culture, history, science & technology and humanistic spirit of Xi'an – the ancient City from multiple aspects. The Expo Site shows the charm of Chinese traditional culture and gathers horticultural products from all over the world. It presents a fashionable and harmonious Xi'an with profound context and advanced development, fully illuminating the theme "Eternal Peace & Harmony between Nature & Mankind, Nurturing the Future Earth - a City for Nature, Co-existing in Peace".

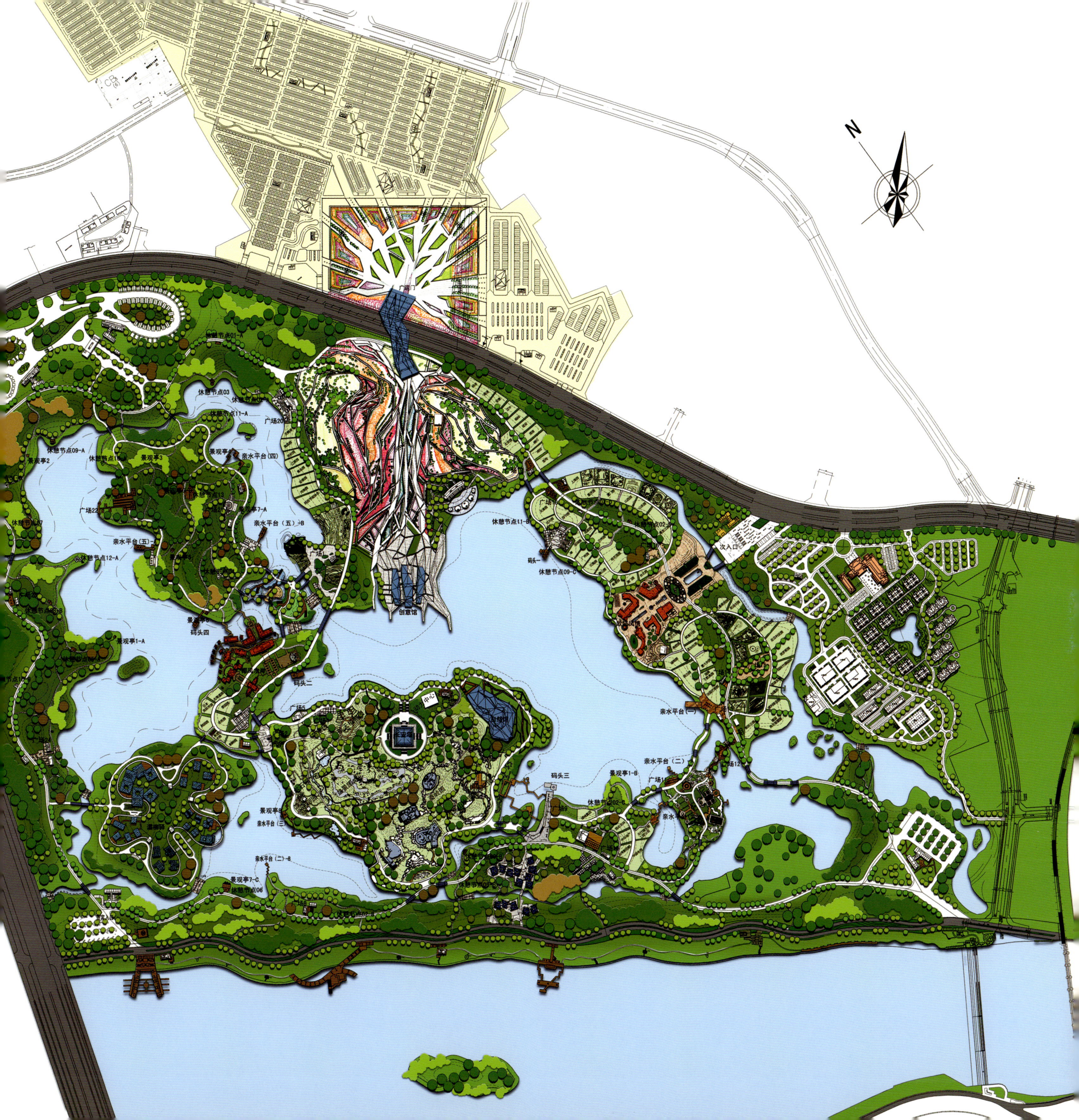

Water System and Waterscape
水系与水景规划

水，作为本届世园会最重要的自然元素，意味着生命的依托，似乎从一开始便与人类生活乃至文化历史形成了一种不解之缘。人类总是逐水而居，滨江临海，繁衍生息。纵观世界文化源流，黄河、恒河、尼罗河和地中海都孕育了灿烂的远古文明。经济建设更以人为本，以水为先。面对山水形胜，孔子说："智者乐水，仁者乐山。"一个"智"字，既反映了先哲对"水"的认知，又破译出"水"所蕴藏的无尽的文化内涵，"水"为"智者"提供了丰富的文化源泉，并且由物质的层面升华到一种精神的境界。

我国古代朴素唯物论把金、木、水、火、土"五行"视为世界的本原。水是生命之源，没有水就没有世间万物生机勃勃的景象。

园区的水系沿用原有广运潭设计水系，总面积为1,880,000 m²（含灞水水域面积），其中核心区水面约800,000 m²，库容约150万立方米。原有广运潭水系进水口最大取水能力为2立方米/秒，每天约18万立方米。正常情况下园区每天取水8.46万立方米，约20天可以更换水体一次。世园会期间是灞河的丰水期，水量充裕可以满足景观用水需要。

As the most important natural element of the Exposition, water, connects to mankind life and cultural history from the beginning, meaning an essential element human beings relying on. The mankind always live and grow near the waterfront, such as river and sea. The Yellow River, Ganges, Nile and Mediterranean all played an important role in human civilization from immemorial time. Water is essential for mankind life and the economy of the waterfront is always highly developed. The Confucius ever said :" The wiser(zhi) delights in water, the kindhearted(ren) likes mountain." The word "Zhi" reflects the sages' first cognition of "water", and interpretes the profound cultural content of water. "Water" provides an abundant cultural source for the "wiser", and the meaning of water is highlighted to a spiritual level.

Metal, wood, water, fire and earth are the five elements in ancient Chinese naive materialism, among which, water is the resource of lives. Without water, the world will never be vigorous.

The design of water system follows the original Guangyuntan, with 1,880,000 m² (including the Ba River) of water area within the Expo Site, 800,000 m² of core area and 1.5 million m³. The original Guangyuntan system can intake 2 cumec at most, which is about 180,000 m³. 84,600 m³ water can be intaken per day on the Expo Site at the normal condition, with 20 days as a cycle to replace water substance. Ba River will see her high flow period during the Expo; the need of water on landscapes can be met.

图例：

水流方向示意

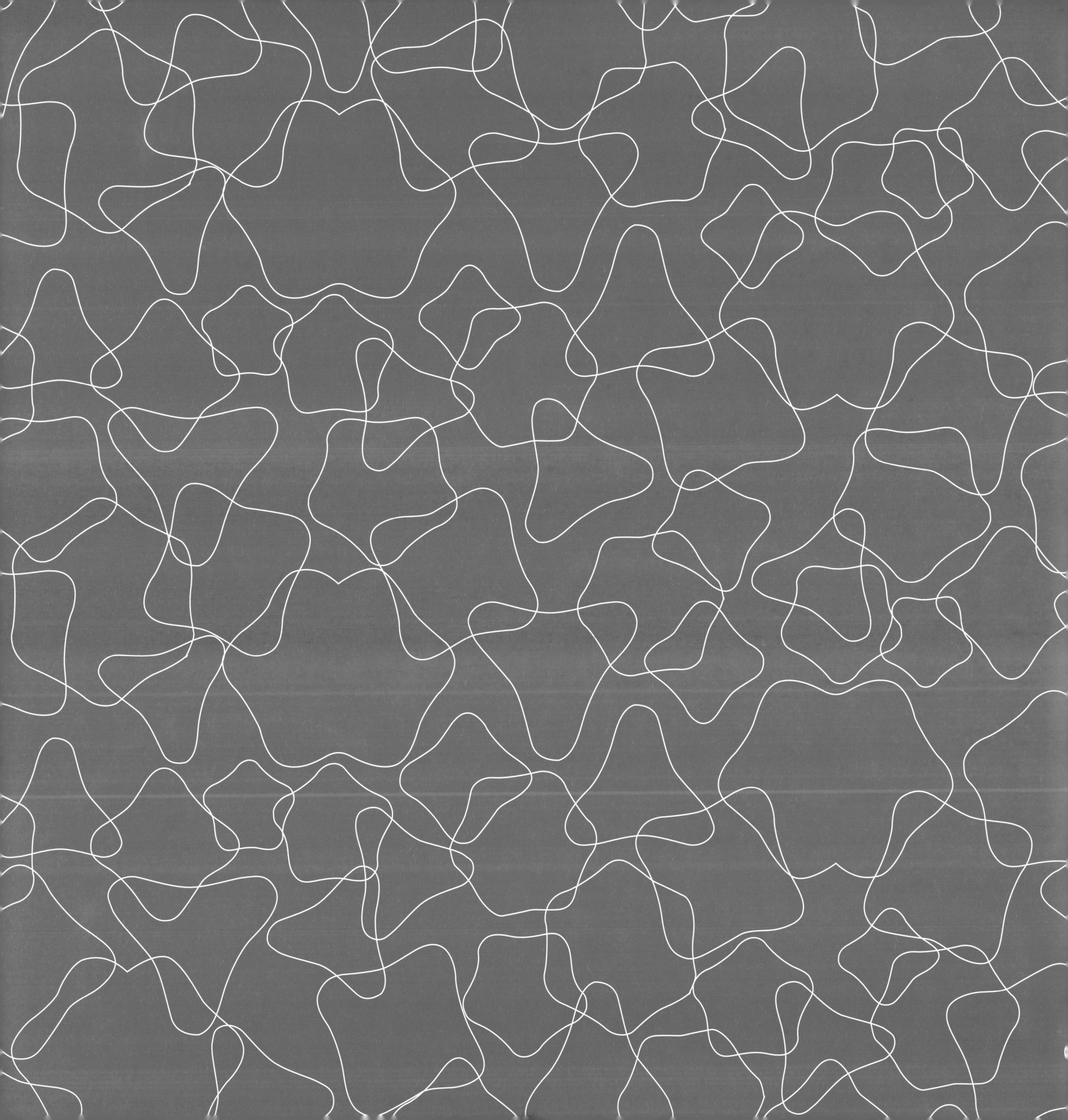

ARCHITECTURE
场址空间

2011西安世界园艺博览会精心策划设计的多组建筑群落,全方位展示了世园会的建筑魅力和主题精神,也可以看做对当代建筑与景观的前沿触探。本章重点介绍了西安世园会的四大标志性建筑、三个特色服务区,及一系列特色景观建筑;这里既有中国建筑大师张锦秋主持设计的气势恢宏的长安塔,也有先锋建筑师马清运设计的精致别趣的灞柳驿,更有国际新一代景观事务所Plasma设计的富含创意的广运门等。东方与西方、传统与新锐、景观与建筑,多元化的设计思维在世园会碰撞出惊艳的火花。

International Horticultural Exposition 2011 Xi'an exhibits multiple groups of architectural complexes and highlights the spirit of its theme, "Eternal Peace& Harmony between Nature & Mankind, Nurturing the Future Earth - a City for Nature, Co-existing in Peace" and the soul of architecture. It represents an exploration to the up-to-the-minute architecture and landscape design.
This chapter features mainly on "Four Landmarks", "Three Special Service Areas" and a series of unique landscape projects: the large-scale Chang'an Tower is designed by Zhang Jinqiu, the renowned Chinese architect, as the lead designer; the exquisite Ba Liu Hotel is designed by Ma Qingyun, a prestigious architect in the forefront of architectural design field; the inspiring Guangyun Entrance is designed by Plasma Studio, a new generation of landscape design team.
Diverse elements – eastern and western, traditional and cutting-edge, landscape and architecture – collide into each other, produce sparks in Xi'an International Horticultural Exposition, and create a dynamic design concept.

Space
空间结构

2011西安世界园艺博览会的总体规划结构呈现"两环、两轴、五组团"的基本特征。其中,"两环"指世园会分为主环和次环:主环为核心展园区,主要的展园和景点均分布在主环内,次环为扩展区。"两轴"指园区内的两条景观轴线,南北为主轴,东西为次轴。"五组团"指主要的展园组团,分别是长安园、五洲园、创意园、科技园、体验园。在规划结构上,这样的设计呈现出层次分明、逻辑性强、利于运营等综合优点。

The Expo Site is based on a pattern of "two circles, two axes, and five groups". "Two circles" refer to the main circle, the core site containing the majority of the exhibition areas and sightseeing spots, and the sub-circle, the extension zone. "Two axes" refer to two landscaping axes, with south-north as the main axis and east-west as the sub-axis. "Five groups" refer to major exhibition gardens, including Chang'an Garden, Five-Continent Gardens, Creativity Gardens, Sci-tech Gardens and Experience Gardens. This kind of structural design is of distinct arrangement and strong logicality, favorable for the operation and reconstruction after the Expo.

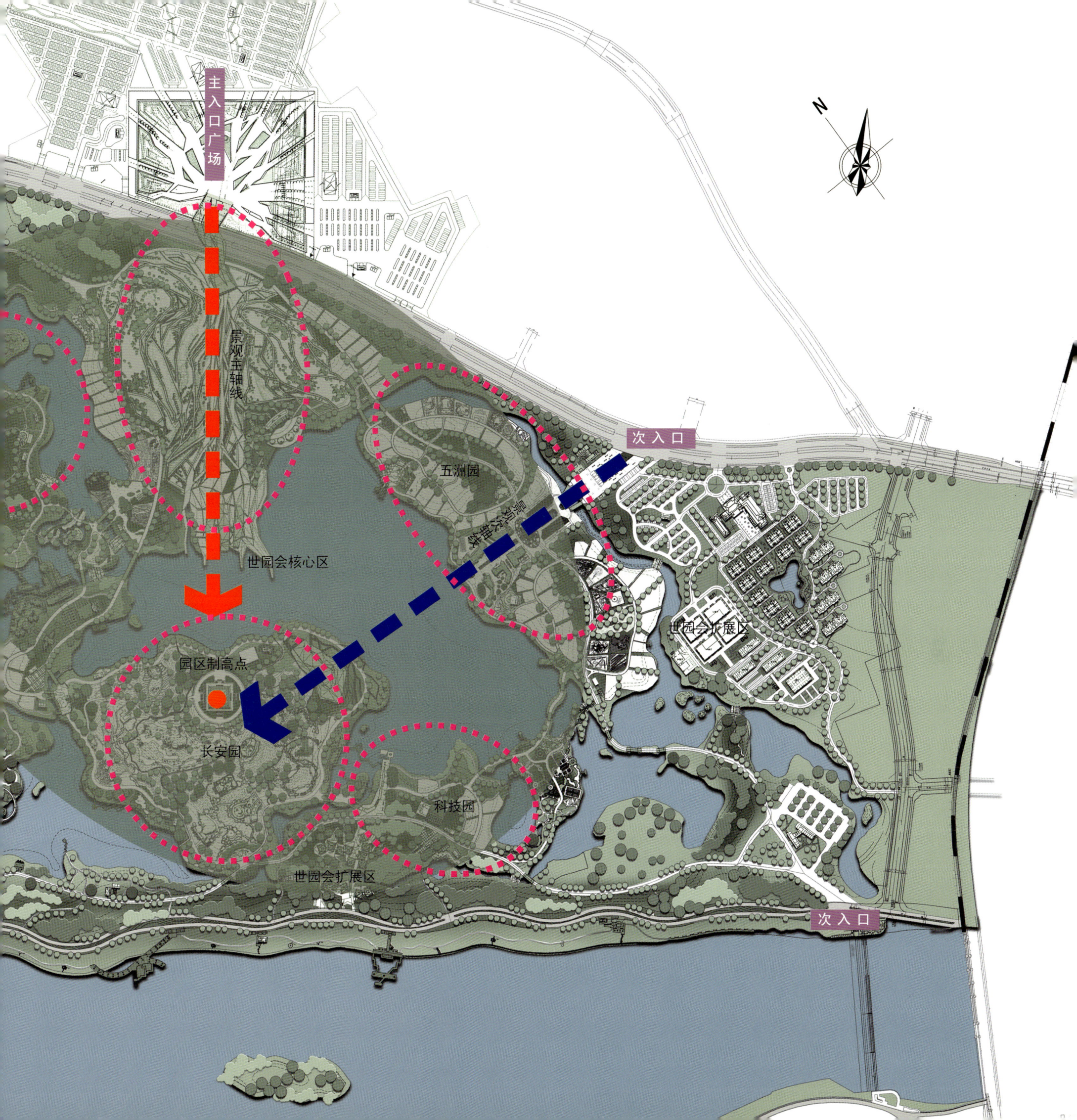

Entrances Zone
入口区

2011西安世界园艺博览会入口包括主入口广运门、团队入口、贵宾入口、次要入口和后勤入口。

There're four entrances for International Horticultural Expo 2011 Xi'an China: Guangyun Entrance as the main entrance, Group Entrance, VIP Entrance, Secondary Entrance and Logistics Entrance.

广运门

广运门位于园区东北部，是一座折线形的步行路桥，横跨了60米宽的世博大道，是2011西安世界园艺博览会的主入口。完备的紧急通道和残疾人通行设施让广运门更添人性光彩。项目整体由踏步、方块式园艺花卉造型组成，延伸至"长安花谷"景区的坡道横跨了进入园区的主要交通道路，形成恢弘气势的同时也将上下之间联系起来。入口前，线形绿化带的布置，将不同功能空间进行区隔的同时，也丰富了景观效果；坡道将进入园区的人车进行了分流，在最高处，设计师也安排了场所为即将进入园区的游客们提供了先睹为快的机会；植物花卉等反映绿色园艺色彩的设计元素贯穿了整座广运门。

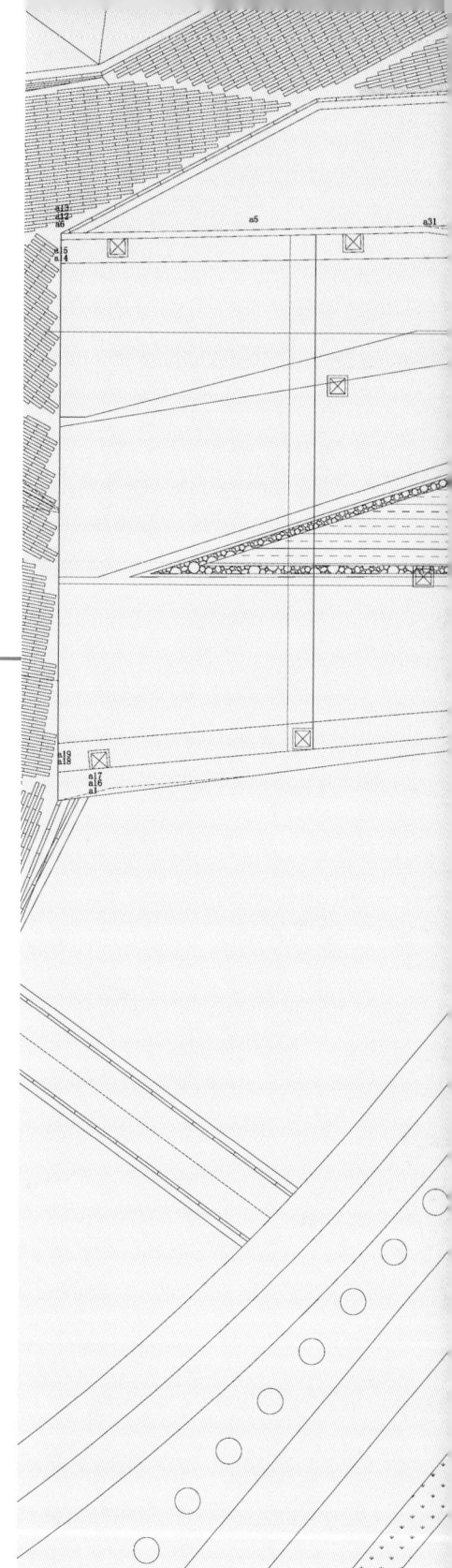

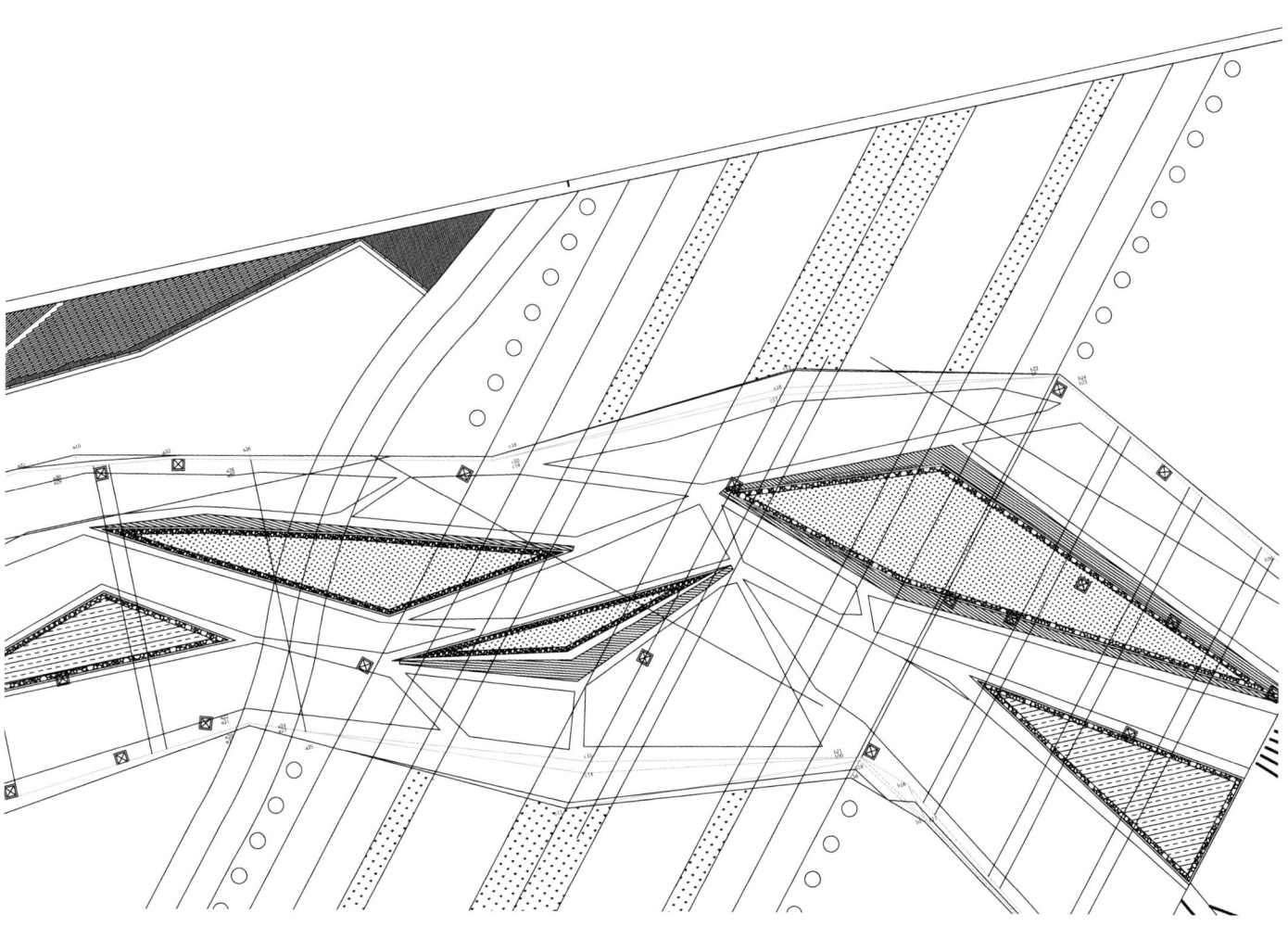

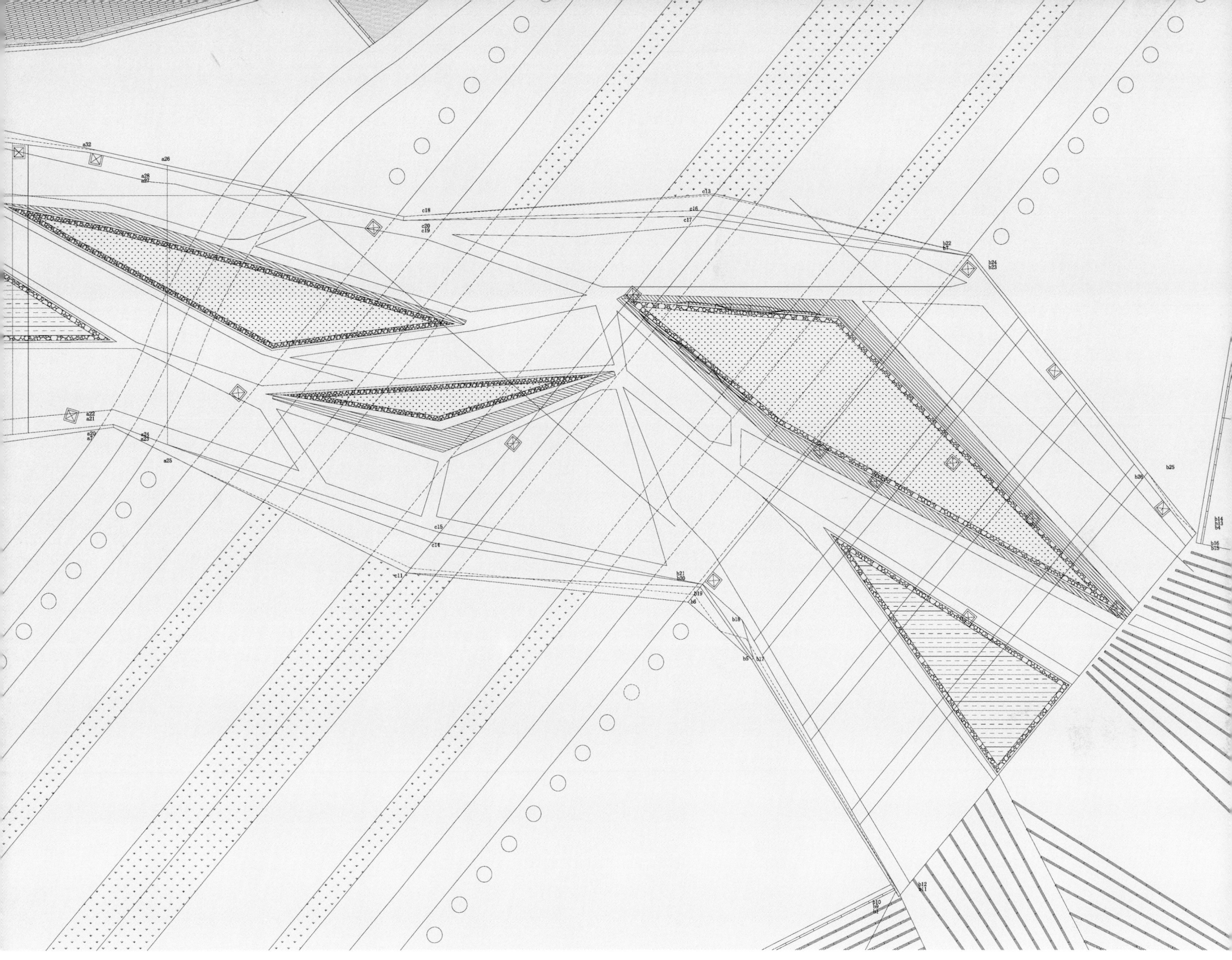

Guangyun Entrance

Guangyun Entrance, the main entrance of the Expo Site located at the northeast of the Expo Site, is a Z-shaped pedestrian bridge stretching over the 60-meter-wide Expo Avenue. The installation of emergency exits and facilities for the disabled brings convenience to tourists. It consists of steps, water features and square flower gardens, extending to Chang'an Flower Valley. This magnificent ramp spans the main road network of the Expo Site. The linear greenbelt defines different functional sub-spaces and adds a layered effect to the landscape. The ramp separates the driveway and the pedestrian. At the highest end, a place is prepared for visitors' pre-appreciation of the landscape. Plants and water features reflect the green aspects of the design. The sunshade pergola of climbing plants reveals vertical greening function.

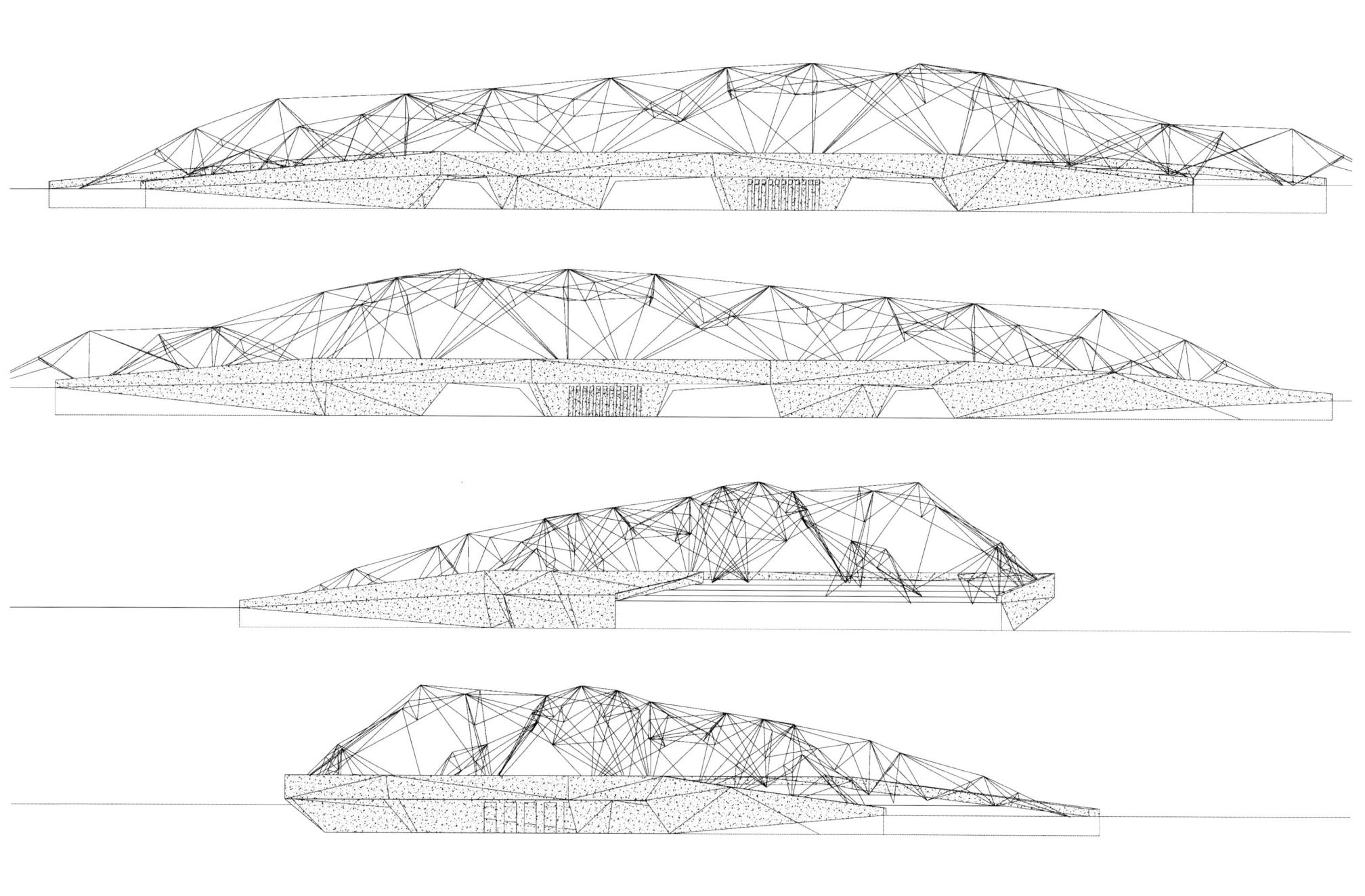

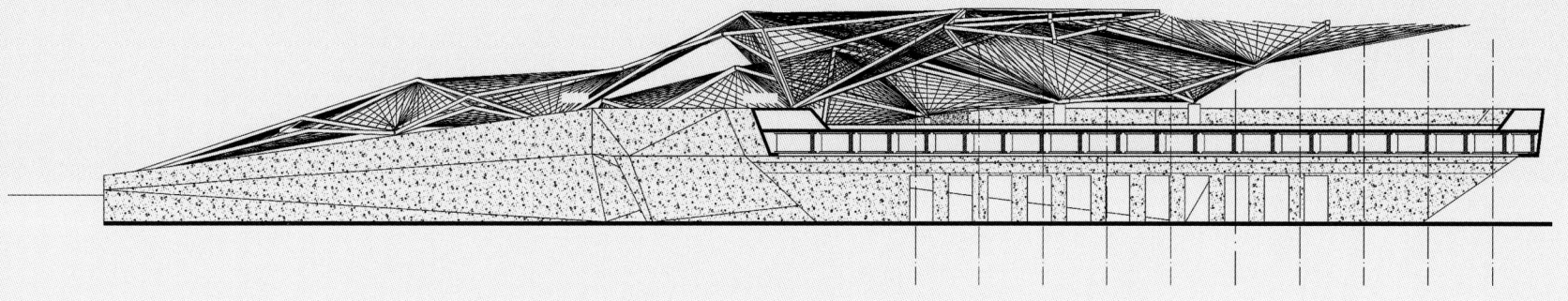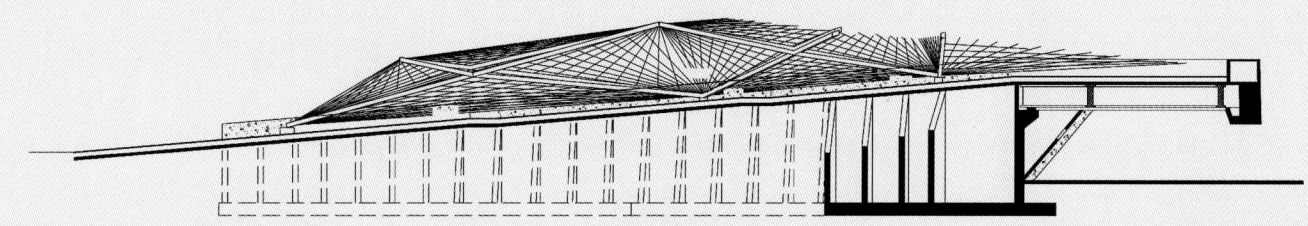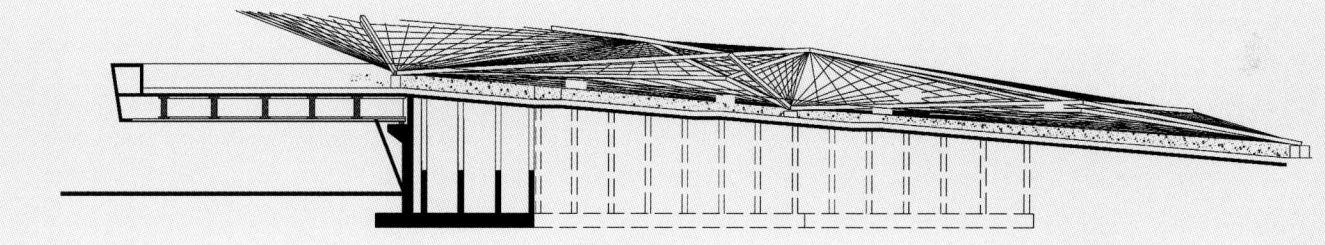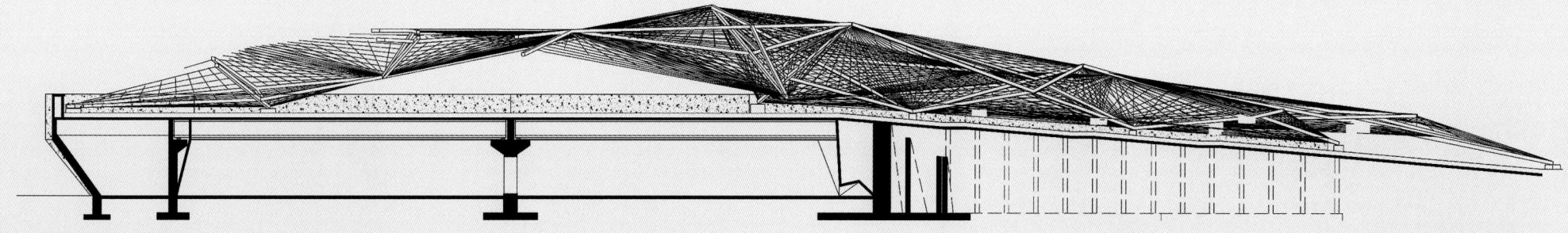

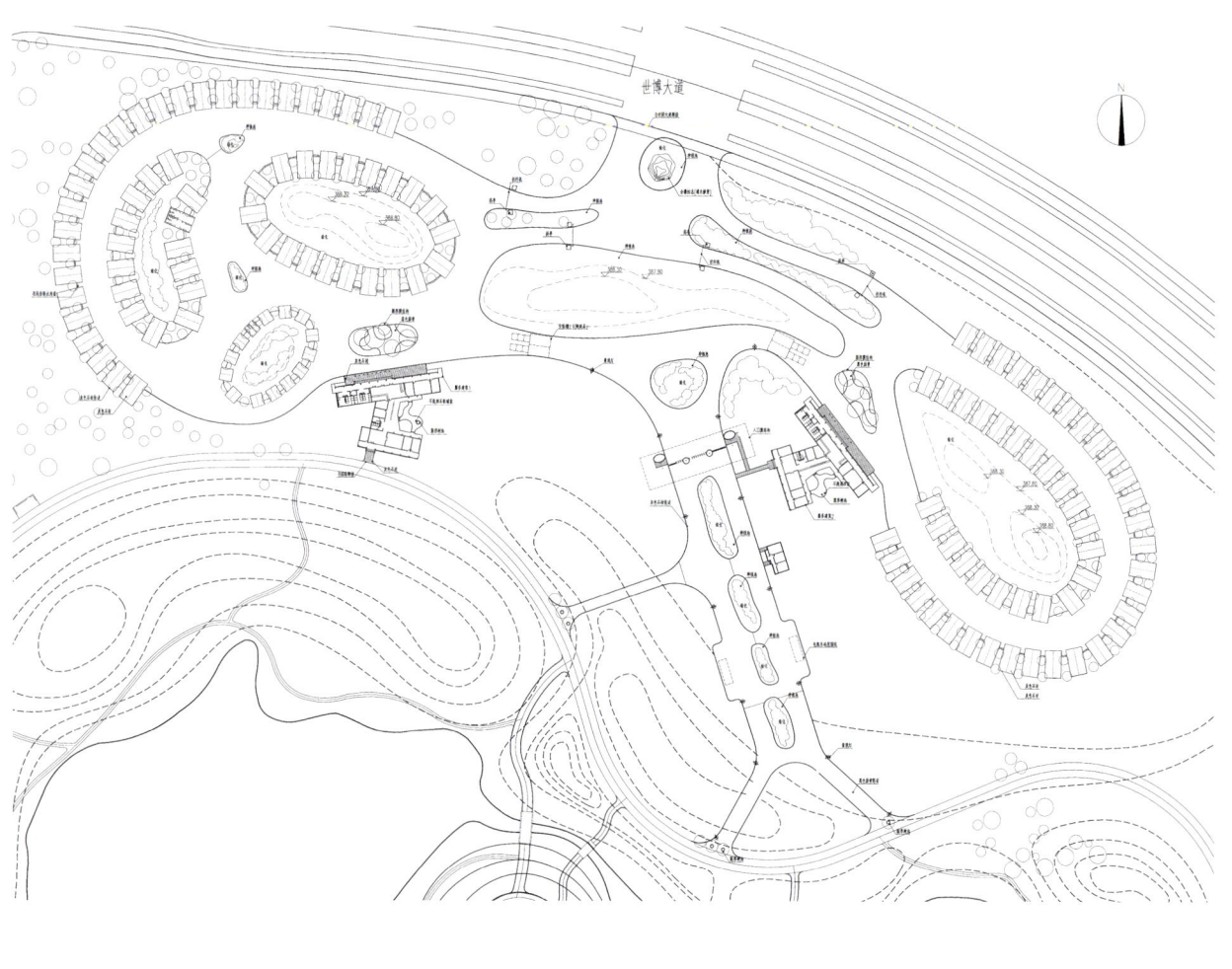

团队入口

团队入口位于园区的西北角,连接世博大道和世园会主道路。团队入口总面积约80,000平方米,主要功能包括停车(以大巴为主),接待,检票,电瓶车站等。入口大门为钢框架膜结构,宽54米,高11米,造型简洁,风格现代。由于场地硬化面积大,所以在停车场和大面积的硬化区域设计绿岛,停车场呈环形分布。地面铺装材料以沥青为主,视觉效果统一,明朗。

Group Entrance

The 80,000 m² Group Entrance Area is located at the northwest end of the Expo Site, connecting the Expo Avenue and the main roads of the Expo Site. Its main functions include parking (focusing on buses), reception, ticket-checking, electric vehicle station, etc. The entrance gate in simple and modern style is built of steel frame and membrane structure, at a height of 11 meters and with a width of 54 meters. Due to the vast hardscape area, green islands are designed and put between the ring-shaped parking lot and the vast hardscape area. The paving material focuses on the asphalt to unify visual effects, bright and clear.

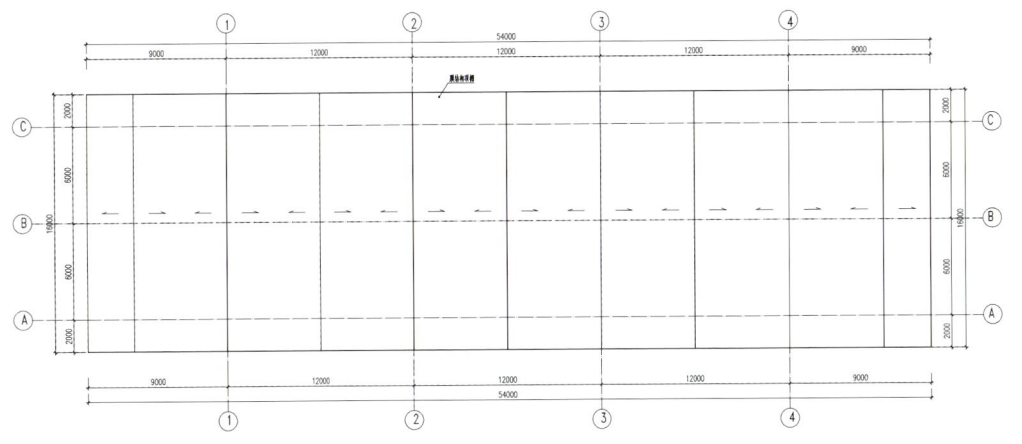

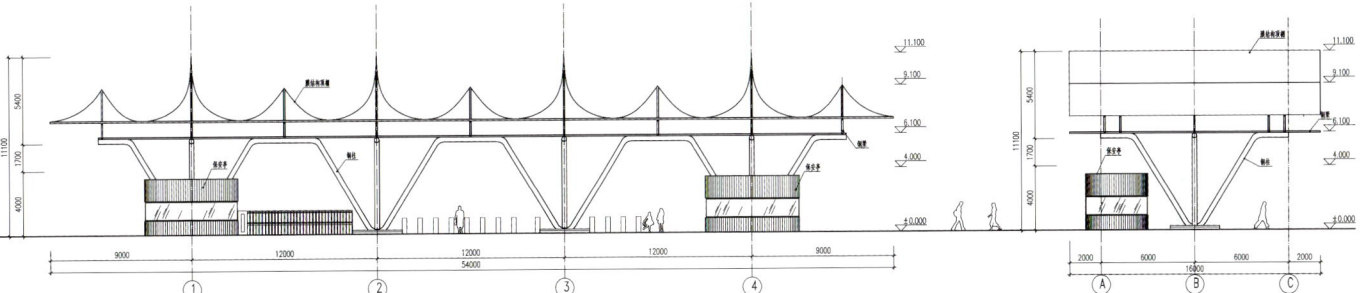

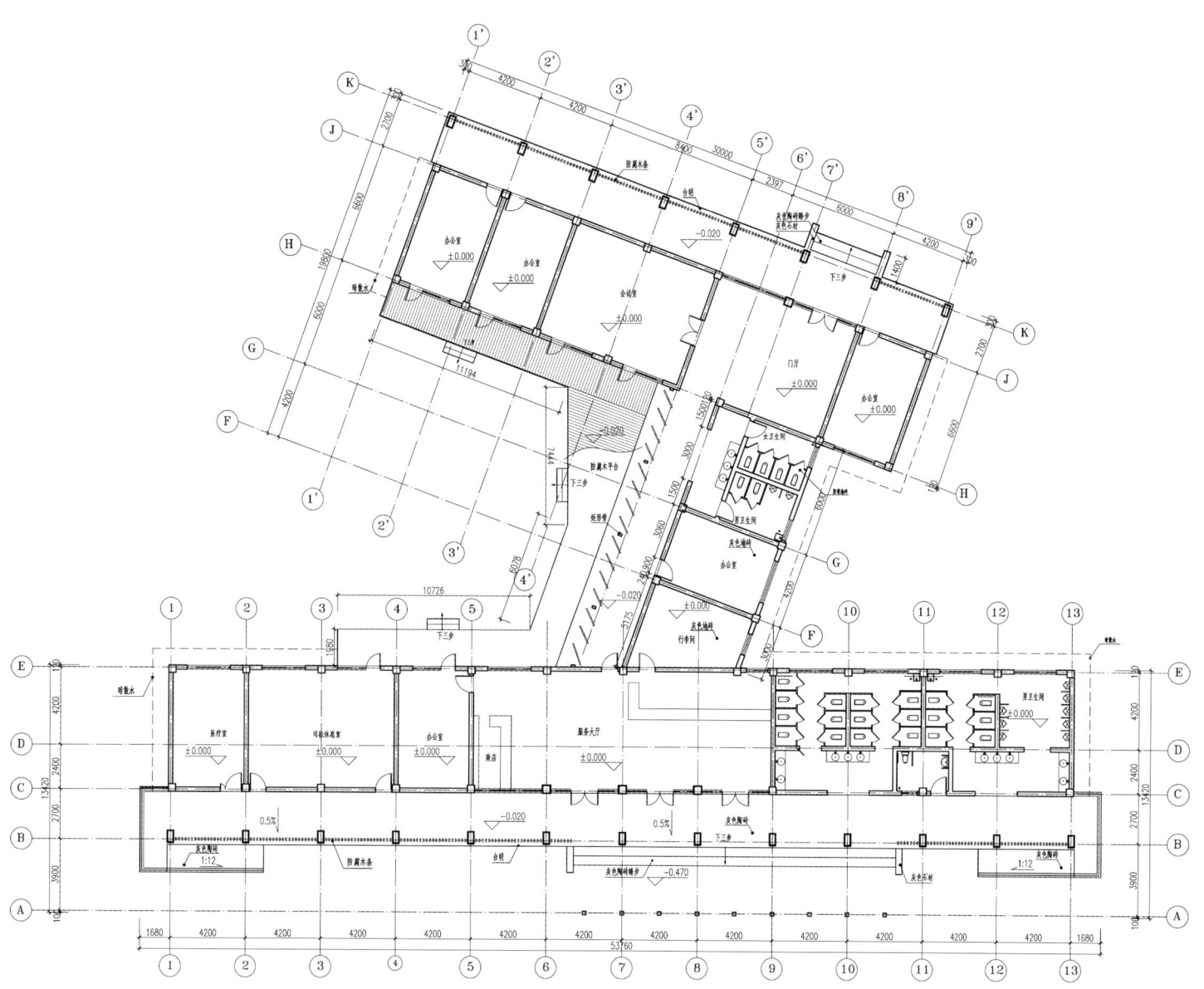

团队入口建筑

团队入口建筑共两幢，位于团队入口大门两侧，单幢建筑面积924平方米，单层框架结构，建筑高度6.65米。建筑风格现代简约，功能包括：服务大厅、游客咨询、会议室、办公室、休息室、卫生间、应急医疗等。

Architectures at Group Entrance

There are two single-storey-frame constructions with an area of 924 m² respectively and the height of 6.65 meters each on either side of the Group Entrance. The style is modern and simple with the function rooms such as service hall, tourist reception area, meeting rooms, rest rooms, toilets and medical emergency, etc.

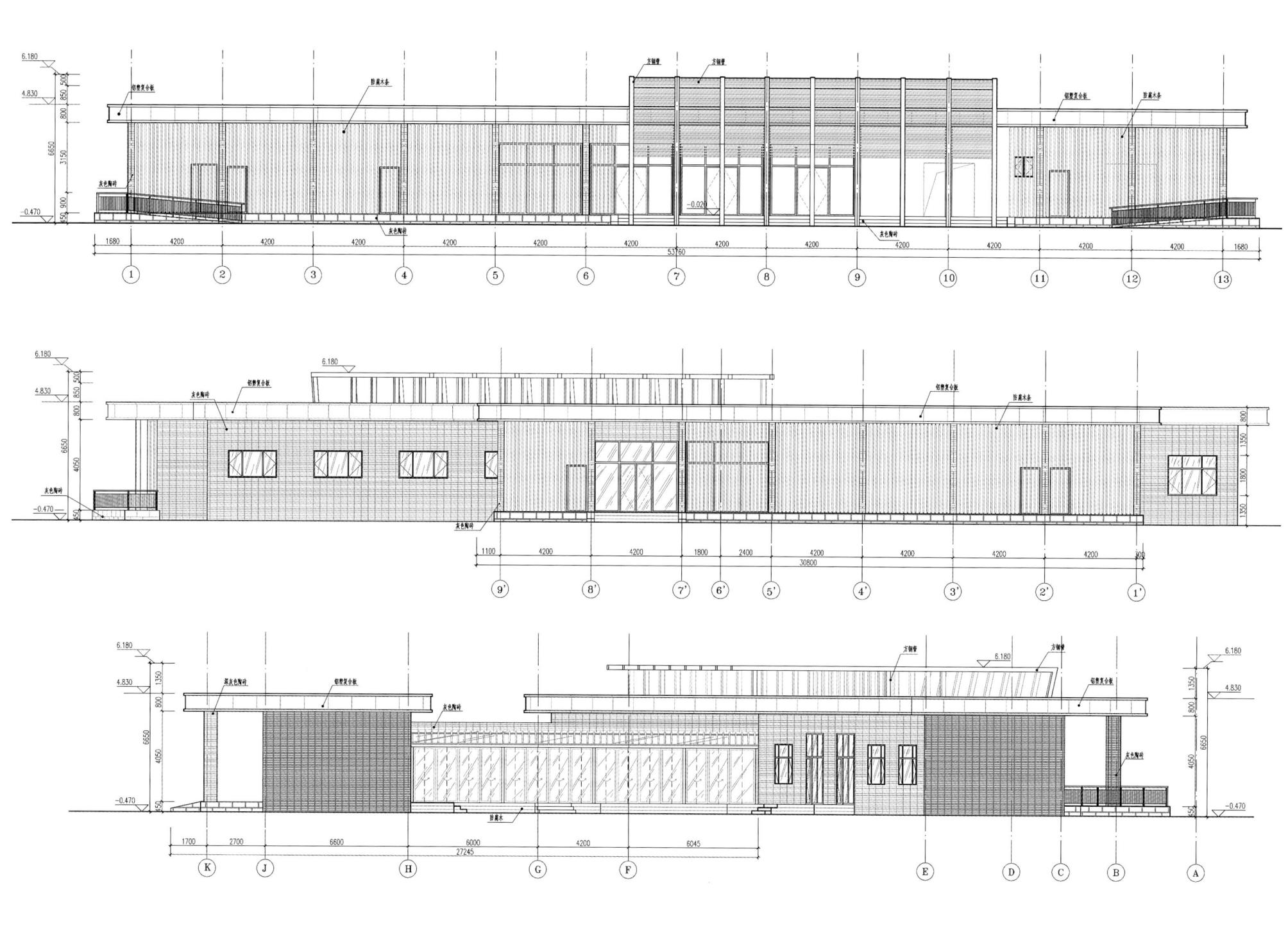

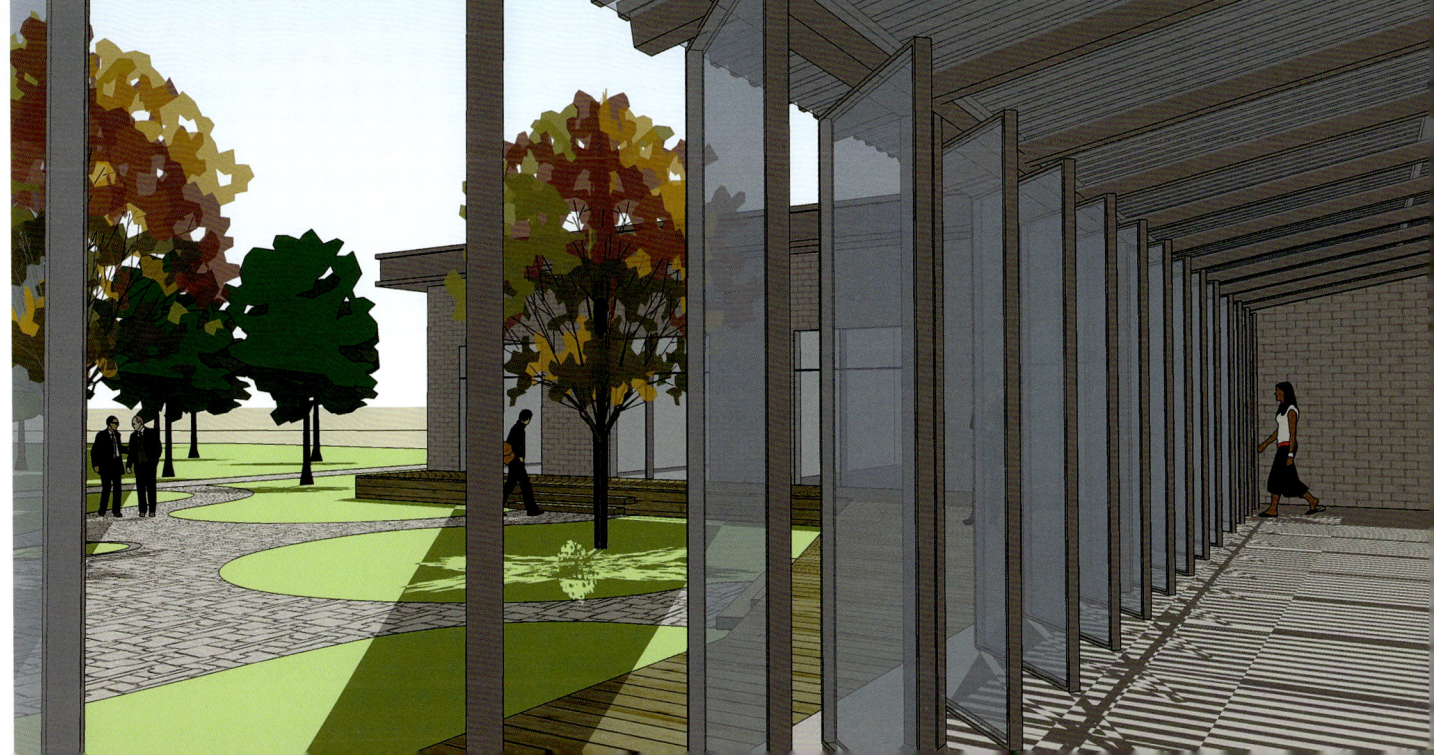

次入口

次入口位于园区东北部,连接世博大道和欧陆风情服务区。总面积32,000平方米,大门为曲面的方块形状膜结构,共7组,高度从7米到12米不等,错落有致。地面铺装为菱形和平行四边形等几何形式,材料为灰色和黑色石材,整体风格统一协调,有较强的视觉冲击力。服务建筑一幢,建筑功能包括游客咨询、办公、休息、卫生间、应急医疗等。

Secondary Entrance

The 32,000 m² Secondary Entrance is located at the northeastern part of the Expo Site, connecting the Expo Avenue and the service area of the European Avenue. Its gate is a curve surface of square membrane structure in 7 groups, with a height varying from 7 meters to 12 meters. The paving material focuses on the gray/black stones in geometrical patterns including rhombus and parallelogram, which produces a strong visual impact and unifies the whole style. A single building is constructed to provide services for the visitors, with tourist reception area, office, rest room, toilets, emergency medical service, etc.

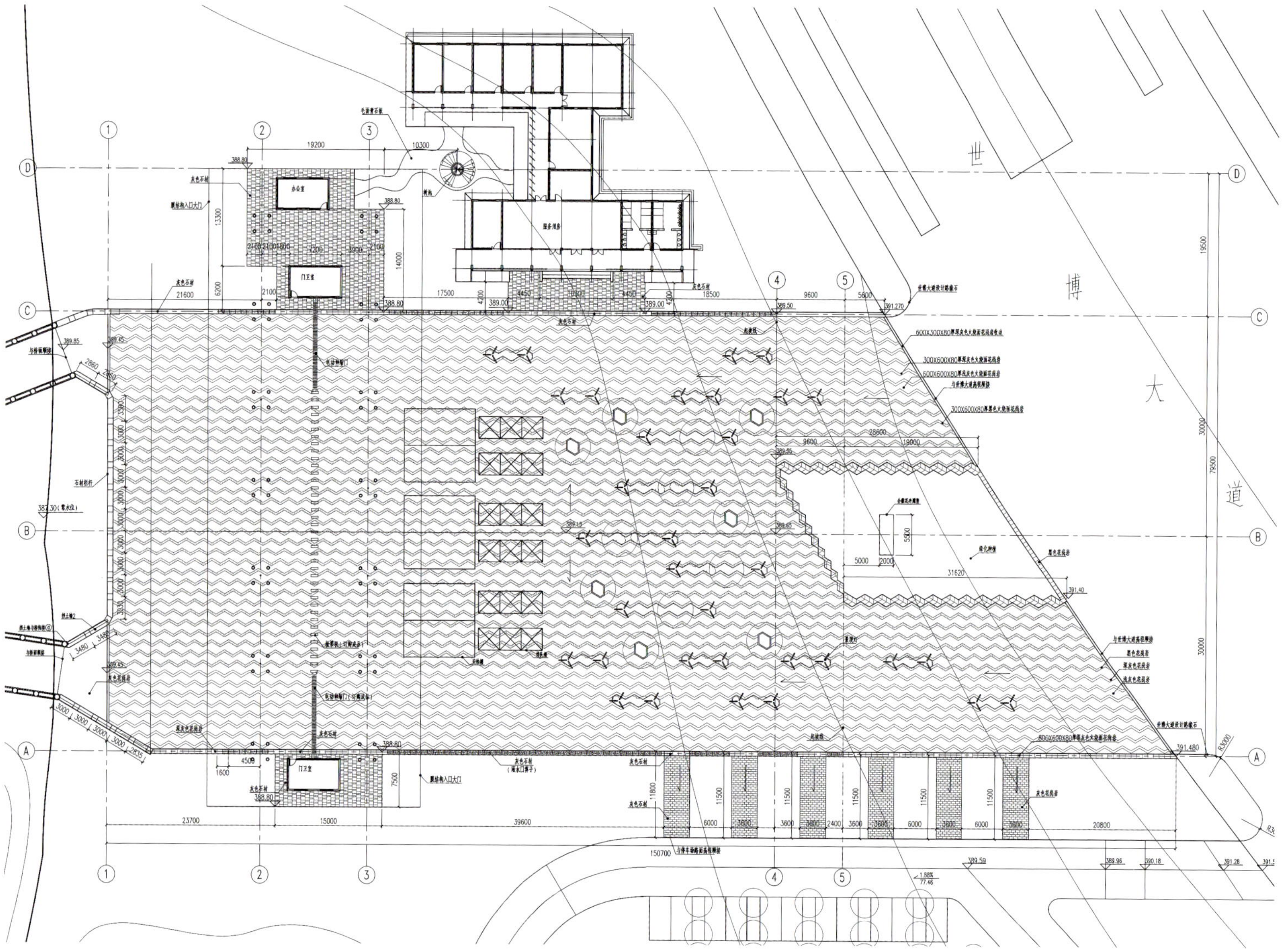

1'
2'
3'
4'
5'
6'
7'
87000
7500
12000
7500
B'
A'
15000
30000

12.000
11.000
9.000
9.500
8.000
7.000
5.000
4.000
±0.000
87000
1'
2'
3'
4'
5'
6'
7'

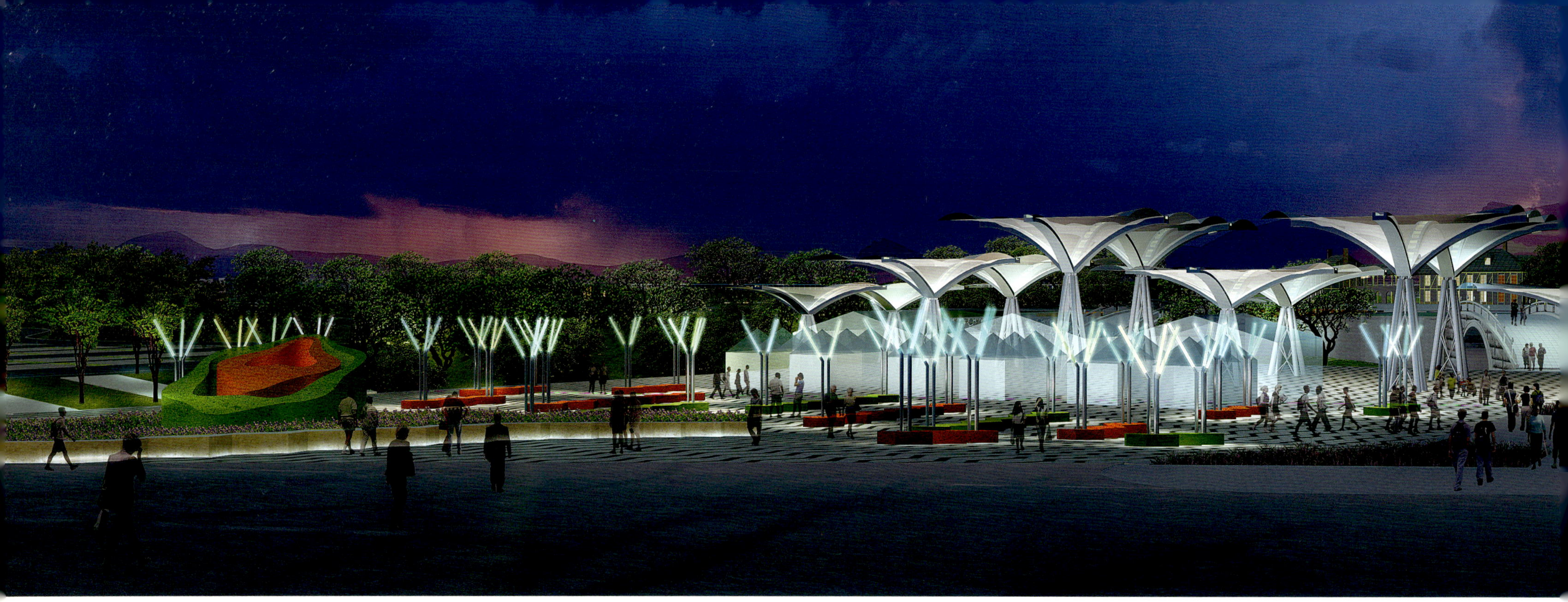

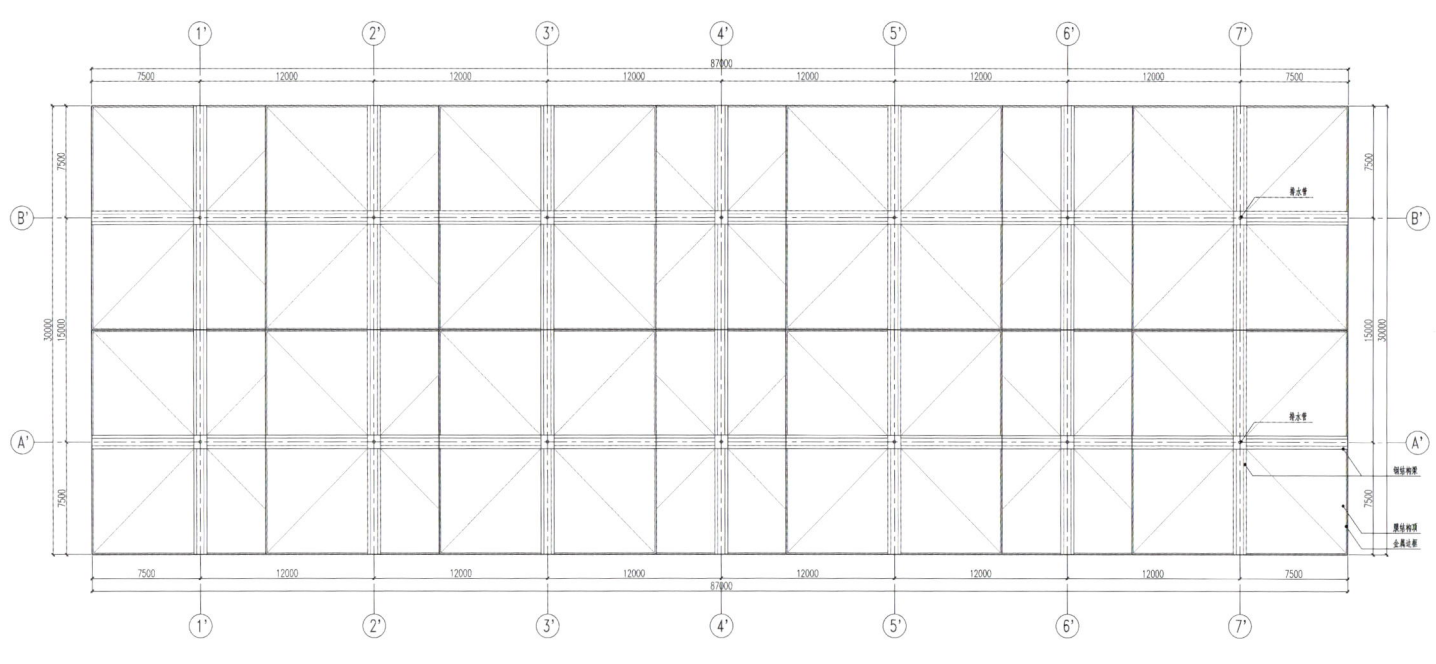

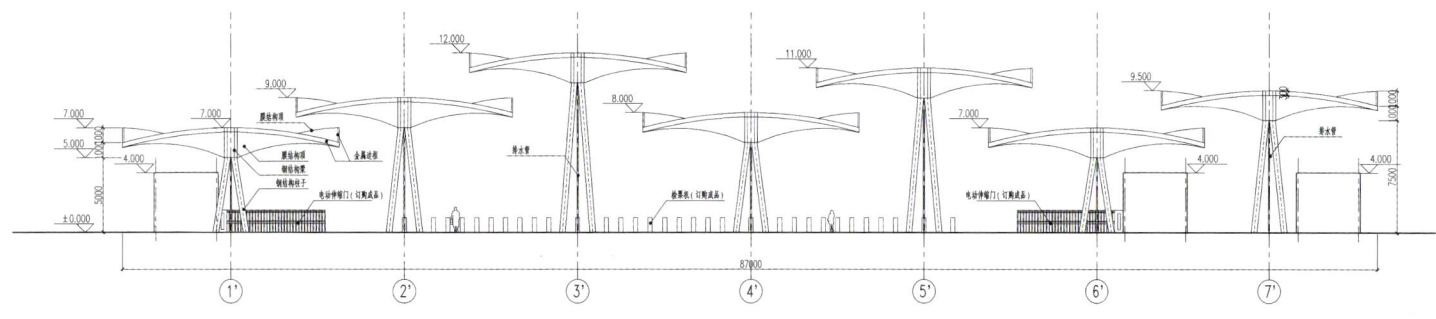

Chang'an Tower

核心区·长安塔

构思立意

项目定位 文脉追寻

项目定位是设计的首要问题。经过深入领会"天人长安"的塔名，逐渐认识到这是一个文化标志性的主题。这里既需要体现中国数千年来的"天人合一"宇宙观，又需要有明显的长安地域特色，作为大型国际博览园的主建筑之一应当充分反映当今的时代特色和审美情趣。

沿着长安城建历史的脉络回顾了隋唐长安建塔情况。隋唐长安塔体林立，大大小小不计其数。总的看来有宝塔、观光塔和风水塔之分。宝塔中供奉舍利、经卷、佛像或其它宝物；观光塔供游人登临、赏景；风水塔则是为取得城市或片区的风水平衡而建，也有一个塔体集三项功能于一身的。如今保存的比较完好的64米高的大雁塔、46米高的小雁塔都属宝塔一类。它们为砖塔，厚重、结实，有永恒感。那些别有风韵的木塔均已荡然无存。长安历史上最高的木结构塔当推位于隋唐长安城西南隅的大总持寺和大庄严寺的双塔。据文献记载，塔总高97米，塔基44米见方，规模与洛阳有名的永宁寺塔相当。之所以在隋长安城建此高塔，首出于城市风水的考虑。因长安市区东南曲江一带地势高爽，而西南隅地势低下，宇文恺等拟以在此建高塔以平衡之。这一对皇家寺院的高塔兼具上述三种类型塔的功能：帝王在此游览，文人百姓来此观光，塔内还曾供奉过佛牙。唐代诗人岑参登塔后留诗"高阁逼诸天，登临近日边"，"槛外低秦岭，窗中小渭川"。宋之问登塔诗曰："梵宇出三天，登临望八川。开襟坐霄汉，挥手拂云烟。"一千多年前登者与自然融合，人与霄汉云烟的亲近赋予设计创作灵感。

园区特色 确定塔高

审视2011西安世界园艺博览会规划所提供的建塔环境，全园主入口广运门、创意自然馆形成的主轴跨过湖面直指人工土山"小终南"上的天人长安塔。塔成为主轴的端景，同时它也在全园副轴与主轴的交叉点上。世博园占地418公顷，其中水域面积188公顷。虽有岗峦起伏，但总的地形比较平坦。中国有句老话说："山地观脉，脉气重于水；平地观水，水神旺于脉"，"山地以山做主"，"平阳以水做主"，由此我们认识到"小终南"上的天人长安塔固然依托于"小终南"的山形地貌，但无论如何它都应该成为具有丰富水系景观特色的世园会的标志性建筑，其高度宜高不宜低。隋代用木结构建起了97米的高塔，现代化的今天，塔理应高出一筹。经推敲塔高定为111米，后因航空管制的要求，不得不降至95米高度。

Background and Intent

Project Orientation and Cultural Context

Project orientation is the first issue to address in design process. It is evident that the Expo theme is a cultural symbol. Based on the deep understanding of the Expo theme, this project should highlight three aspects: "Harmony between Nature & Mankind" is a universal concept in China for thousands of years; local features should be stressed in this ancient city – Chang'an (the former name of Xi'an in the Sui and Tang dynasties); as a main building in a large-scale international exposition, Chang'an Tower should reflet features and aesthetics of contemporary times.

We trace back to the construction of pagodas in the Sui and Tang dynasties upon the urban history of Chang'an. In the Sui and Tang dynasties (581-907), there are numerous pagodas in various sizes in Chang'an. In general, they could be classified into three categories: Buddhist pagoda, sightseeing tower, Feng Shui tower. Buddhist pagoda is used for keeping sacred relics safe and venerated, holding sutras and figurines of the Buddha and other jewels; Sightseeing tower functions as a scenic spot for visitors to climb up and enjoy the panorama around; Feng Shui tower is constructed for the purpose of melting a city or district into the nature. Besides, there are also pagodas that integrate three functions mentioned above. For instance, the well-protected 64-metre-tall (210 ft) Giant Wild Goose Pagoda and the 46-metre-tall (151 ft) Small Wild Goose Pagoda are Buddhist pagodas, built of brick, massive and solid, with the feeling of eternity. Those exquisite wooden pagodas have not survived. The tallest wooden pagoda in Chang'an history was the Twin Pagodas in Zongchi Temple and Zhuangyan Temple, built in the southwest of Chang'an City in the Sui and Tang dynasties (581-907), both standing at a total height of 97 meters (318 ft) with a 1,936 m² base (20,839 sq. ft). Their scale was similar to that of the well-known pagoda of the Yongning Temple in Luoyang. The royal Twin Pagodas' purpose was to win the balance between the highland of Qujiang in the southeast and the lowland in the southwest of Chang'an City. They integrated three functions of the three categories of pagodas: offered spectacular views to the Emperor and its people, once housed Buddha tooth relic. Many famous poems were composed upon climbing the pagodas. It is this intimacy with nature one thousand years ago that provides us with inspiration for creative design.

Special Features and Tower Height

Expo Site planning is based on the context of constructing tower. The main entrance of Guangyun Entrance, Theme Pavilion and Greenhouse form the

main axis. Chang'an Tower is located on a man-made mound – the Young Zhongnan Mountain. It is the meeting point of the main and auxiliary axes of the landscape as well as a focal point for the whole Expo Site and an ideal location for a panoramic view. The Expo Site is 4,180,000 m^2, of which water area is 1,880,000 m^2. The landform is flat in general though meandering in some places. Chang'an Tower shall be a landmark in Expo Site and integrated with waterscape because it is located on a mound – the Young Zhongnan Mountain and that it complies with the traditional Chinese concept in architecture – mountains and water shall be main elements in architecture. There was a wooden pagoda standing at the height of 97 meters (318 ft) in the Sui Dynasty, so is today's tower which shall be taller. The height of tower was calculated originally to be 111 meters (364 ft), but it was altered to 95 meters (312 ft) due to air traffic control.

远观塔势 近赏细形

"千尺为势,百尺为形"是指建筑设计时首先要从全局把握建筑形体和山水环境大的态势。从近距离观看时,则应有精彩独到的建筑形象细部。出于前述项目定位,决定采用唐风木结构塔的基本造型:方形的塔体,稳健的逐层收分,悬挑的平座,深远的出檐。从远处任一角度都可以识别这是长安之塔。由于设计采用传统建筑的革新,选用了钢结构外框内筒的方案,屋顶,挑檐,明层的外墙一律采用超白玻璃,外露结构构件和檐下创新构件一律采用沙光不锈钢色。这样使天人长安塔具有闪亮、透明的"水晶塔"的风韵。即使行至距塔50-100米近处时,塔体细部简明的韵律,构件鲜明的节奏在明媚的阳光下无异于呈现出一组超大型的现代工艺品。人们由远及近会对塔有一个感悟过程。

塔中菩提 绿色永恒

天人合一要求这个现代高层建筑融入园区甚至更远的山水环境,同时也要让塔中游人有身临山水之间的感受。塔周的玻璃幕墙提供了这样的条件。曾设想在塔顶层高敞的玻璃屋顶下做一个花卉大厅,以形成登塔的高潮,后由于经营管理等原因未能实现。进而试图创造一个永恒的绿色环境,这个设想在室内设计师,画家和建筑师的密切合作下终于变成现实。把塔的七个明层的塔心筒墙面视作一幅巨画,用油画的手法绘出一组菩提树林。菩提象征着圣洁,和平,永恒,这是园中塔,塔中树的生动畅想。行走在这样的观光塔中,无论塔外四周作何变化,都能感受到万古长青,绿色永恒的意境。

Appreciating the Architecture in Appropriate Distance

"A distance of a thousand feet represents the general configuration and a distance of a hundred feet represents the concrete form." This sentence stresses the macro frame and the unique detailing in architectural design. Based on the project orientation, we decided to use main elements of the wooden pagoda of the Tang Dynasty: square body, stories reducing the size gradually to the top, eaves stretching out above each storey, eaves gallery set on every storey. Hence, Chang'an Tower is recognizable beyond everywhere. The architectural design is traditional and innovative: frame and inner column are built of steel; roof cover, roof overhangs and the exterior façade of the bright stories are built of ultra clear glass; the exposure parts and the innovated parts below eaves are built of stainless steel. All these techniques highlight Chang'an Tower as a "crystal tower" of brightness and transparency. At a distance of 50-100 meters, the exquisite details and the succinct parts is a metaphor of a large-scale modern artwork shining in the sun. This get-closer experience from a distance is the essence of appreciating the architecture.

Ficus Religiosa in Tower, Green for Eternity

According to the concept of "Harmony between Nature & Mankind", the modern architecture Chang'an Tower should melt into its surroundings – mountains and waterscape, at the same time, offering visitors a unique experience in nature. The glass curtain wall provides such condition. Our former proposal was to create a flower hall under the tall glass roof on the top floor for the climax of climbing, but it failed due to operation and management. Then, a green environment for eternity is proposed and implemented upon the collaboration of interior designers, painters and architects. Oil painting techniques are used for sketching Ficus Religiosa onto the inner tube wall of the seven stories of Chang'an Tower. Ficus Religiosa is the symbol of sanctity, peace and eternity. The interesting idea of tower in Expo Site and Ficus Religiosa in tower infuses a feeling of eternity into sightseeing tours no matter what changes outside the tower through the seasons.

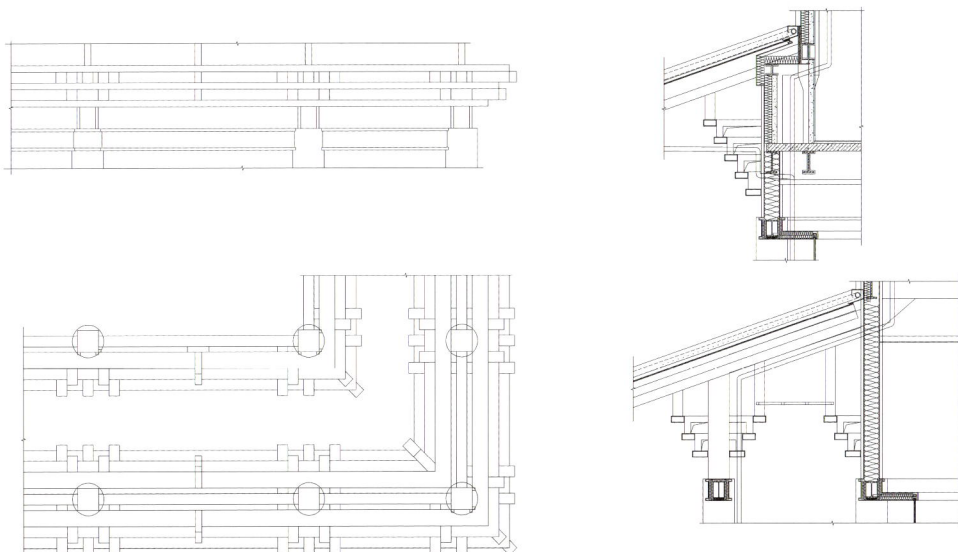

建筑设计

长安塔总建筑面积13,060平方米，地上十三层共计12,066平方米，地下一层994平方米。具有唐代传统木塔的基本造型，由于采用现代钢框架结构，玻璃幕墙，因而具有鲜明的时代感。

建筑平面设计

整栋塔体为正方形，塔座3.9米高，55米见方，台座之上是1.2米高42米见方的台明。台明之上是塔的一层明层，周围有2.9米宽开敞的檐廊，檐廊柱间设可供游人休憩的石凳。

每层正中五开间宽度均为4.5米，尽端梢间从一层的4.2米宽，逐层递减0.7米，至顶层梢间为0米，顶层正好总面宽5×4.5米。这样的柱网安排保证了塔体的收分符合古塔塔身收分的韵律，又保证了结构垂直传递荷载的合理性。核心筒设有能容纳水、暖、电管道的竖井，能方便通达各层。暗层在核心筒的东西两侧设男女卫生间，还可安排小型展览和纪念品商店及部门管理用房为上下相邻的明层服务。楼上各明层外檐柱外有平座及栏杆，为安全起见，平座不对游人开放，仅供工作人员之需。玻璃幕墙设在外檐柱之内，使玻璃面连贯，增加空间的现代感。核心筒与玻璃幕墙之间是游客主要活动的"回"字形空间。七明层是塔的顶层，顶层结合核心筒设3.4米高的观景平台。站在平台之上，上，可以透过玻璃屋面望天上风云卷舒，下，可以俯瞰苍翠大地春花秋实。

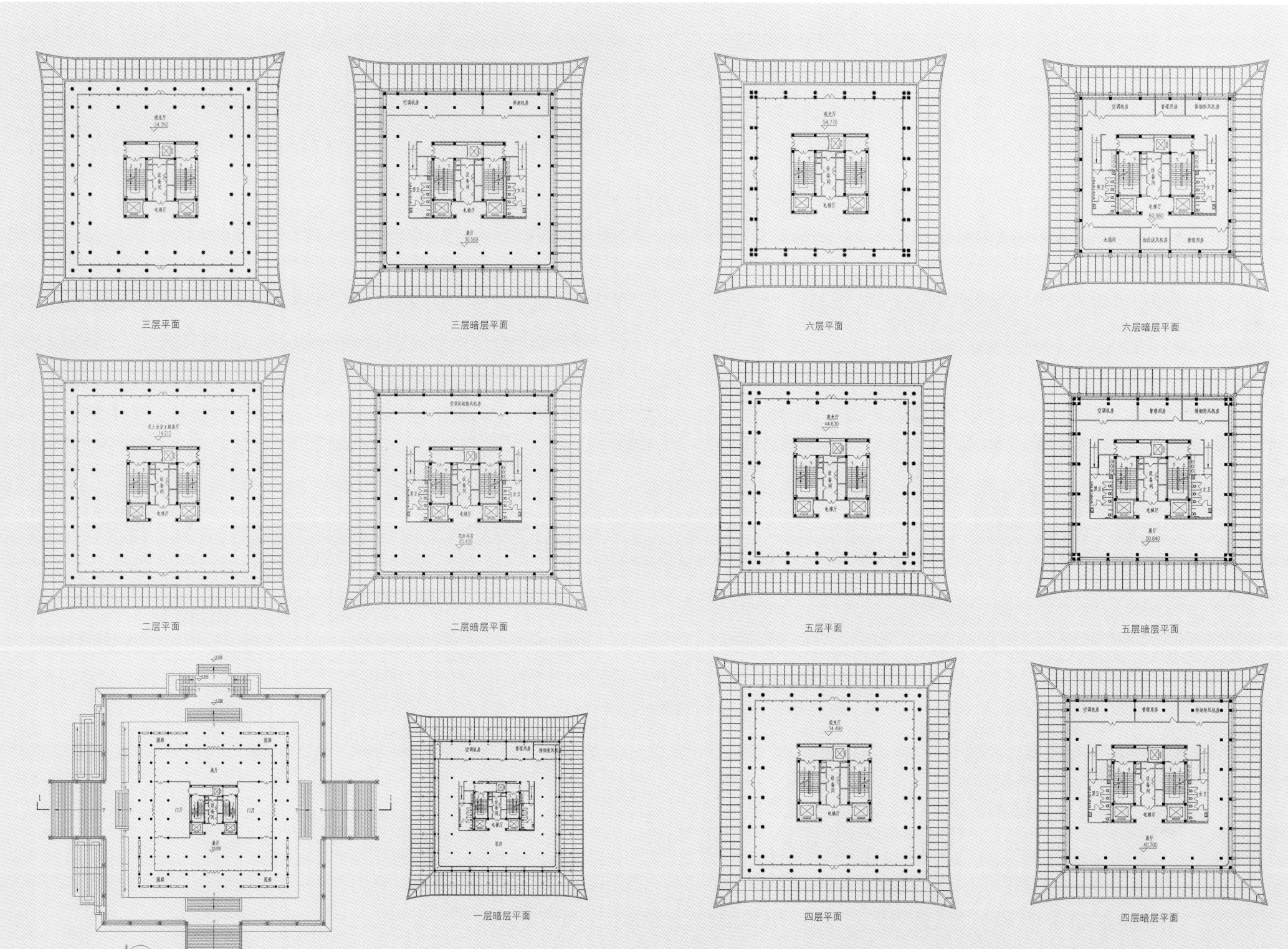

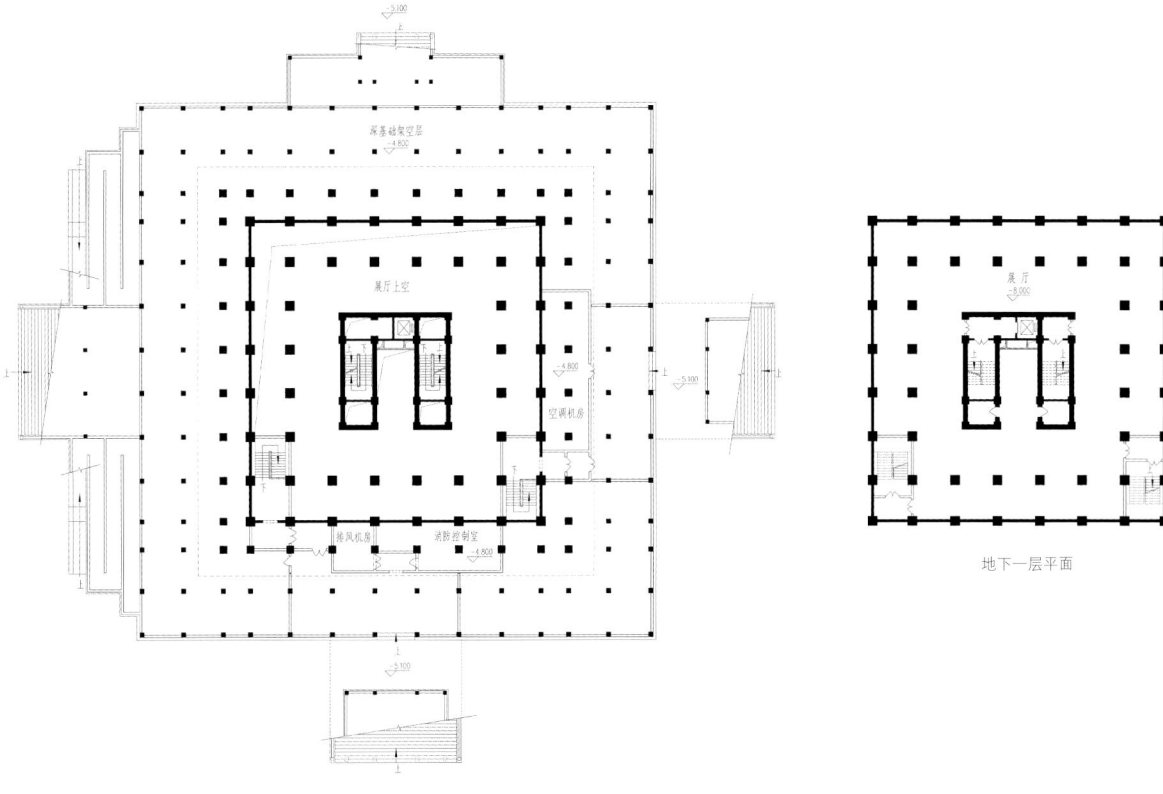

Architectural Design

The overall building area of Chang'an Tower is 13,060 m², of which the total area of thirteen stories is 12,066 m² and the underground floor area is 994 m². It mimics the traditional wooden pagoda of the Tang Dynasty, while its modern style is evident due to the usage of modern steel frame and glass curtain wall.

Architectural Plan

The tower is square shaped. Its base stands at a height of 3.9 meters with an area of 3,025 m² (32,561 sq. ft). The substructure on the top of the base develops into two parts – the upper part is the pedestal, and the lower part is the platform. The pedestal is 1.2 meters high, covering an area of 1,764 m² (18,988 sq. ft). Above is the first bright storey with an open eave gallery of 2.9 meters in width. The space between eave galleries and columns is set with stone benches for visitors to relax.

Each storey has five rooms of 4.5 meters in width. The rest space of stories reduces the size gradually from 4.2 meters in width within the first storey to the top by 0.7 meters to 0 meter in width within the top storey. Then, the top floor area is precisely 22.5 m². Such arrangement is in accordance with rules of the tiered tower design and meets the requirement of vertical carrying structure. Water pipes, heating and electrical facilities are inserted into the vertical shaft of the inner tube connecting every storey. In the east and west sides of the inner tube of the blind stories are male and female toilets as well as small exhibition area, souvenir shops and office rooms for the staff. All these facilities of the blind stories are convenient for providing services for the bright stories above and below. The upper bright stories are including eave galleries and balustrades, and only the staff could use the eave galleries in consideration of the visitors' safety. The glass curtain wall is installed within the external eaves column, its continuity lending a modern tone to this space. The concentric-circle-shaped space between the inner tube and the glass curtain wall is the main activity place for the visitors. The seventh storey is the top floor of the tower which has a 3.4-meter-tall viewing terrace beside the inner tube. Stand on the roof terrace, looking up through glass plane is the sky with clouds curling and stretching, looking down is the aerial view of green land with glorious flowers and solid fruits.

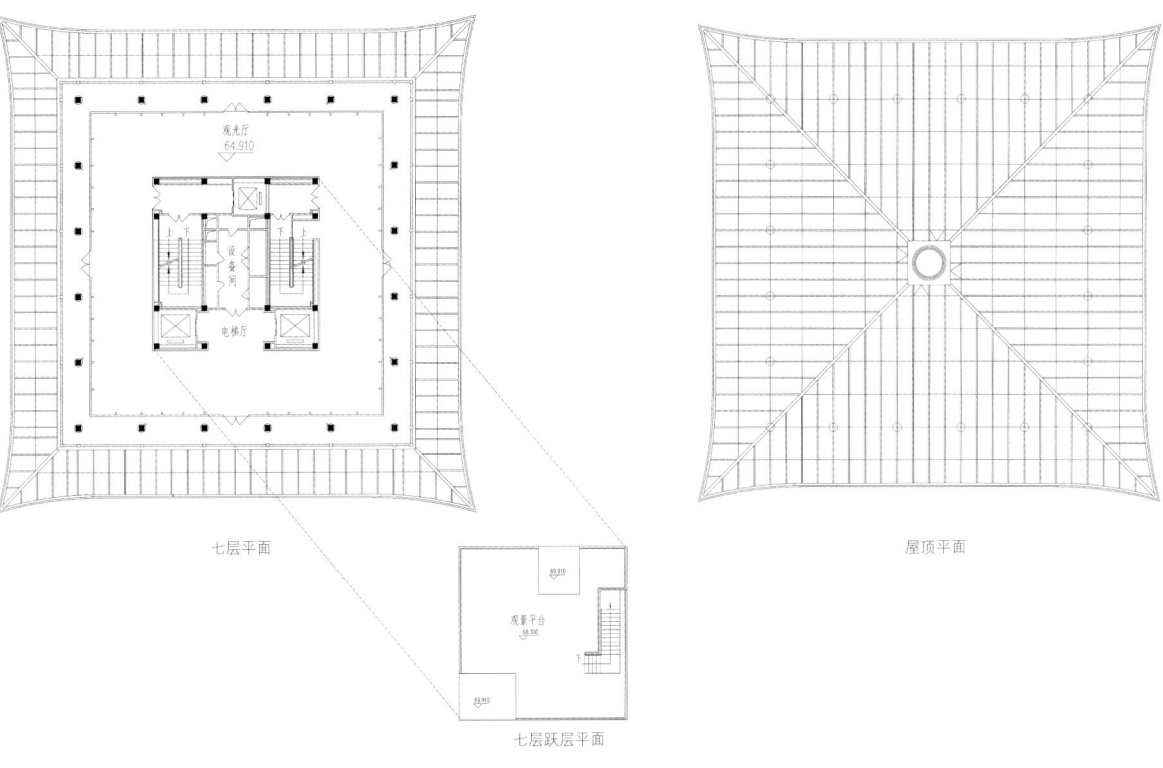

建筑立面

为了使天人长安塔远望之有唐代传统木塔的造型特色,剖面设计按照一层挑檐上面有一层平座的做法,逐层收分,这样设计符合古塔塔身收分的韵律。内部也就分别形成七明、六暗共十三层,挑檐尺寸较大达4.6米,体现唐代木结构建筑出檐深远的造型特色。

楼上各明层四周外围是平座栏杆,平座与栏杆均为沙光不锈钢的金属质感,栏杆造型简约,是对唐代建筑栏杆造型的抽象概括。

平座栏杆内侧是圆形的金属檐柱,檐柱之间在柱顶的高度上通过同样简洁的金属梁连接,金属梁之上的檐柱截面为正方形,这样处理是对传统建筑柱头和栌斗的高度概括。柱头与檐下之间层层出挑的金属构件相互搭接组合,是中国传统木结构建筑檐下构件系统的溯源和创新,比传统的斗栱系统更简洁,造型也由于更真实地反映结构的力学特性而显得更具有现代感。

挑檐采用钢结构上铺装夹层玻璃,玻璃分块尺寸结合开间模数,屋面玻璃采用竖明横隐玻璃幕墙的做法,竖向的明框在屋面上形成顺着屋面坡道方向的线条,仿佛传统建筑屋面瓦的节理。

挑檐玻璃下设遮阳百叶,所有百叶与屋面"瓦"的走向相同,遮阳百叶在透明的玻璃挑檐下面形成一个半透明的层次,给挑檐的透明玻璃与钢结构梁强烈的质感对比增加了一个细腻的中间层次,使气势恢宏,雄浑大气的唐风建筑多了些许柔和、隽秀。顶层屋面为钢框架结构梁之上安装中空夹层玻璃,玻璃之下在结构梁间的铝合金百叶可电动调整角度,冬天遮阳百叶全部打开,使阳光透过玻璃照进室内,在顶层形成温暖的阳光房。夏季遮阳百叶全部关闭,攒尖顶部的天窗打开,利用热空气上升产生的"烟囱效应",在建筑室内形成自然气流,有效地降低建筑室内温度。

建筑剖面

明层供游人观光使用,所有明层外围护结构为无框落地玻璃幕墙,玻璃幕墙整体退后圆形金属檐柱,在立面上形成具有强烈立体感的光影效果。人在其中视野开阔,全园景色尽收眼底,具有现代感。明层室内塔心筒墙上满布象征圣洁和谐的菩提树林的油画。总高约31米,总面积达1,400平方米的大型油彩壁画分为七段分别敷设在一至七层的塔心筒墙面上,蓬蓬勃勃,郁郁葱葱体现了天人合一的主题思想。

暗层立面在上一层平座与下一层挑檐之间的部分,这个部分外围护结构全部为金属幕墙,不开窗。在平座底部挑出部分采用叠涩的做法做成圆弧形的线脚,使建筑在局部形成层次丰富具有韵律感的光影效果。

明层玲珑剔透的"虚"与暗层的整体统一的"实"形成鲜明的对比,在立面上随着楼层的升高虚实相间,从大到小渐变,形成对比鲜明,具有渐变韵律,经典传世的建筑立面。

在天人长安塔最高处,是用沙光不锈钢板制成的塔刹,塔刹底部方形的须弥座,象征我们栖居的世界,也暗合天圆地方的中国传统宇宙观;正方形的须弥座上做莲花台,莲花又称荷花,与"和"谐音,象征和谐、和平。在莲花台之上是十层从大到小渐变的圆环,圆环之上金属球象征太阳,太阳之上是月牙,象征日月同辉、天人合一。

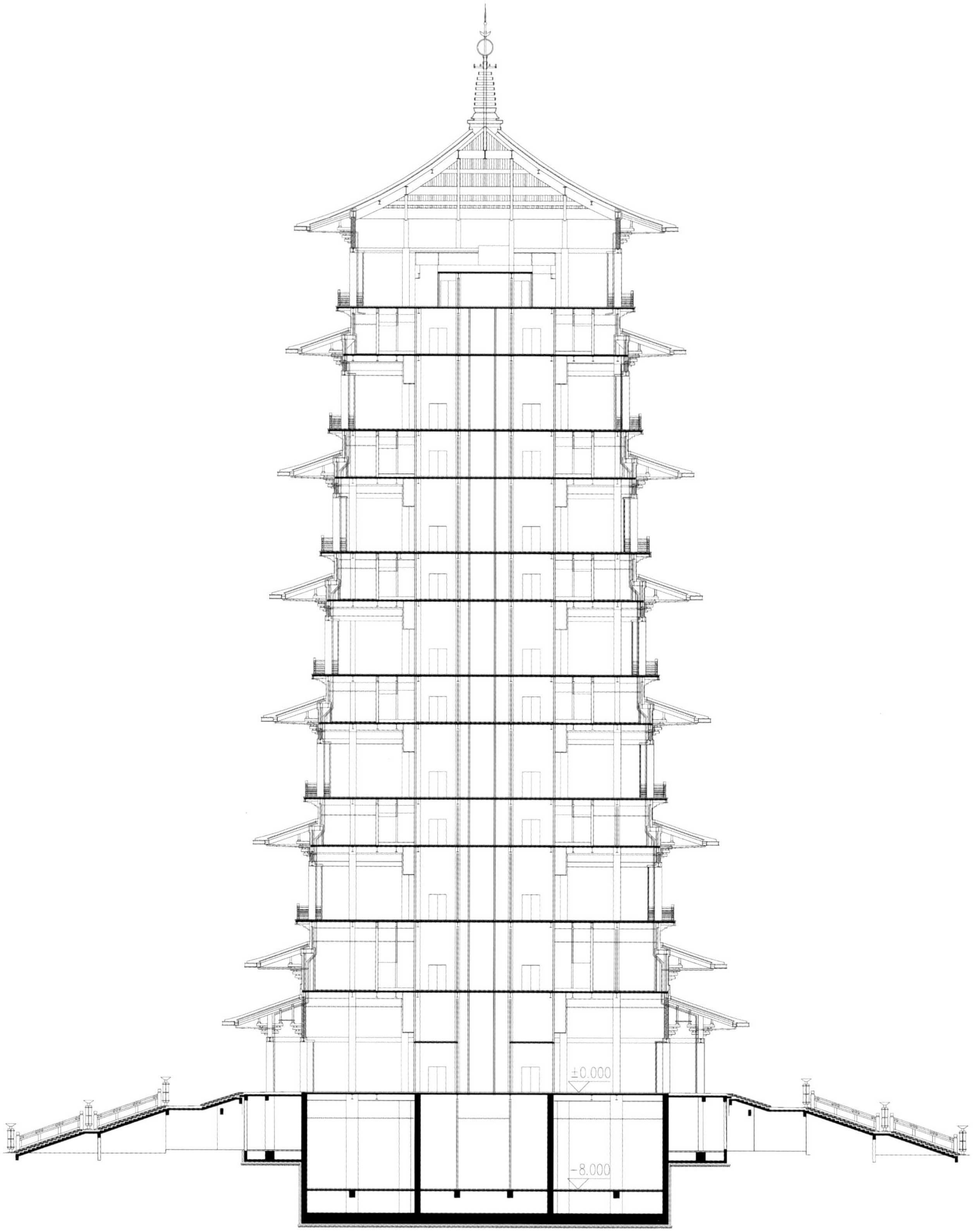

Architectural Elevation

According to rules of the ancient tiered tower design, the section is reducing the size gradually to the top with an eave gallery above an overhang eave. Thirteen stories consist of seven bright stories and six blind stories. Overhang eaves are large in size with the largest one of 4.6 meters, accentuating the special character of the wooden architecture of the Tang Dynasty – eaves stretching far out above each storey.

All bright stories are full of eave galleries and balustrades built of stainless steel. These metal balustrades mimic the simple style of the architectural design in the Tang Dynasty.

Inside the balustrades of the eave galleries are metal eave columns. Metal beams joint the eave columns on the top. The parts of the eave columns above the metal beams are square in section. This solution's purpose is to reproduce the traditional design of column caps and cap blocks. Column caps and the metal parts stretching out layer by layer below eaves are jointed imitating the traditional design of wooden structure, while the style of bracket set is simpler than the traditional one, and its modern shape reflects the essence of structure mechanics. Overhang eaves are steel structure inserted with glass planes in varying size according to room modules. Glass curtain wall is horizontal invisible and vertical visible. The vertical visible glass curtain wall forms a parallel line to the direction of ramps like the joint of traditional roof tiles. Louvers are placed below the glass curtain wall of the overhang eaves, with the same orientation as that of "tiles". The translucent sunshade louvers below the transparent glass curtain wall add a feminine beauty to this magnificent architecture of steel structure and metal beams. Roof cover of the top storey is built of steel structure beams inserted with hollow glass planes. Below the glass, aluminum louvers among beams could change their angles on the basis of electric power. Therefore, in winter, all sunshade louvers are opened allowing natural light, creating a sunlit room on the top floor; in summer, all sunshade louvers are closed, and the skylight of the steeple is opened for cooling through "Stack effect".

Architectural Section

The bright stories are used for the visitors' sightseeing, so the building envelope is frameless glass curtain wall. The whole glass curtain wall draws back, along with circular metal eave columns, encouraging a three-dimensional effect of light and shadow. The visitors will have a broad view of all parts of the Expo Site. The inner tube wall of the bright stories is a 1,400 m² oil painting at the height of about 31 meters sketching the sacred Ficus Religiosa. It consists of seven parts which are produced on seven stories' inner tube respectively. Its abundance and vitality represent the Expo theme "Harmony between Nature & Mankind".

The blind stories are located between the upper eave gallery and the lower overhang eave. The building envelope is built of metal curtain wall without windows. Curve moldings are realized through corbelled coursing, encouraging a layered effect of light and shadow.

The transparent bright stories is "unreal" space, while by contrast, the blind stories is "real" space. This vivid unreal/real distinction gives the architectural façade an everlasting feature that could be handed down from generation to generation.

The tallest component of Chang'an Tower is the steeple built of stainless steel plane. The base of the steeple was square Sumeru pedestal, symbolizing the world we are residing and complying with the universal concept in Chinese history – "Sky is round and ground is square." Blooming lotus petals are built on the Sumeru pedestal, rendering peace and harmony. Above the blooming lotus are discs of ten layers reducing the size to the top. A metal ball on the discs is a symbol of the sun. A crescent moon is put on the sun. The crisscrossing light of the sun and the crescent moon is a metaphor of "Harmony between Nature & Mankind".

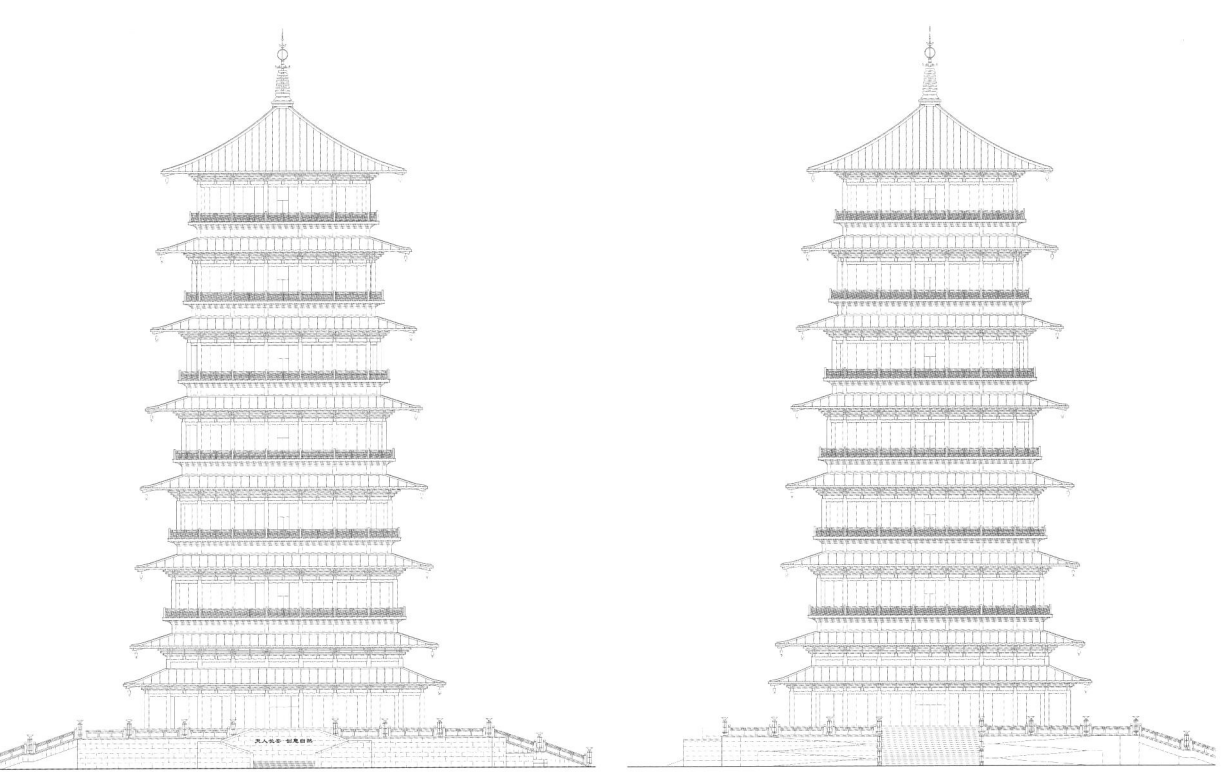

Theme Pavilion
核心区·创意馆

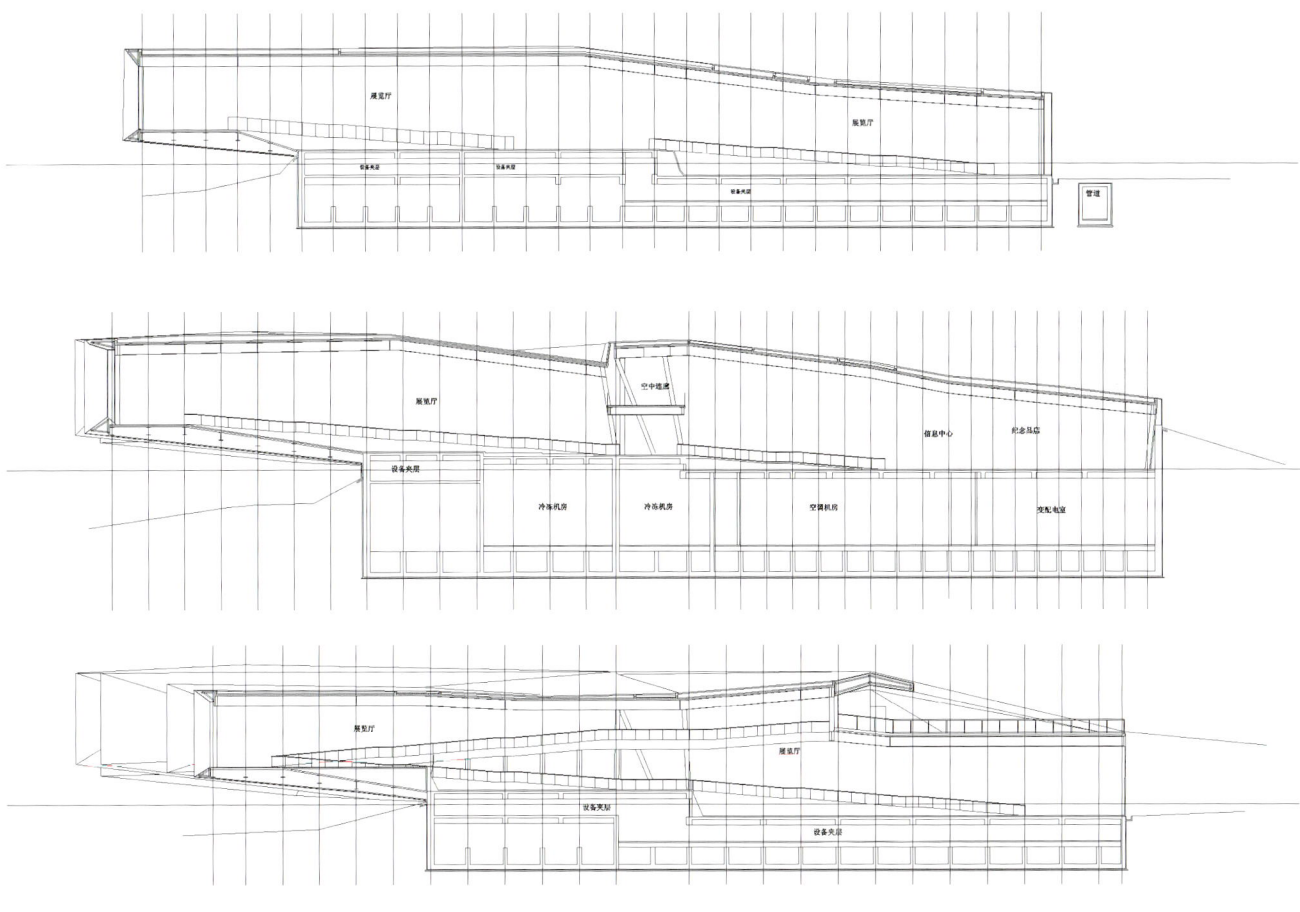

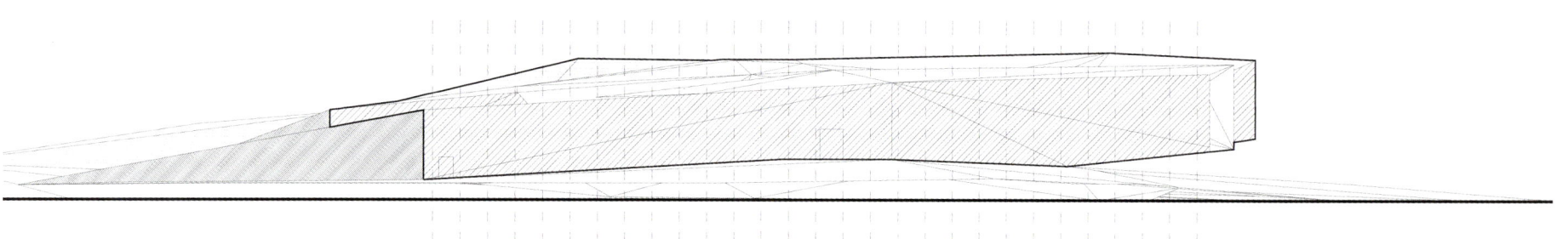
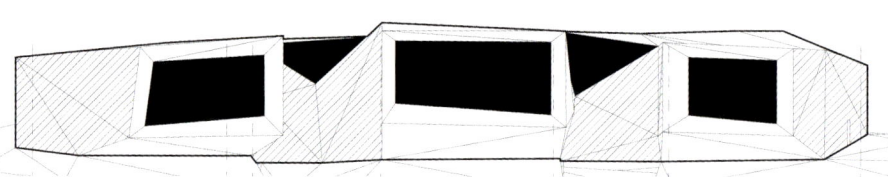

创意馆位于2011西安世界园艺博览会主轴线上，整个展馆结合码头和周边场地进行设计。展馆之前的广场通过着色与地景进行区分，延续了步道，有效引入了人流。展馆布局由三翼不规则几何体组成，自然地被分割成三个相对独立，并向水面延展的展览展示区域。游客可以经由一系列的室内展示坡道上达屋顶观景平台。建筑体块的连接处婉约细腻，与主体有明显区别；屋顶所使用的混凝土褶皱板结构满足了建筑对大跨度空间的需求，体现了优雅的结构形式；尺度宜人的扩张与收分完美诠释了功能与形式相结合的建筑艺术。从入口到展馆之间的过渡地块着色鲜艳，为了与过渡空间热闹的气质进行区别，建筑整体大面积使用纯色材质，营造出整体简洁大方的气质，也使得参观流线具有了韵律。在细部处理上，展馆通过运用青铜金属、石材及花园式种植屋面等不同饰面的无规则衔接处理，形成了错落有致、内涵丰富的艺术效果。

Theme Pavilion is located at the main axis of the Expo Site. Designers fully consider the landform of the waterfront site, utilize colors to separate space of the front plaza, introduce pedestrian element in access control. The Theme Pavilion consists of three irregular geometrical bodies, independently extending to the waterfront, so that the visitors could reach the observation platform on the rooftop through a series of interior display ramps. Connections between bodies are graceful and restrained, showing different characteristics from the main body. The folded plate concrete roof is elegant and meets the demand of architectural design of large space. The integration of style and function reveals architectural art. The bustling front plaza is of multiple colors. By contrast, the architecture is built of the same color materials in succinct style, lending the visiting route at a rhythmical element. Architectural details highlight layered art effects of abundant elements including bronze, stone, green roof etc.

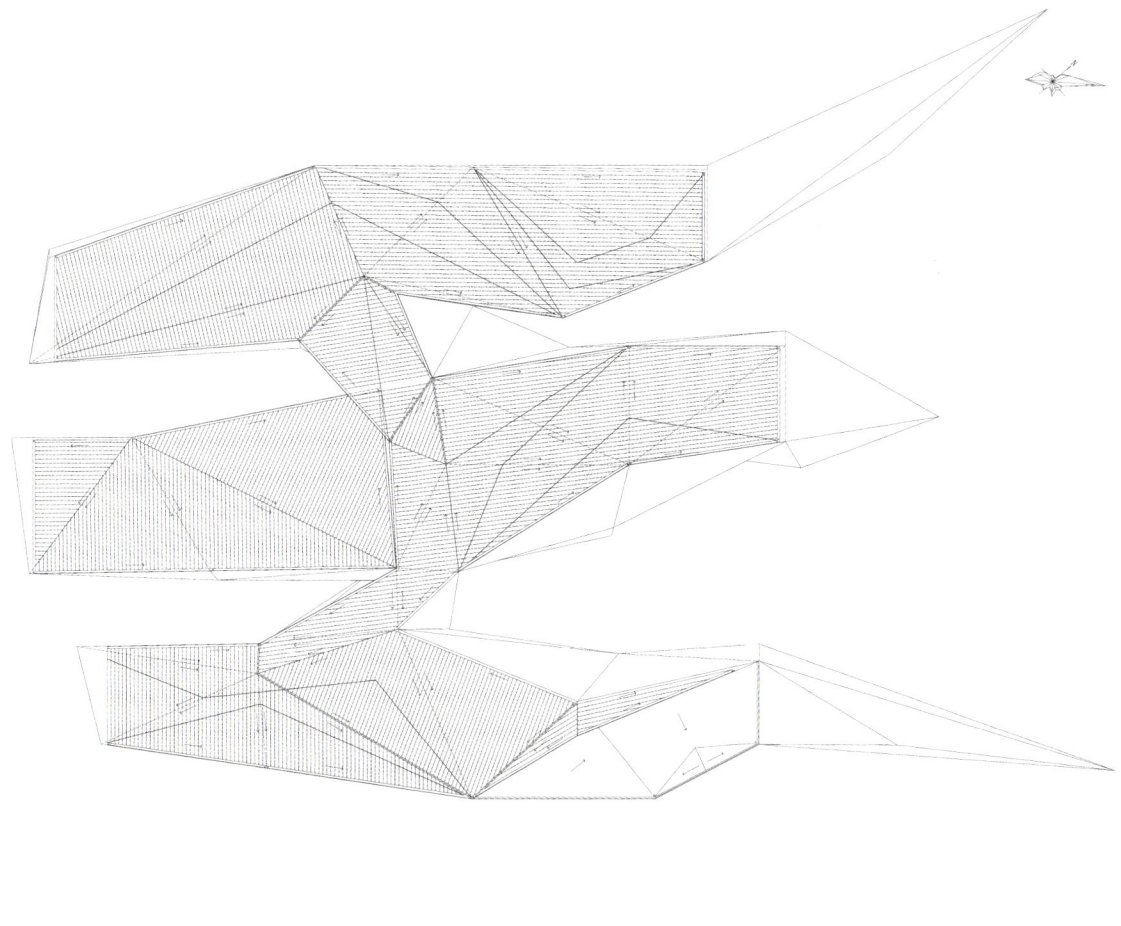

Greenhouse

核心区·自然馆

自然馆是2011西安世界园艺博览会的植物温室，用以展示多种类植物及其生态景观，以及不同气候带下的典型植物景观。设计参考了当地传统的建筑形式——窑洞，通过对窑洞居住环境和地方文脉的分析，进行了现代化的转译，展馆大部分位于地下，从高度上，视觉上弱化了建筑的体量，保证在建筑室内可以从不同标高领略湖面和对面花园的美景。建筑立面选用玻璃，木材与少量混凝土结合，朴素的材质让建筑与地形进行结合，而丰富的切面让建筑显现出梦幻般的氛围。

从广运门到自然馆，Plasma缔造了一条富有深刻内涵的轴线而非孤立的节点。他们设计的不仅仅是三个重要建筑，更将周边的场地联系了起来，与建筑之间形成富有逻辑关系的整体。Plasma将"流动的花园"作为设计的主体概念，通过人造景观与自然景观有机结合的手段，将整体设计融于场地的肌理与界面，并且从"线"的连贯性出发，解决了人流疏通的问题，与设计中包含的水资源回收系统一起，立体地阐释了"连贯"的概念。

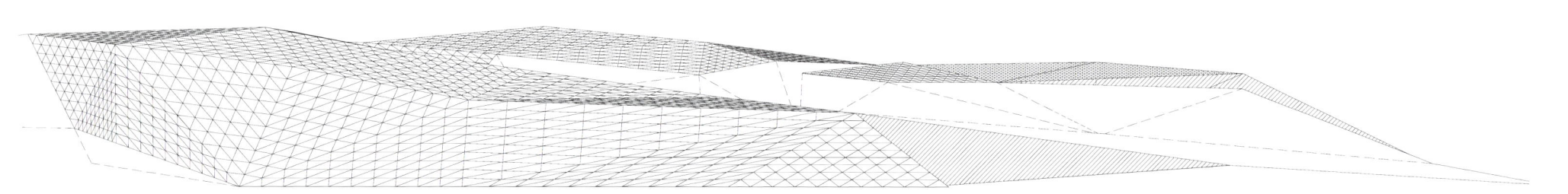

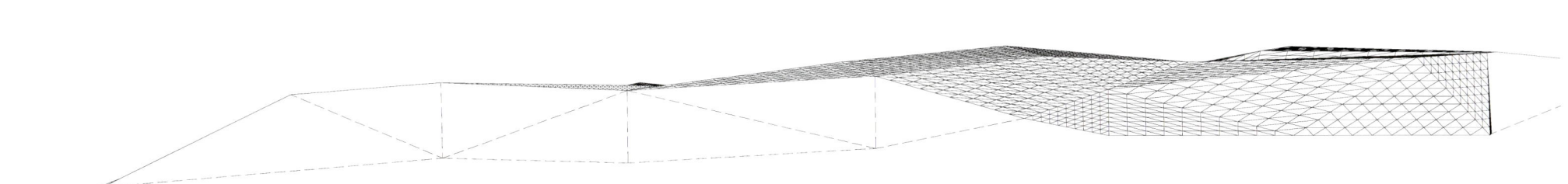

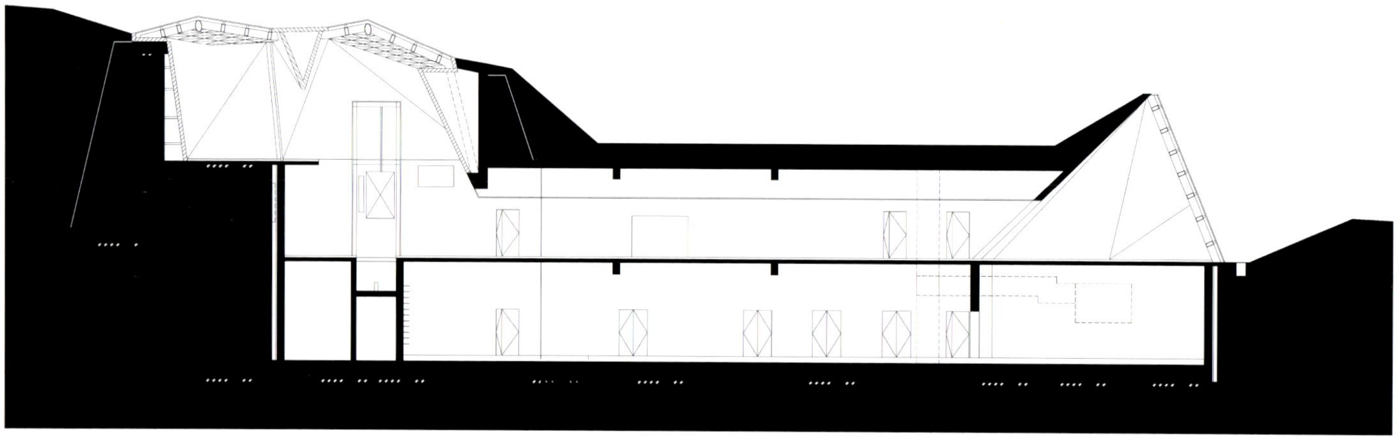

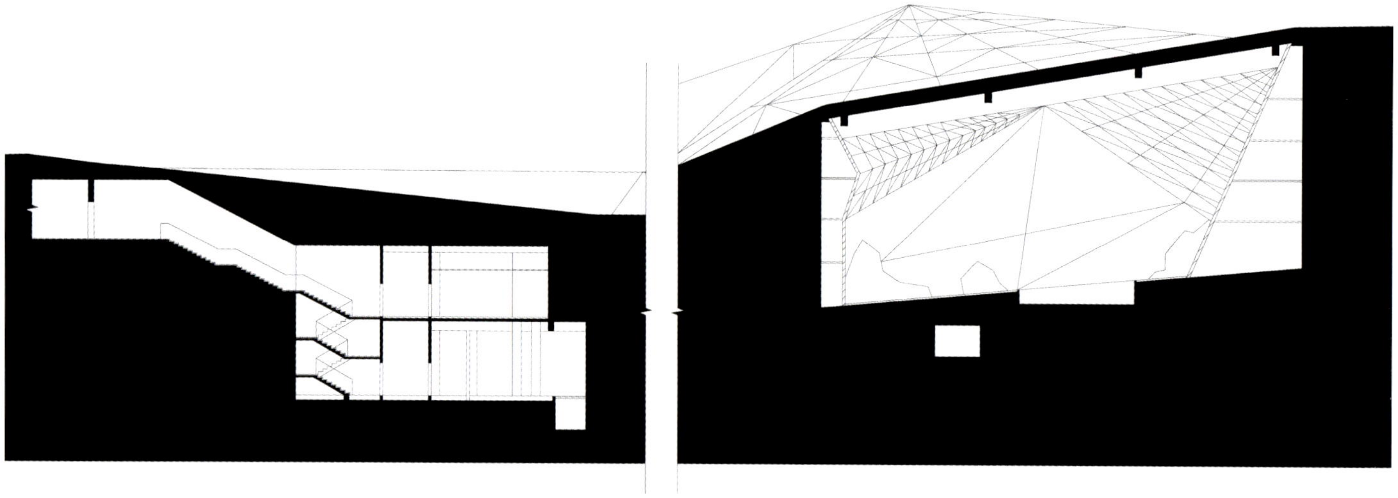

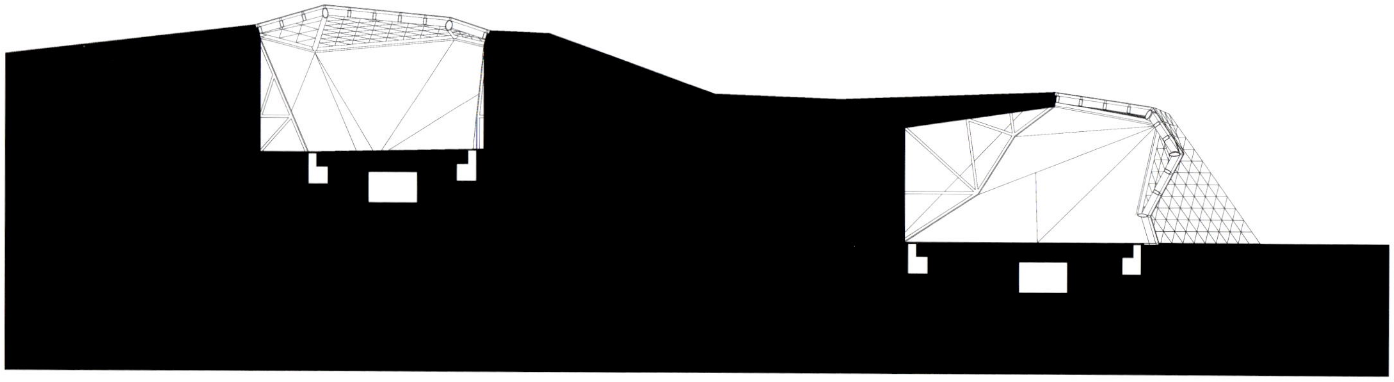

Greenhouse, the botanical garden of International Horticultural Exposition 2011 Xi'an displays various species of vegetation, the eco-landscape and typical plants in specific climatic conditions. Based on the analysis of cave dwelling and local culture, local architectural style – cave residence – is translated into modernity. Most parts of the Greenhouse are underground, and the architectural volume becomes small in height and visual effects, ensures a wonderful view of the opposite garden and lake outside. The architectural façade is built of glass, timber and a small amount of concrete. These simple materials incorporate architecture into landform, while varying façade features a dreamlike world.

Plasma Studio creates a rich axis of connectivity with nodes like Guangyun Entrance and Greenhouse. Its design is not only related to three landmarks, but also combines with the site topography forming a logistical relation between buildings. With the theme "Flowing Garden", the design is incorporated into the texture and interface of the site through the combination of man-made and natural landscapes. In perspective of the "Line", the "Connectivity" concept is interpreted in water recycling system and traffic orientation.

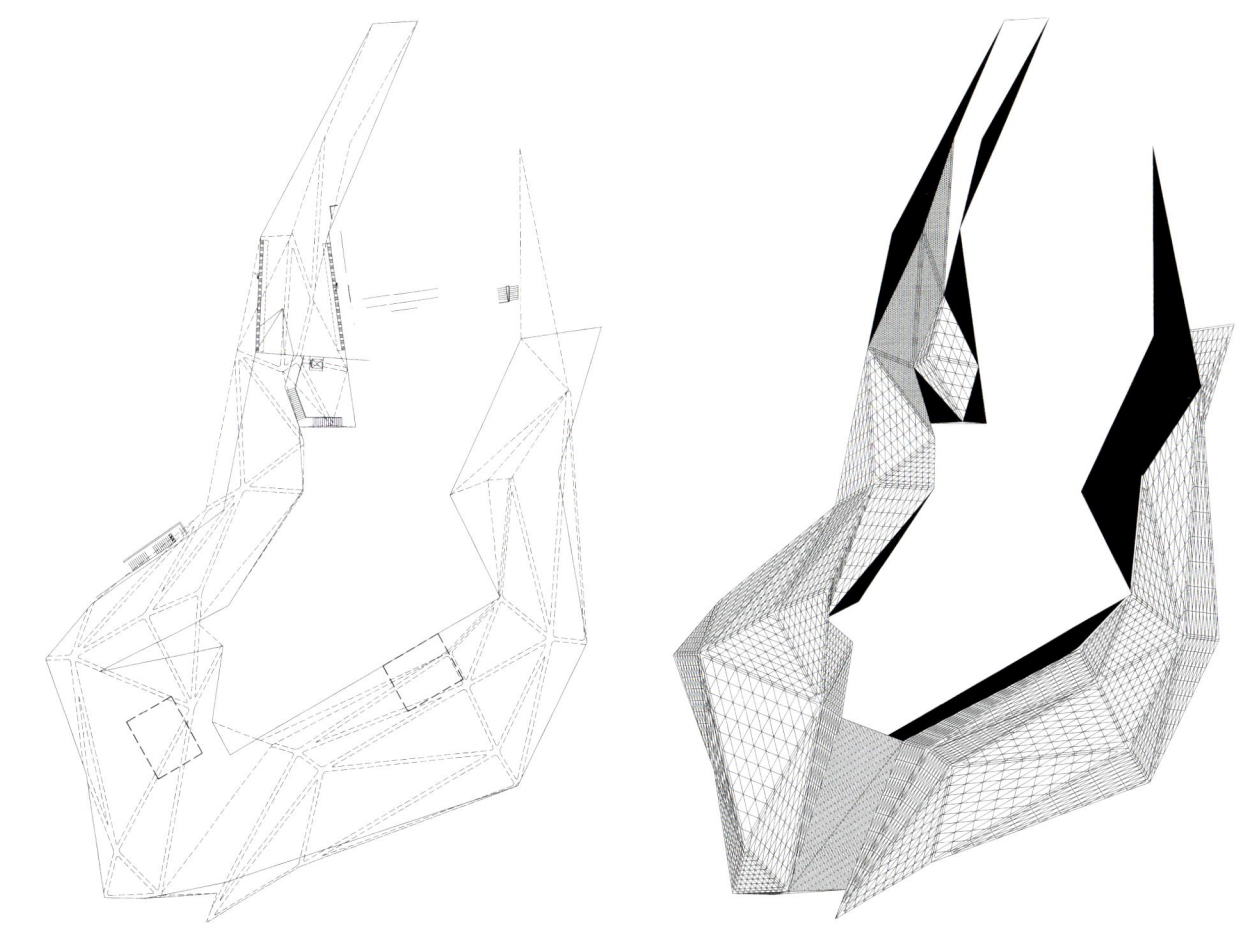

Southeast Asian Street

滨水区·椰风水岸

椰风水岸位于园区北部，为典型东南亚风情文化特征的服务建筑群。总占地面积33,874平方米，总建筑面积4,106.6平方米。

椰风水岸营造出自然、充满活力和风情的异国区域。采用华丽的东南亚传统——多层屋顶作为建筑表达形式，也有典型的东南亚建筑风格，简朴、浪漫，尽显其特有的地域性符号。椰风水岸秉承了自然、健康和休闲的特质，椰子树、棕榈树等高大的热带植物配以低矮的灌木丛，营造出充满生机的热带氛围。静谧的凉亭、曲折的栈桥、围绕着楼体的蜿蜒水街，造型多样的绿色植物与建筑在荡漾的碧波中交相辉映，各个元素之间浑然天成。 椰风水岸总体设计提取东南亚地区建筑规划及园林风格特点，结合东南亚乡村及城市典型的肌理特征和能反映东南亚建筑形式的符号（轴线、广场、曲线与直线街区组合、港口、内庭院空间等），置身其中，能充分感受到东南亚异国文化风情的氛围。

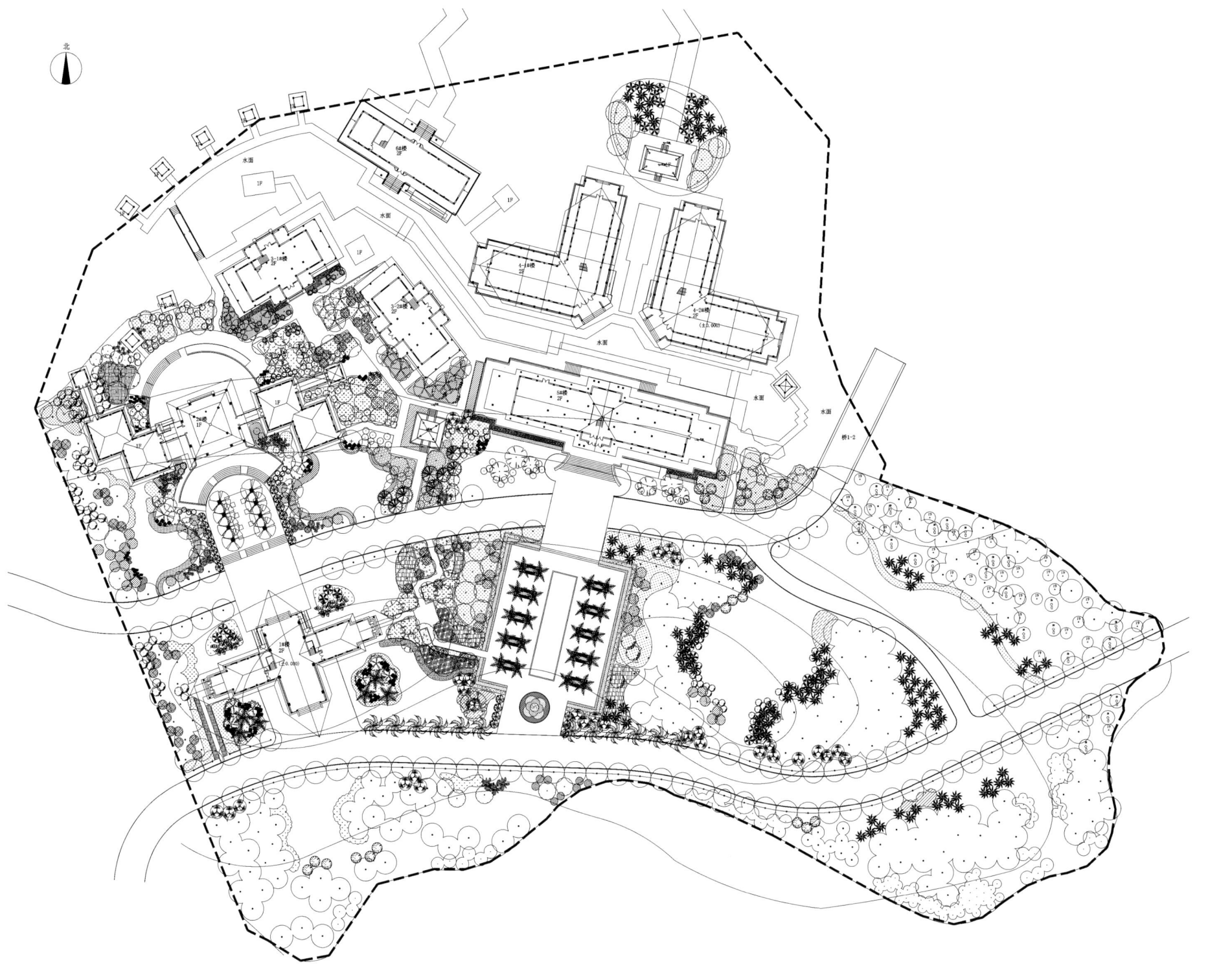

The Southeast Asian Street, located in the northern Expo Site, is a service-oriented architectural complex of typical Southeast Asian style. It covers an area of 33,874 m² and the total building area is 4,106.6 m². The natural Southeast Asian Street is alien and vigorous. There are multi-layered roof buildings that mimic the traditional architecture of Southeast Asia. Some other buildings are of typical Southeast Asian architectural style – simple and romantic, fully showing the particular local feature.
The plant of tall tropical Coconut and Palmae and low shrubs brings lively tropical atmosphere to the natural, healthy and recreational Southeast Asian Street. All elements, including tranquil pergola, zigzag bridge, undulating building-oriented water street, multi-shape plants and architecture, are interwoven naturally and reflected in the water. The general design of Southeast Asian Street extracts the characteristics of architectural planning and landscape style in Southeast Asia. Associating with typical rural and urban textures, and architectural details of Southeast Asia (axis, plaza, the combination of curved and straight lines, harbor, inner courtyard etc.), it presents tourists strong exotic culture atmosphere of Southeast Asia.

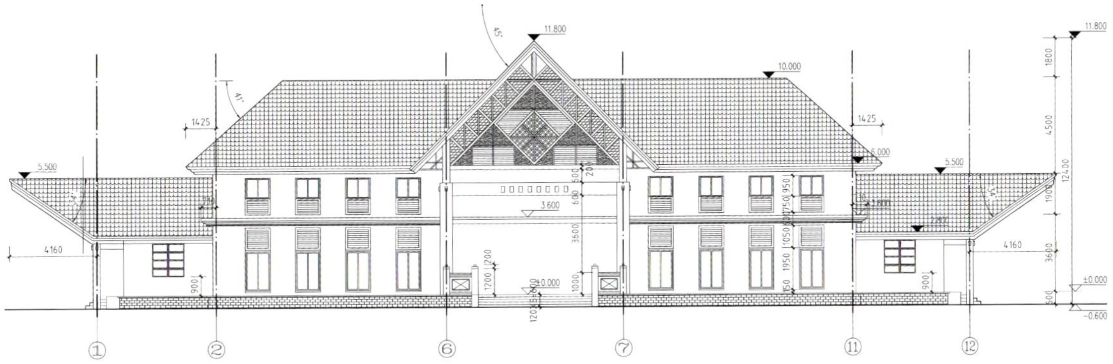

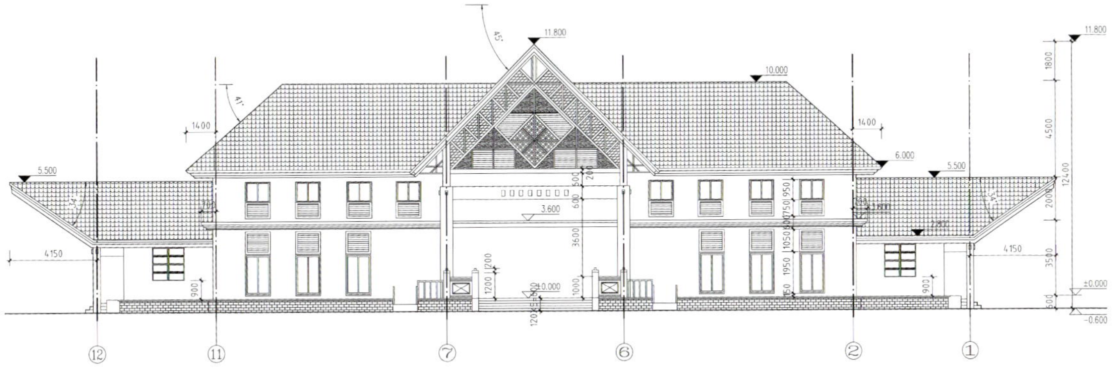

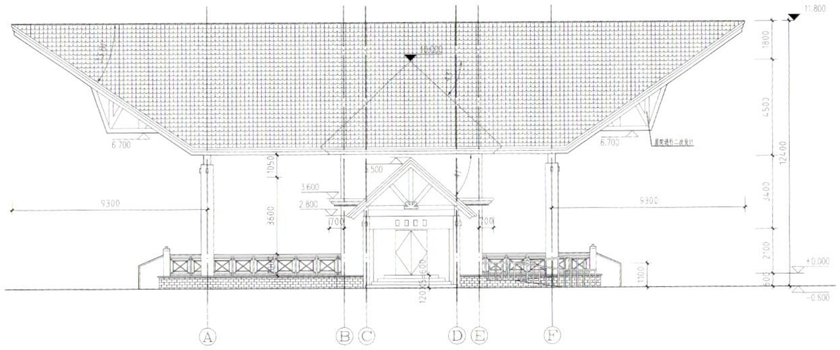

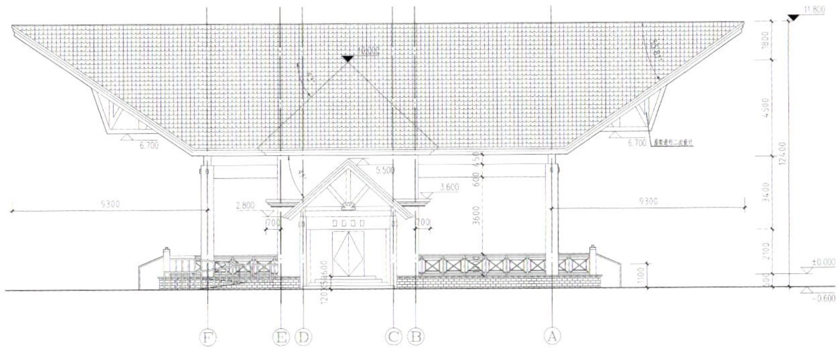

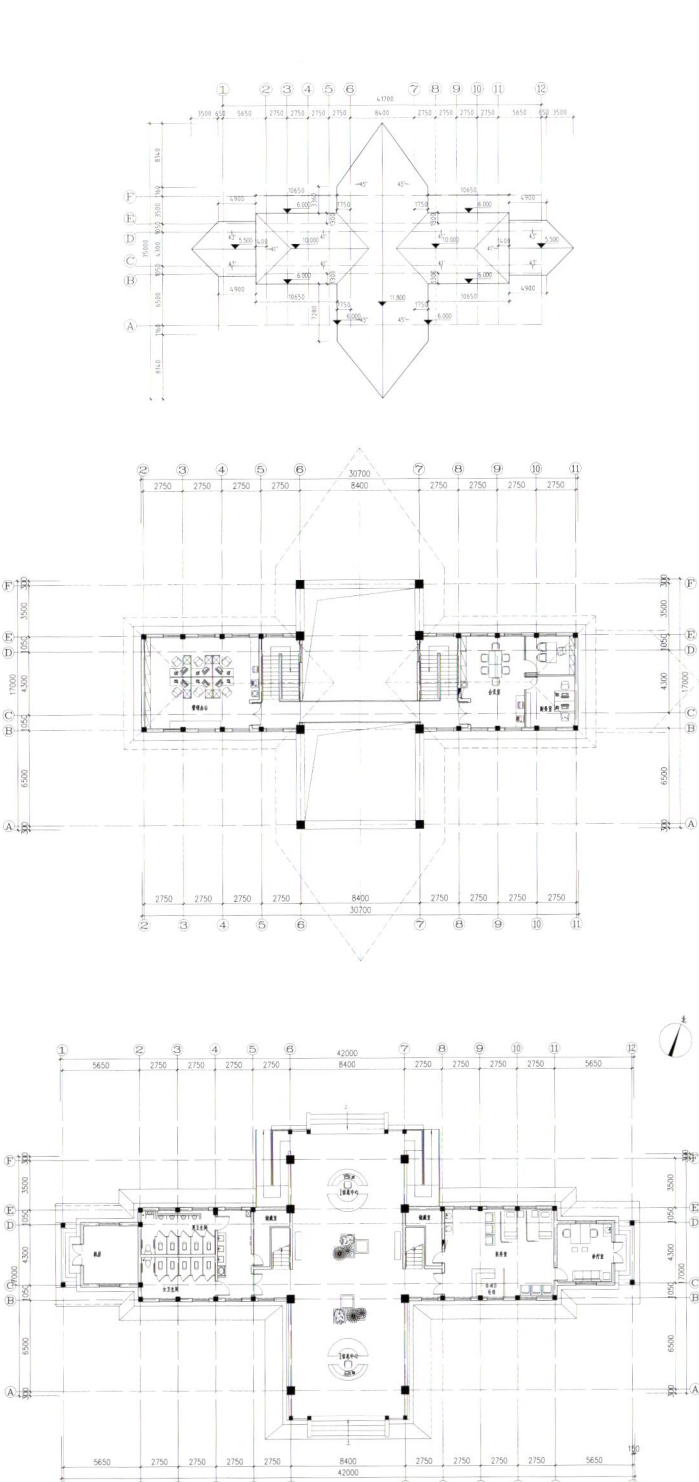

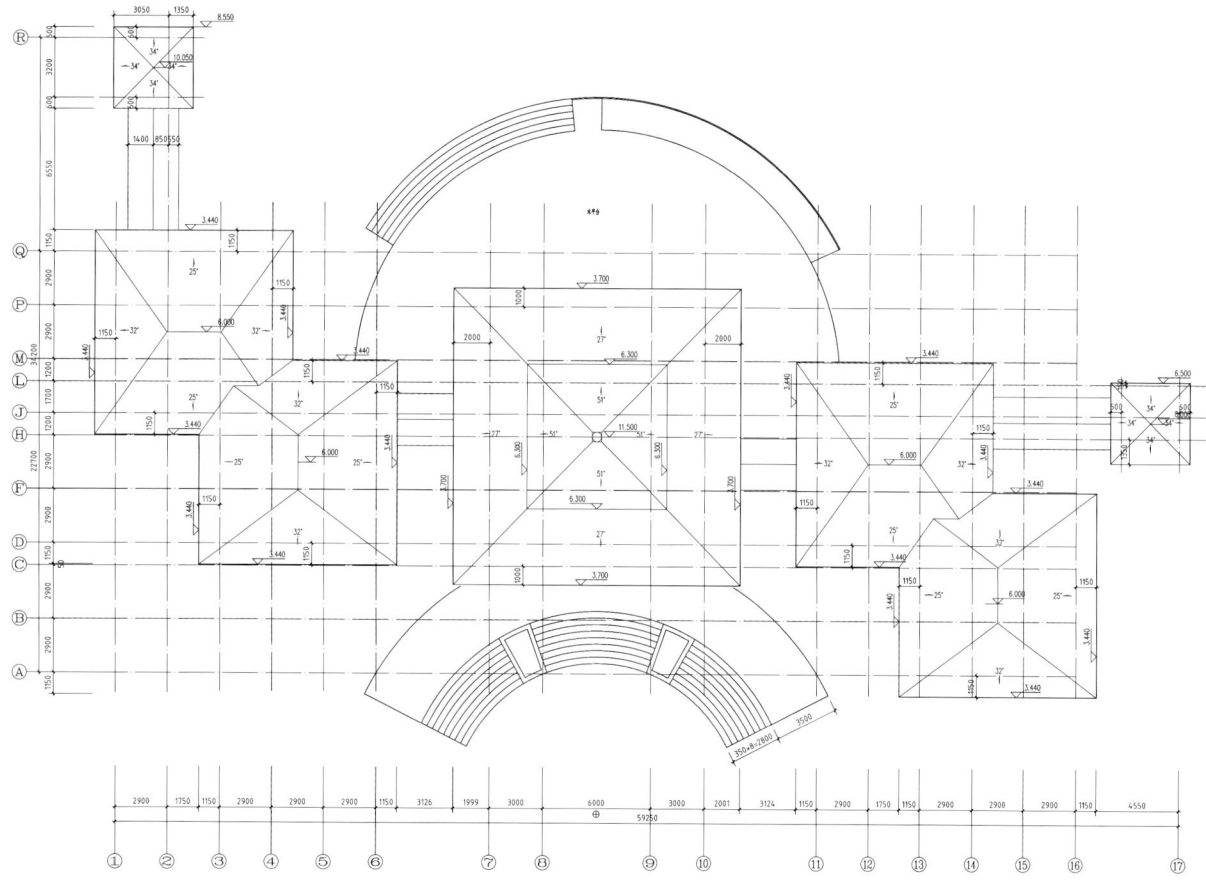
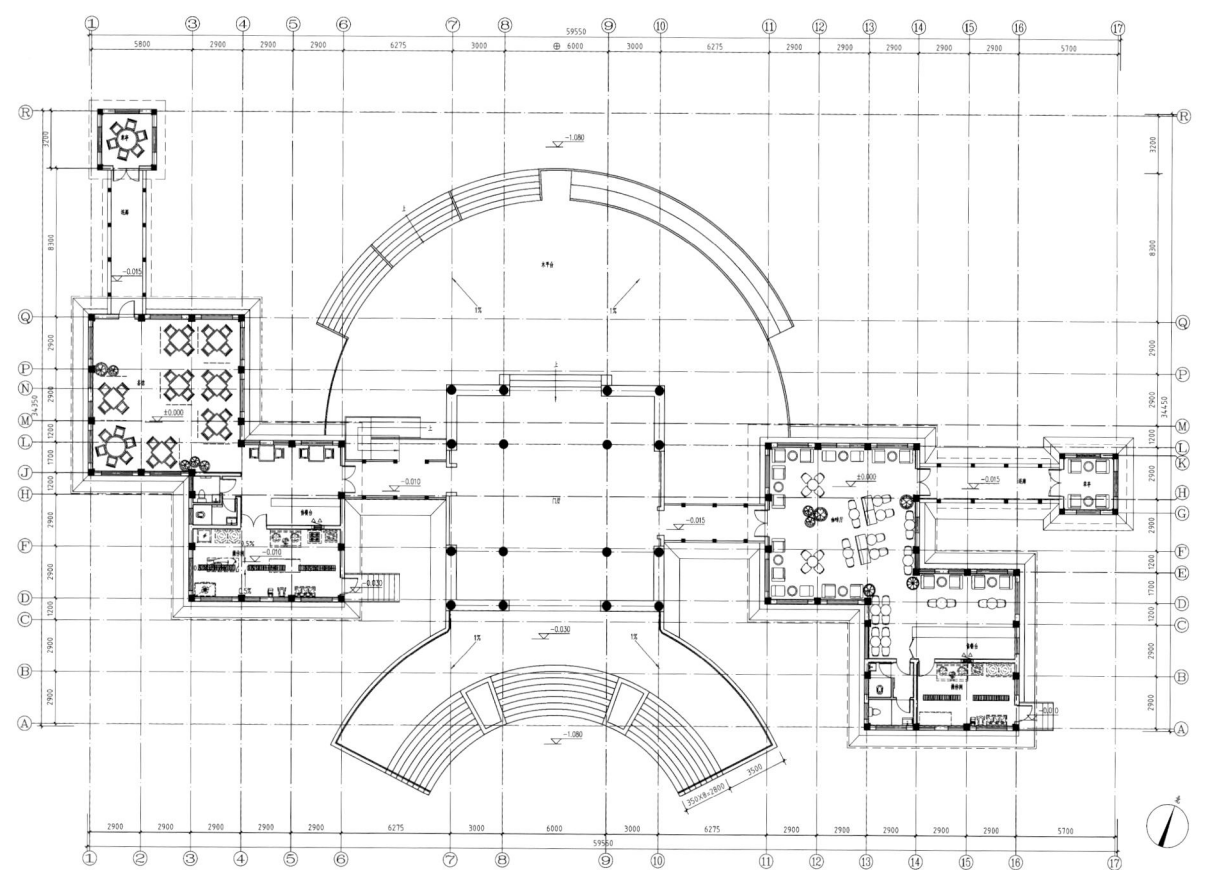

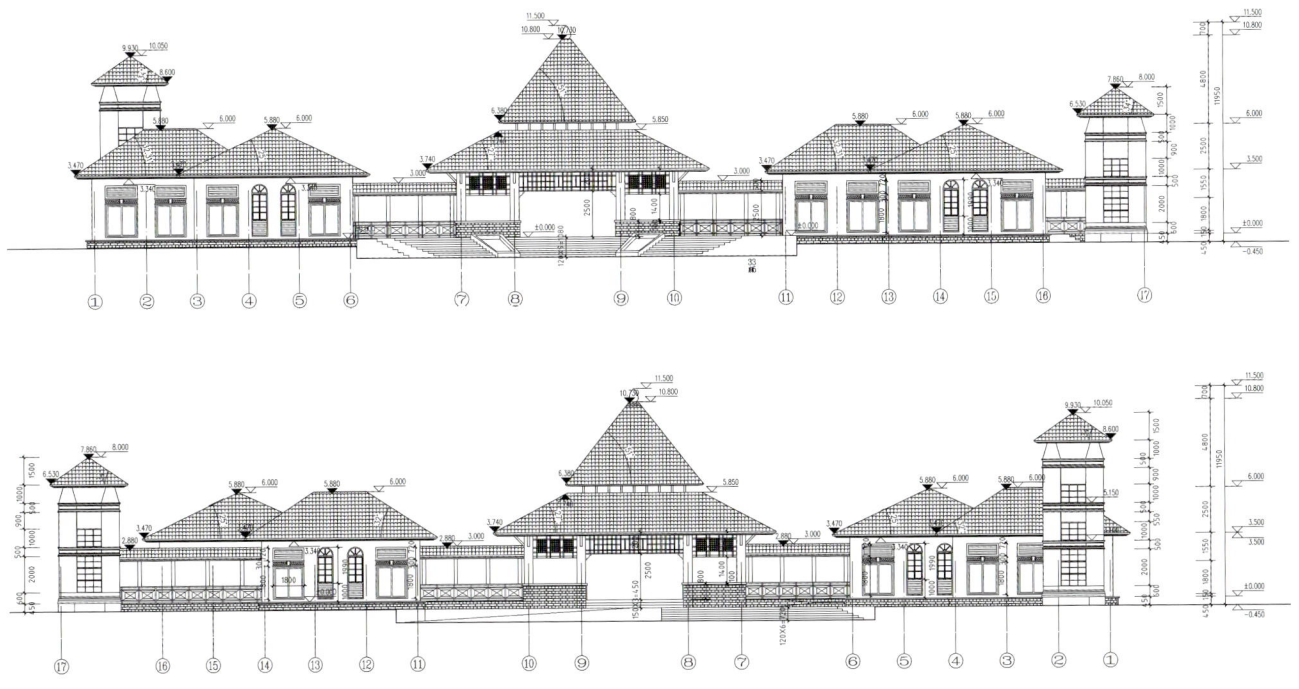

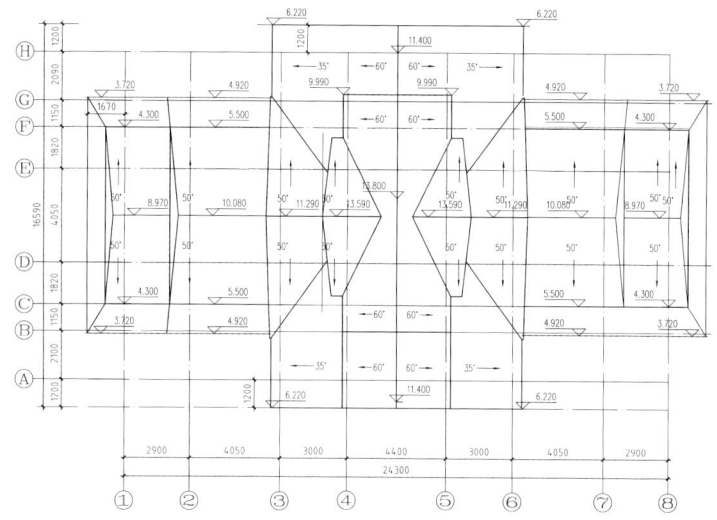

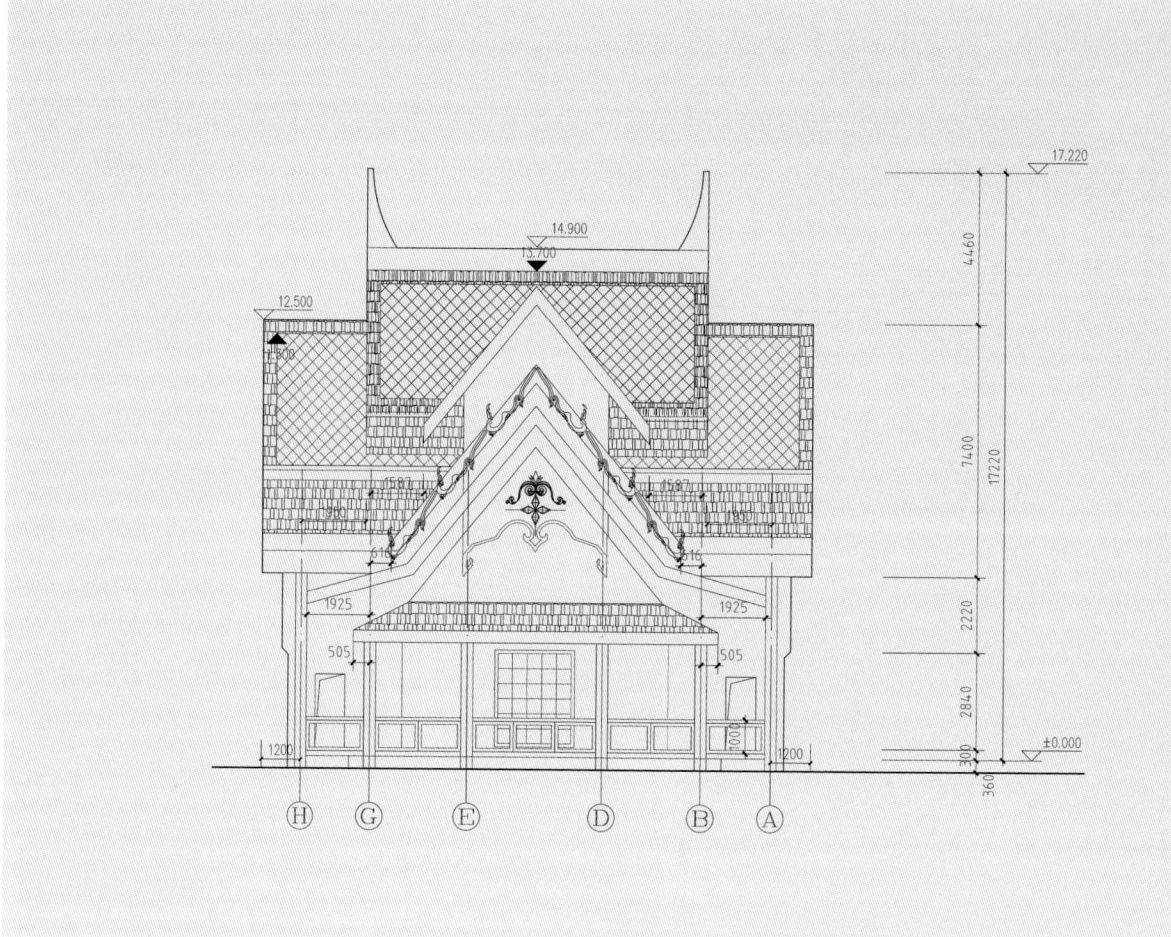

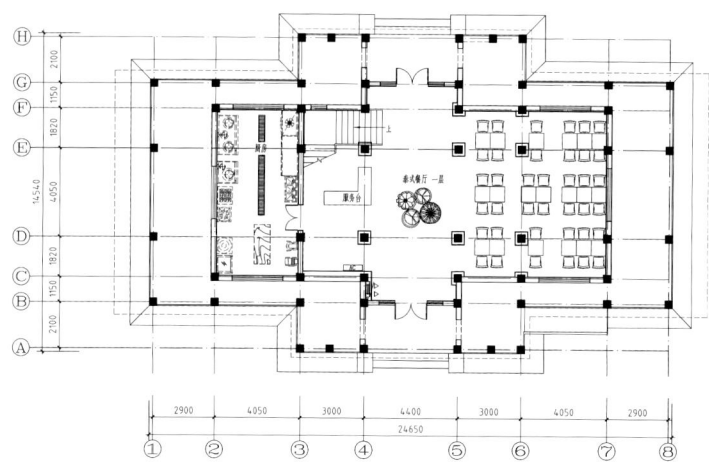

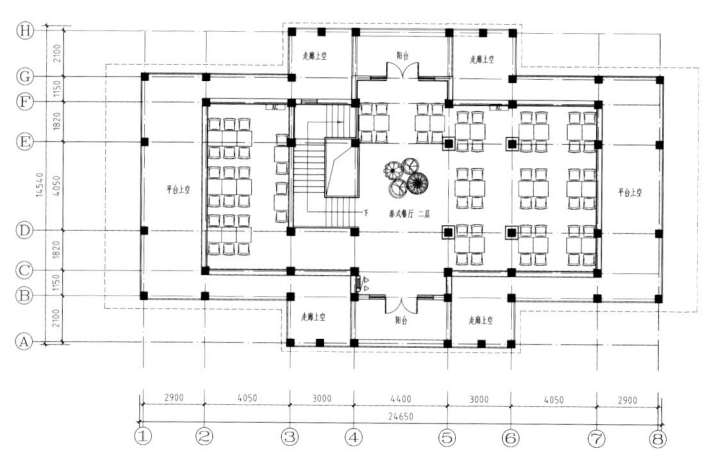

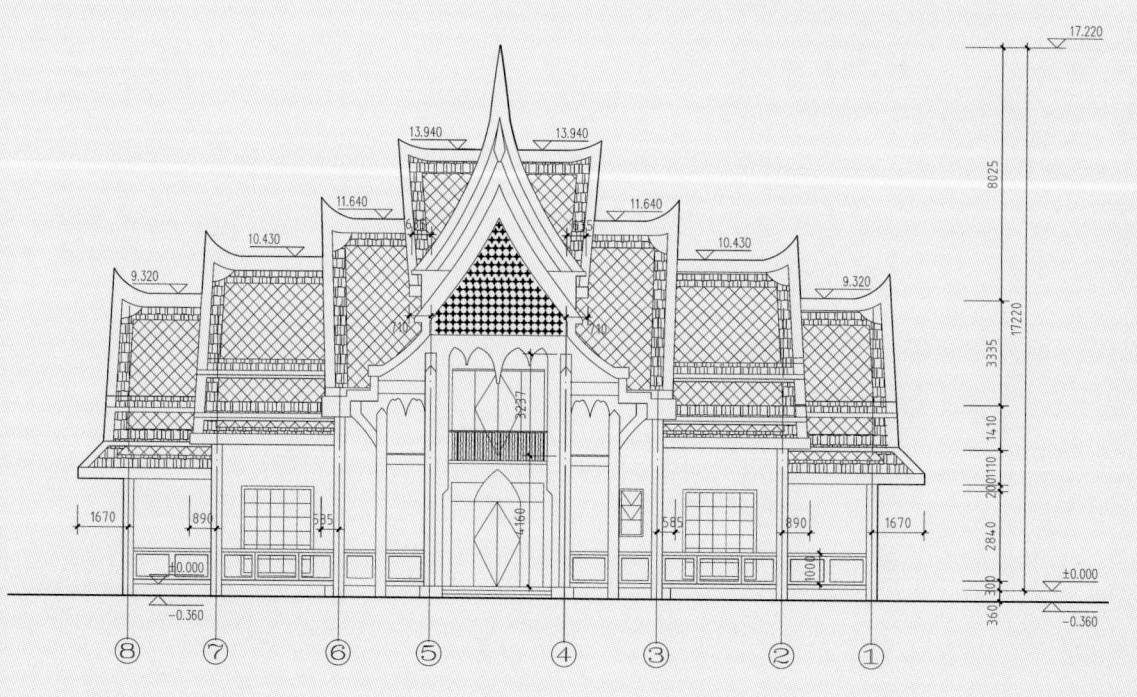

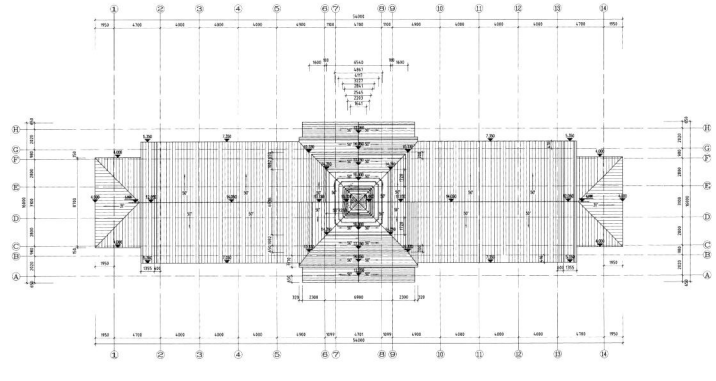

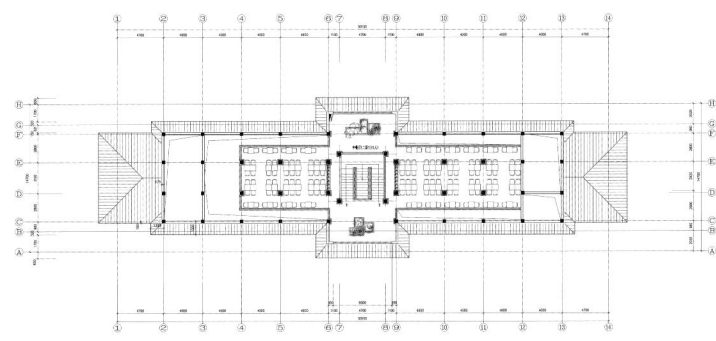

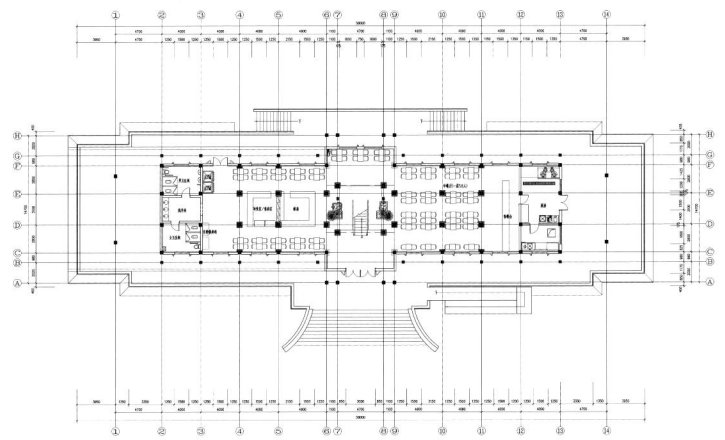

European Avenue

滨水区 · 欧陆风情

欧陆风情地块位于园区东部，紧邻次入口。总占地面积38,444平方米，总建筑面积6,081.7平方米。

欧陆风情1#楼的建筑形体造型多样，烟囱、圆塔、钟楼的设计结合了欧洲典型的建筑特点，建筑饰面材料的表现要求纯朴自然，返璞归真，也是本服务区内建筑形式最为多样的建筑单体。

欧陆风情3#楼毗邻水岸，两座建筑由室外游廊连接。姹紫嫣红的花圃点缀在前庭后院，别有一番景致。以游廊为界一分为二，前为中餐厅，后为西点咖啡屋。而游廊则是您闲庭漫步休憩的最佳场所。

景观设计结合了欧洲传统皇家园林风格特点及欧洲城镇的规划典型特征，重新将欧洲传统建筑形式进行拼贴混搭、组合，形成亲切近人的建筑空间尺度，营造出一个典型欧洲小镇建筑群，游客不仅可以享受餐饮休息接待等服务，同时还可以身临其境地感受欧洲异域的传统文化风情。

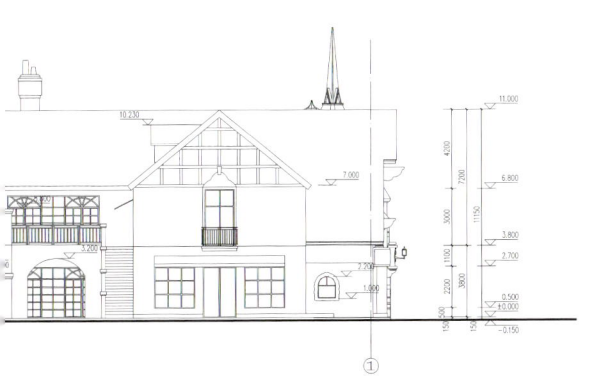
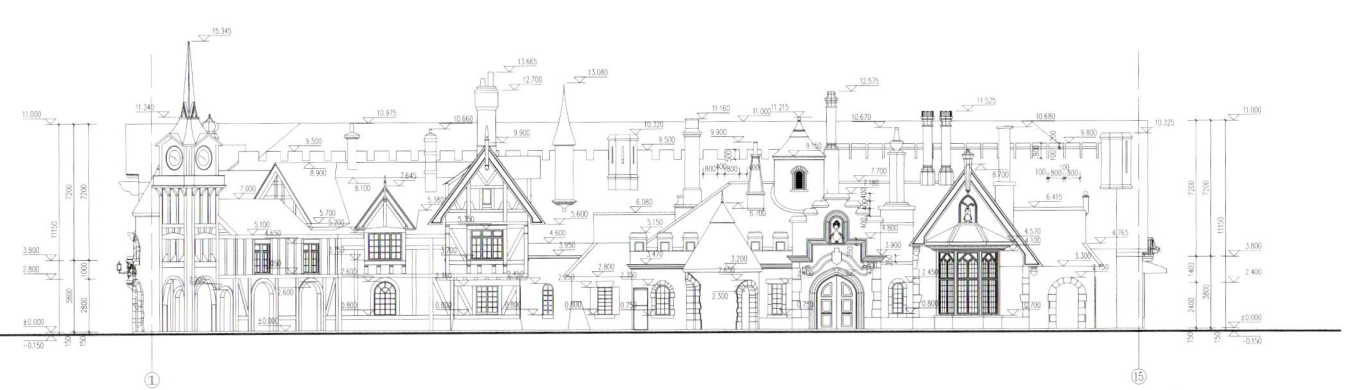

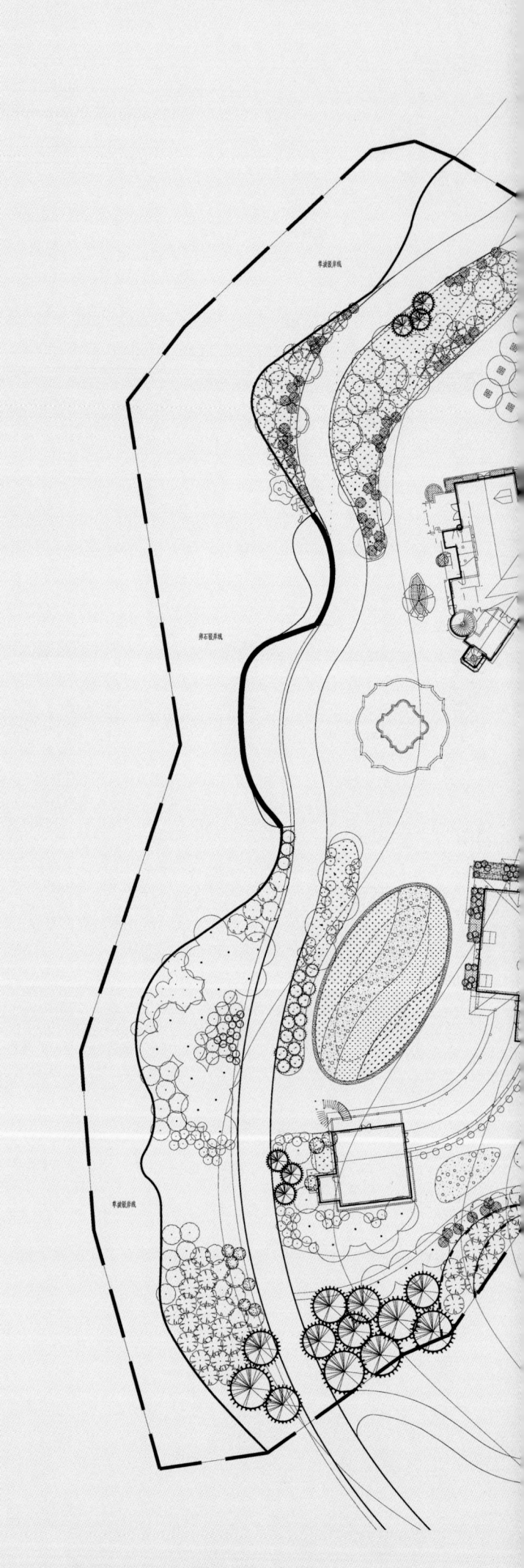

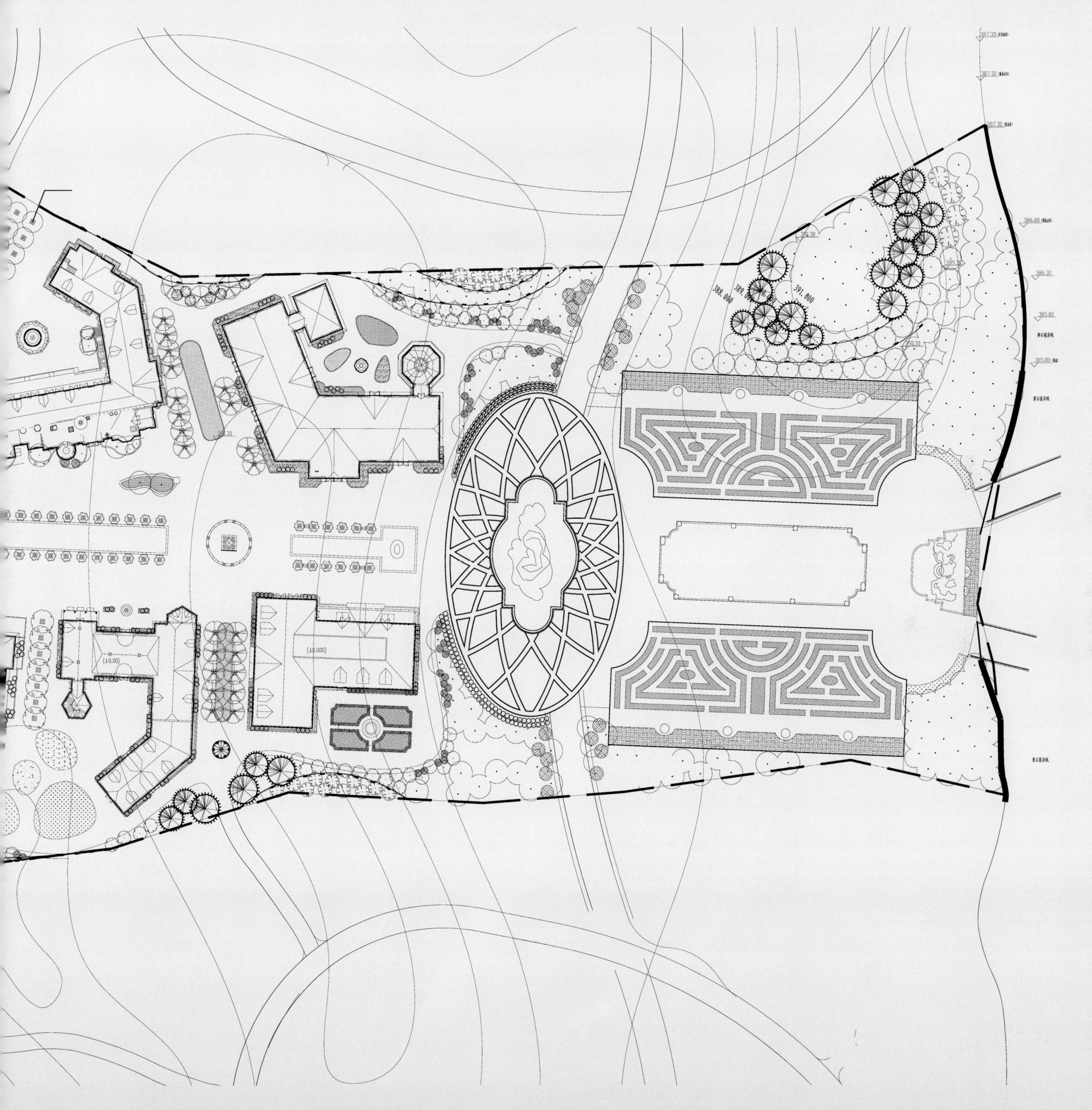

The European Avenue is located in the east of the Expo Site, closely adjacent to the Secondary Entrance, with a total area of 38,444m² and construction area of 6,081.7 m². With natural materials, designers melt typical European architectural elements into chimney, round tower and bell tower of the 1# building. The whole building is presented with diversified architectural styles, which freshens visitors' eyes in this Service Zone. The waterfront 3# building is two interventions connected by a veranda. The veranda divides it into two parts— the front courtyard with Chinese restaurant and the backyard with café, colorful nurseries dotting throughout yards. The visitors are invited to walk through the veranda and overlook the various gardens.

The landscape is incorporated with the elements of traditional royal gardens and typical lawn planning of Europe, which translates the design with a new architectural appearance of combination. The visitors not only get close to the traditional European culture, but also dine and relax among buildings.

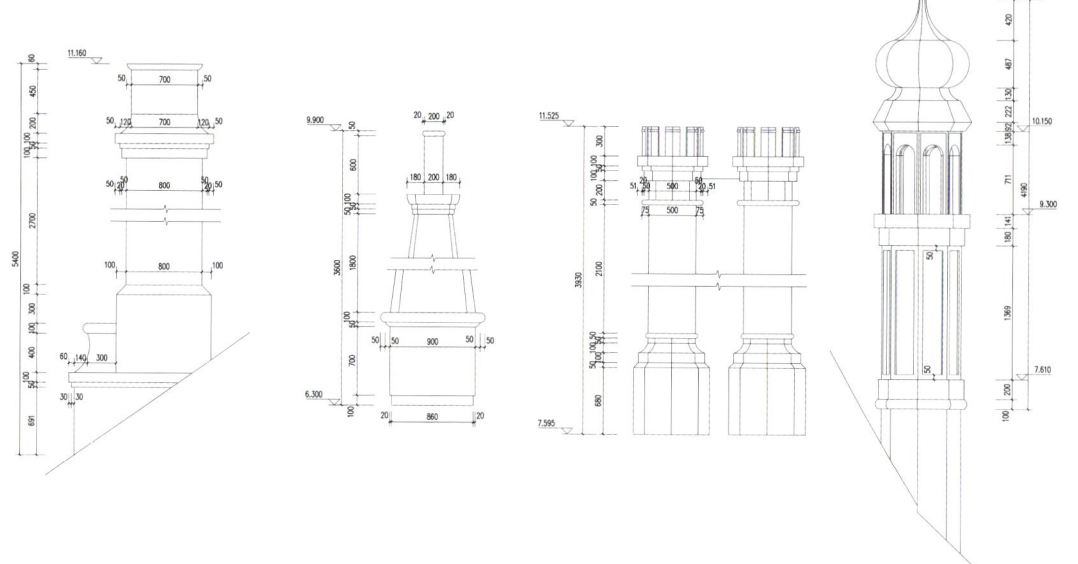

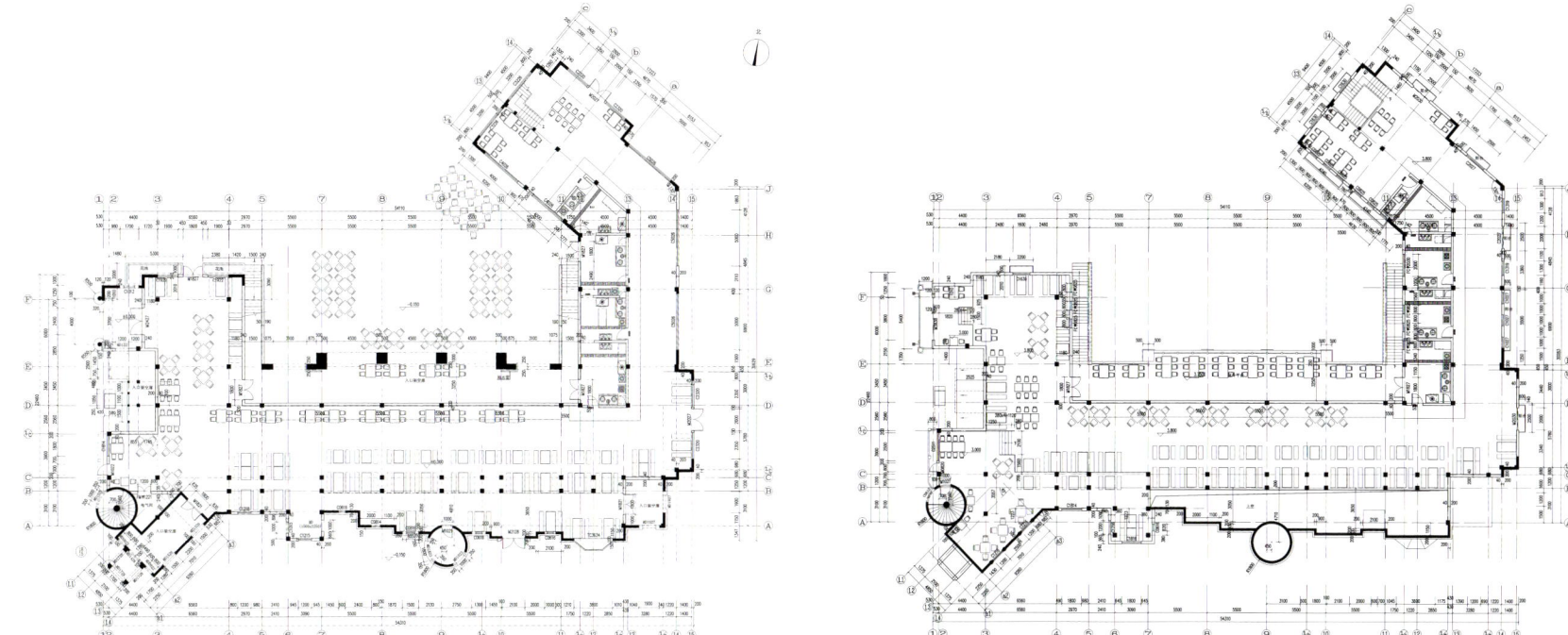

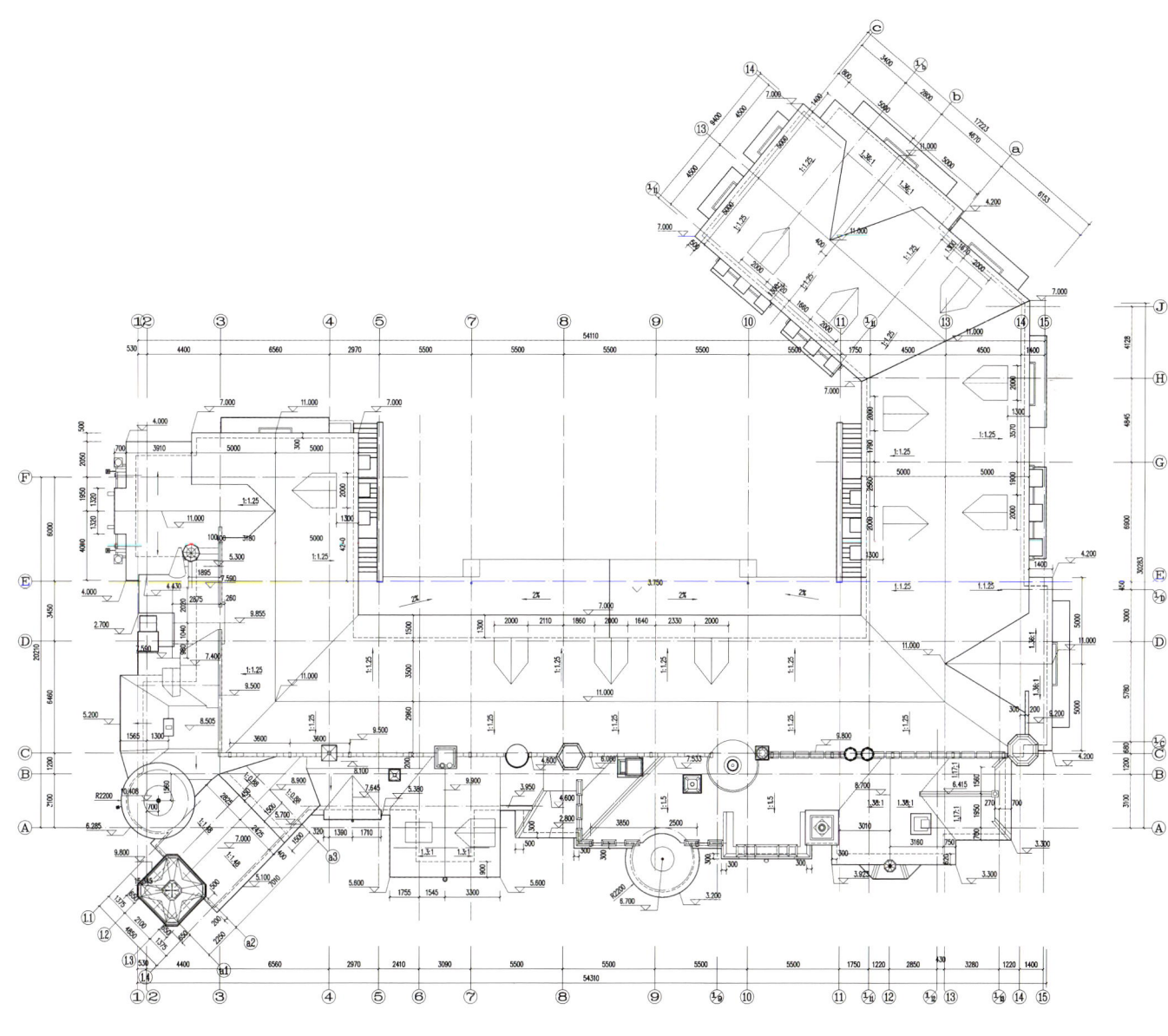

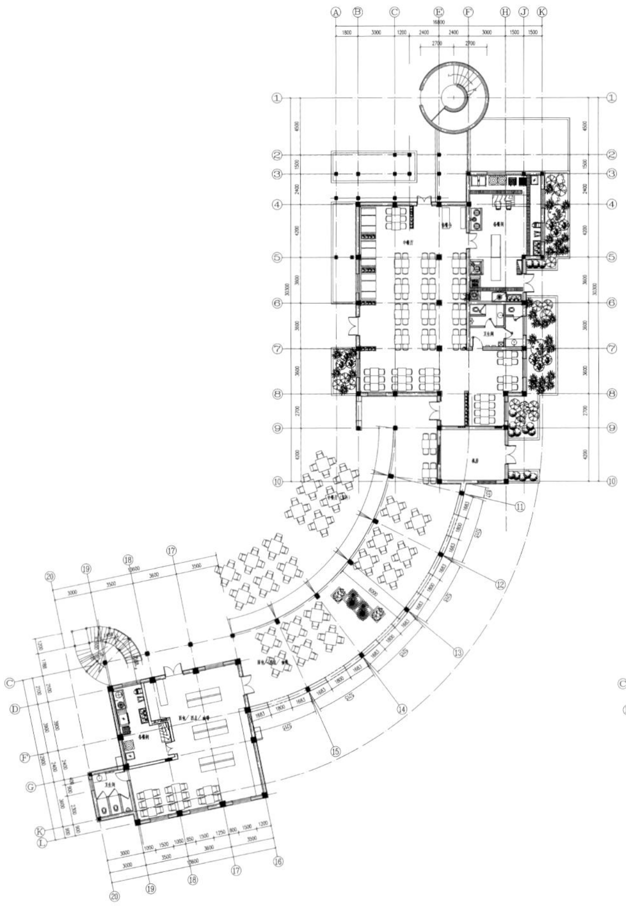

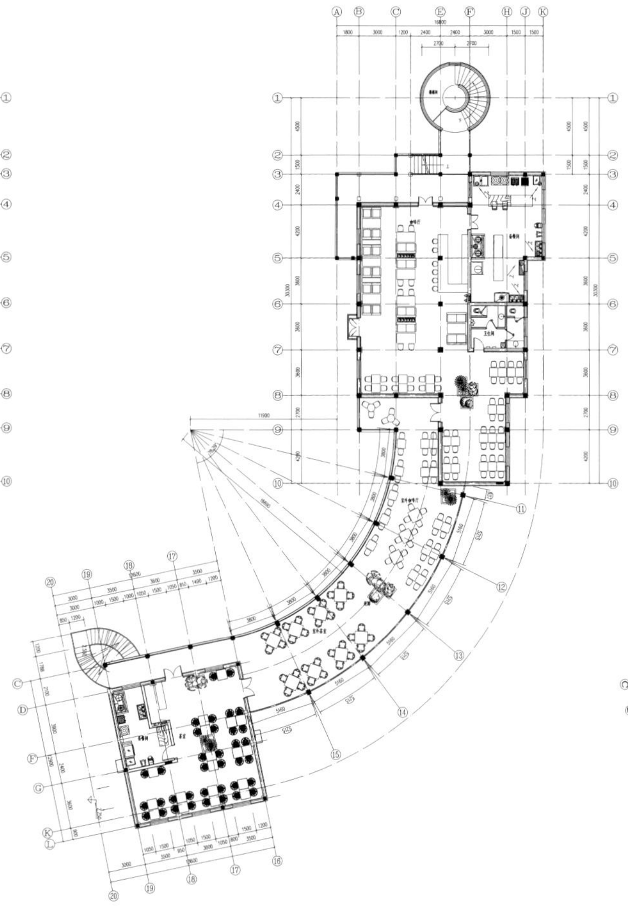

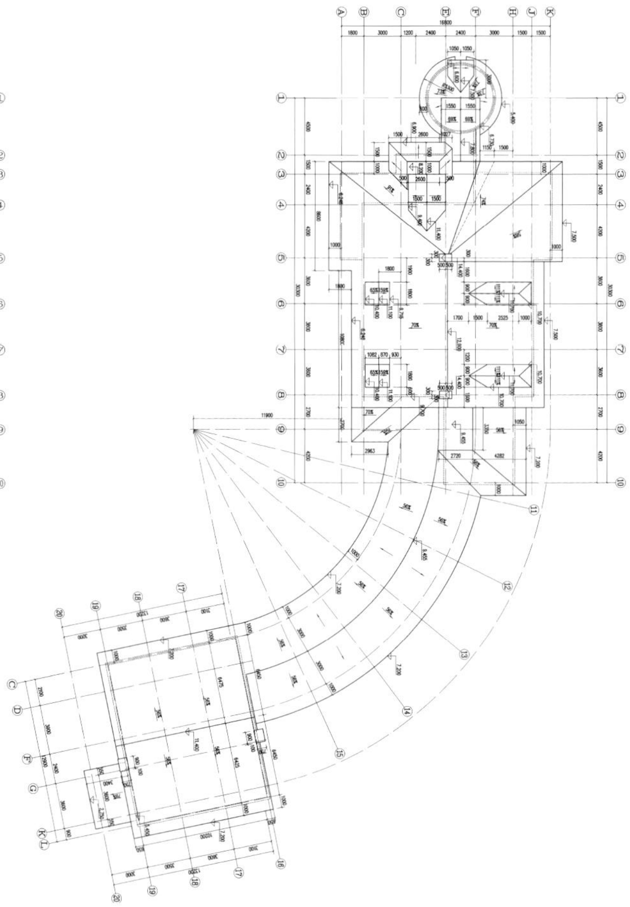

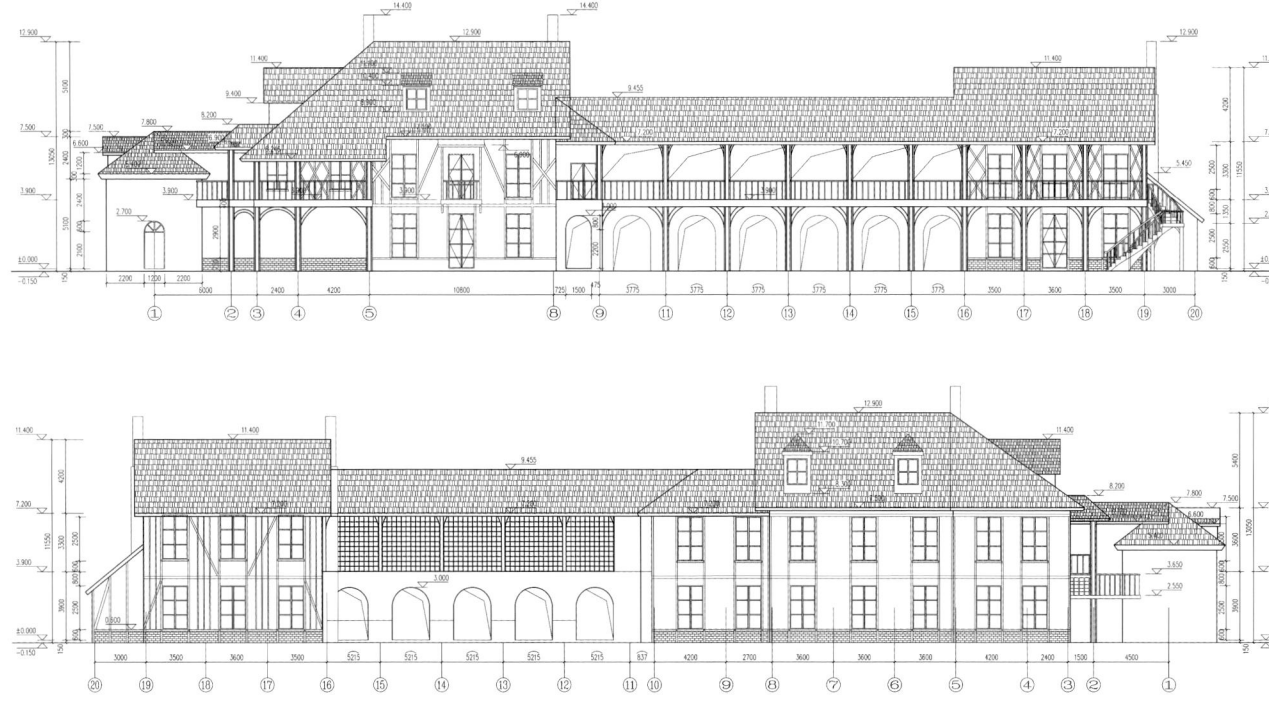

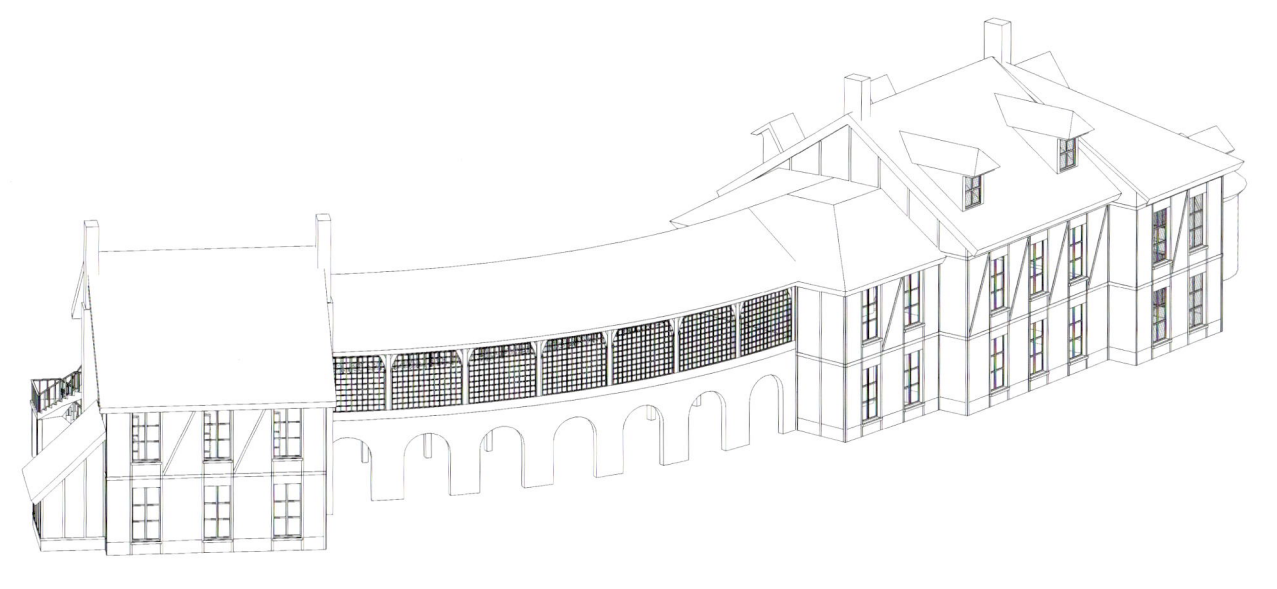

Romance by the Ba River

滨水区·灞上人家

"灞上"是当年刘邦项羽楚汉争霸中鸿门宴的所在地，唐玄宗曾在此举办规模盛大的水运博览会。可以说，这块宝地既有丰厚的历史与文化，又有极佳的自然景观。作为世界园艺博览会的三大服务区之一，本项目试图提供给人以特别的建筑与景观体验，让游人更多地体验到陕西关中的人文精神和自然风貌。

"灞上人家"位于园区南部，南邻灞河堤顶路，北接园区一级路，定位为具有本土文化建筑特色的服务建筑群，总用地面积约2万平方米，建筑以中国传统四合院为蓝本，采取四水归堂的形式，并从建筑尺度的延展，表皮的变异与功能的变通作为切入点，力图使建筑摆脱简单的文化标识，使之成为一个个容纳新生活，催生新故事的盒子。在群体组织上，采取簇群式的布局方式，让建筑最大程度的亲近自然，一条宽约8-9米的溪流从用地中部穿过，营造出自然的、田园的、充满野趣的景观环境。景观种植品种纯净，主题明确，并以地方乡土植物为主。乔木种植方式模仿村落宅前路旁的自然种植，灌木种植成片、成簇，模仿河岸两侧慢坡优美的自然景象。

看起来丰富多变的建筑群体实际上由一个单元通过翻转、镜像组合而成。因此，看似变化很多，实则高度重复，这意味着建筑的构造和节点都是相同的，这给设计和施工都带来了很多的方便。再者，施工时可以先做好一个单元，推敲其细部，成熟后再同时加工其他的单元，从而形成了高效科学的施工方式。这样的方式，体现了低碳建造的建筑观，呼应了世园会关注生态环境资源的精神。

The site by the Ba River was the location where Xiang Yu invited Liu Bang to "Hongmen Banquet" (a historical event memorialized in Chinese history) during the Chu-Han Contention (206 - 202 B.C.). Emperor Xuanzong of Tang once held a large-scale water transport exposition on th site in the Tang Dynasty (618 - 907). The promising land is of rich history, culture, providing fine natural landscape. Romance by the Ba River is one of the Three Special Service Areas in International Horticultural Exposition 2011 Xi'an. It offers the visitors a unique experience of architecture and landscape as well as the spirit of residents and the natural scenery in the Guanzhong region.

The 20,000 m² Romance by the Ba River is located in the southern Expo Site, to the south by Diding Road along the Ba River, to the north by the arterial road of the Expo Site. This architectural complex aims at displaying local culture and architectural styles. Siheyuan (Chinese quadrangles) – a courtyard surrounded by four buildings – and the "All Water into Hall" layout of Jiang Nan residential houses (the south region of the Yangtze River) are rendered and extended to a new architecture. Being a cultural symbol, it also incorporates transformed surface and changed function to embrace new life and things. The cluster layout makes it get closer to nature. A creek, between 8 - 9 meter wide, runs through the middle part and creates a natural, pastoral and wild landscape. The plant palette focuses on native vegetation. Designers imitate the natural plant pattern on the roadside/ in front of the countryside houses in planting arbor, while mimic the natural scenery on the riverside, clusters of shrubs growing throughout the gentle slope.

The architectural complex seems to be changing, but it is in fact composed of similar units in inverted or mirror imaged. These units are identical in height, structure and node, providing convenience for design and construction. The effective method in construction – other units are constructed based on the previous one – complies with the concept of low-carbon architecture and the spirit of the horticultural expo focusing on ecological environment.

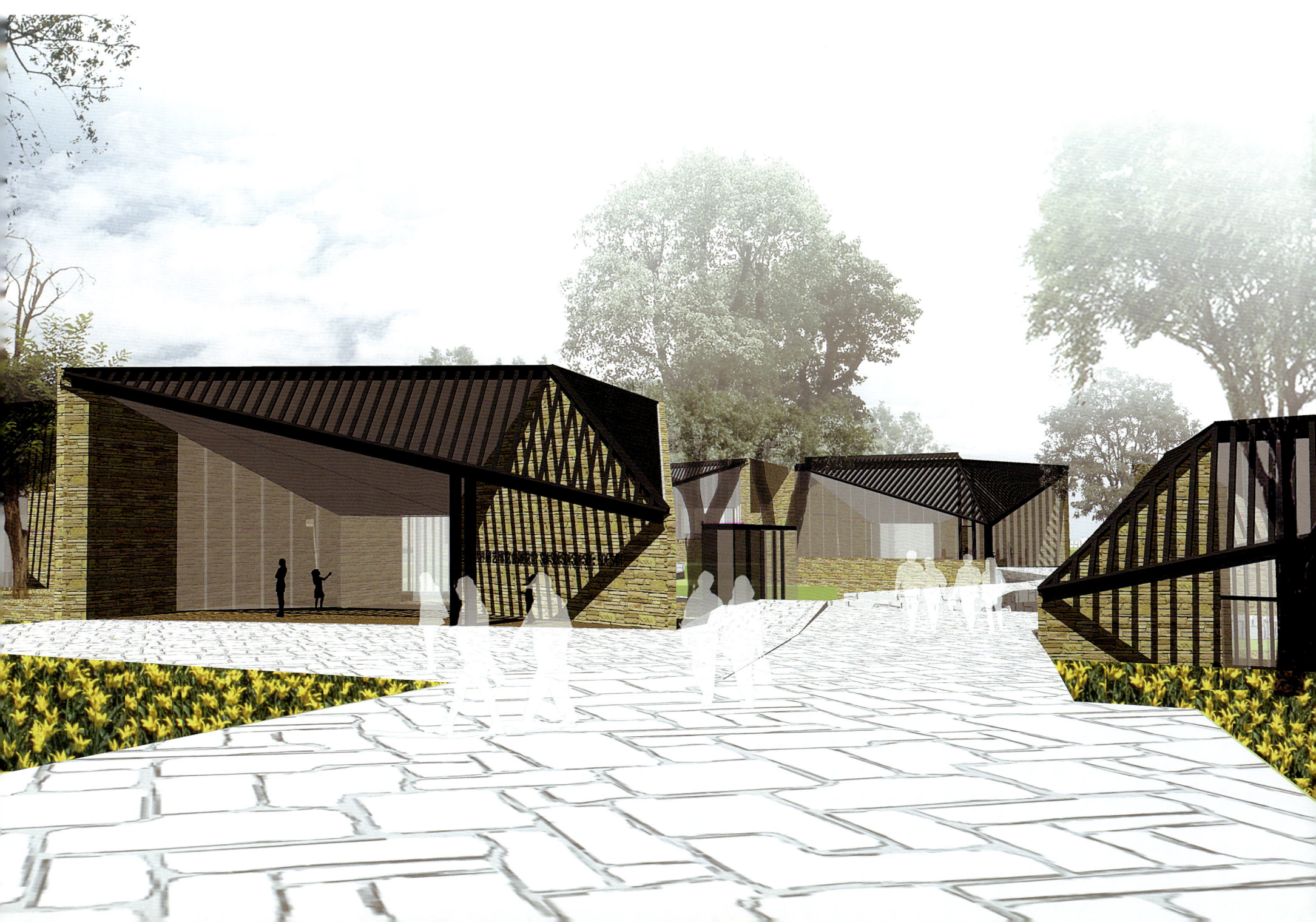

Ba Liu Hotel

服务区·灞柳驿

灞柳驿位于西安世界园艺博览会中心区的西南角,为一座四面环水的独立小岛,占地面积93,770平方米。精品酒店的设计来源于对水流的观察思考,浐灞新区的独特区位和周边河流、岛屿、山脉的自然景观,以及西北大地独特的气候植于其中,被引发出无限可能。随小岛地形流动的闭合曲线通过青砖墙勾勒出酒店的边界,内侧则是连续变化的木条边墙,两道围墙形成了私密又富于趣味变化的廊道空间。

实墙包围着酒店和水疗中心,效仿了池塘上漂浮的荷花形式,以石墙为中心的酒店和水疗设施形成了5个"花瓣"区,每片花瓣代表了独特的景观特色。这5个"花瓣"区面向水面的方向延伸出去,分别是公共接待区、高端会所区及3个客房区。每个客房区又都设置了单独的接待区域,既保证了私密性,又满足了不同的接待需要。5个区分据小岛景色最为优美的5处水岸,既相对独立,又能通过中心庭院互相联系,中心庭院形成了一个大的公众聚集的文化活动和公共艺术展示空间。根据岛屿地形环境差异,5个区采用5种不同的建筑形态,并由此产生了无论是内部使用体验,还是外部视觉观察来看都各有特色的系列空间。

酒店建筑材料选择尽可能考虑当地传统材料(包括大面积粗砾黄土堆积而成的夯土墙;精致细长木片织缀而成的木条边墙;以及无数三角形玻璃折面构成的内庭接待长廊),以实现低能耗和被动式节能,大量采用天然材料,并结合了地源热泵、智能灯控温控等一系列技术,并在建筑、景观、室内采用LEED认证标准进行控制,整片建筑群扎根在基地所处的环境之中,通过对当地文脉与材料的表述,激发出场地特有的气质与力量。花园的体验贯穿了整个中心,植物种类繁多,它们巧妙地组成了一年四季多样化的景观艺术体验,在基地的开发阶段,许多树木种类被保存,以此倡议和促进园艺博览会对植物种类的重用,表达对2011西安园艺博览会的美好愿望。

The 93,770 m² Ba Liu Hotel is located at the southwest of the Expo Site. It resides in water like an isolated island. The observation and reflection of flowing water inspires the design concept. It integrates various elements of Chanba ecological zone and adapts to the unique climate of northwestern China, and creeks, islands, mountains etc invoke incredible imaginations. The closed curves of blue brick walls along the island's landform define the border of the hotel. The inner continuous changing wall is built of wood strips. A corridor between the two enclosing walls is intimate and interesting.

The hotel and spa is enclosed by a solid wall like a floating lotus in a pond. It is divided into five "petal" areas surrounding the central stone wall. Each petal denotes a special landscape feature, stretching to the water area. The five "petal" areas are Public Reception Area, High-end Club and three Guest Room Areas. Each Guest Room Area has a private reception zone to meet the visitors' different demands. The central courtyard not only connects the five separate waterfront areas, but also provides a large public space for activities and exhibiting public art installations. According to different landforms and surroundings, the five areas separately own five different architecture forms which stimulate special experience inside and unique views outside.

The building materials mainly include traditional local ones (for instance, large rammed earth wall using coarse gravel and loess, sidewall built of exquisite wood strips, reception corridor built of numerous triangle glasses within the inner courtyard) so as to meet the requirements of low energy consumption and passive energy saving. Other materials and techniques are a large number of natural materials, ground source heat pump, intelligent lighting control system, temperature control system, and the introduction of LEED rating system (utilization in architectural design, landscape design and interior design). The whole architectural complex is rooted in the site, reflecting the local history, culture, materials and site specific characters. The palette of various plant species offers an art tour of appreciating seasonal phenomena. During the stage of developing the site, trees are largely preserved to stress the important role plant species play in this horticultural exposition and extend best wishes for International Horticultural Exposition 2011 Xi'an.

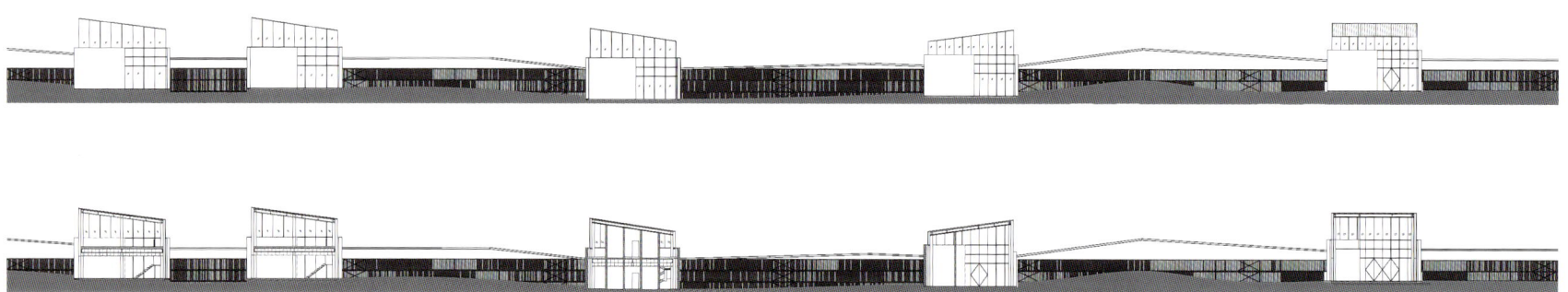

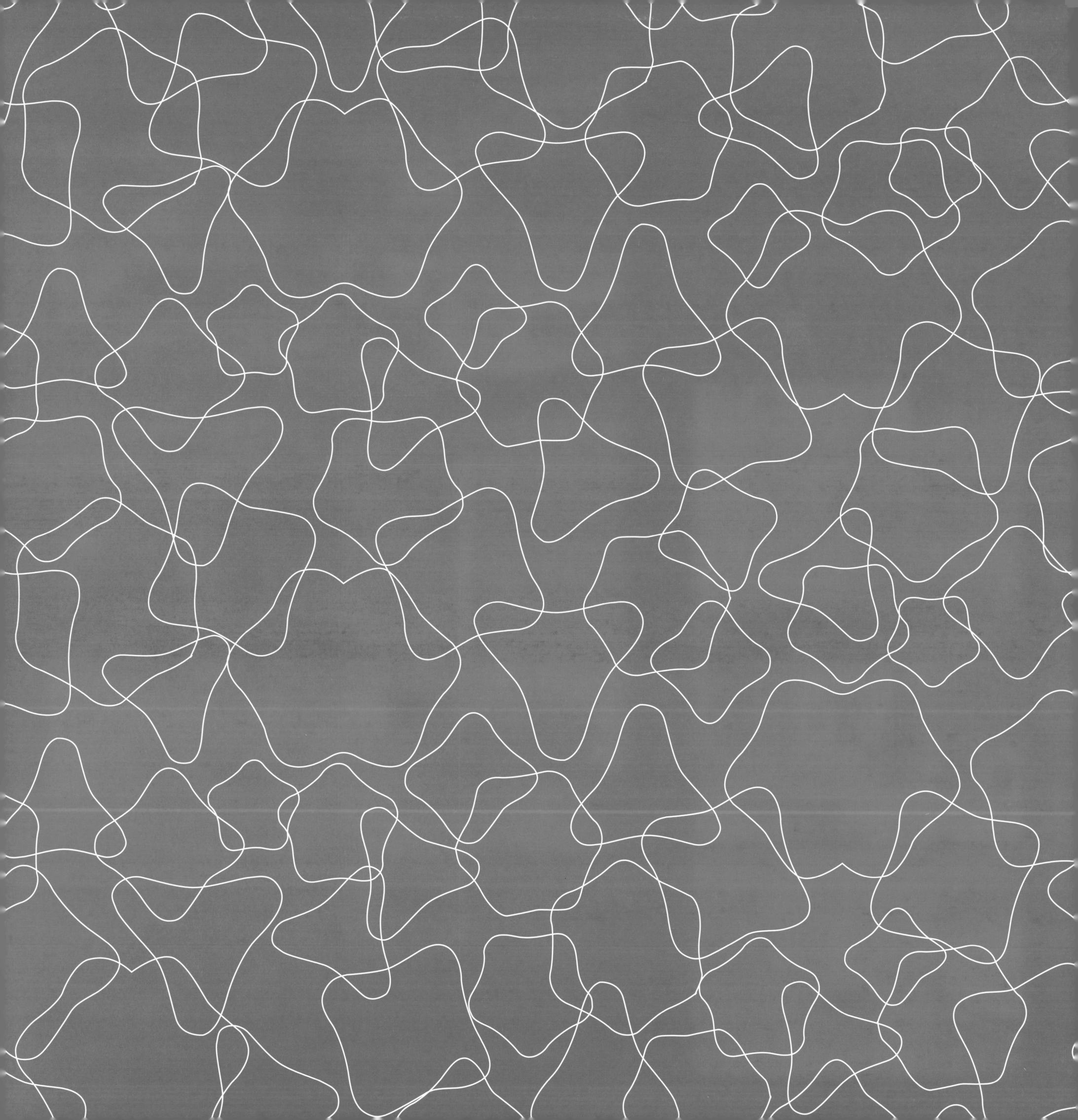

Ecological Gardens
生态园艺

缤纷多姿的各类展园是2011西安世界园艺博览会的主角。以参展国国家命名的国际展园是五洲风情的缩影；由国际一流景观建筑设计师创作的大师园可谓设计盛宴的一道大餐；来自世界著名大学的创意园透视出当代学者和学生们对当代城市景观环境的思索。以省份城市命名的港澳台北园、内地园、省内园，以及以地域特色命名的专类园，是中国特色景观智慧和园艺设计的集中展示。此外，以生态新材料为主题的优秀代表企业的展园则充满着对未来的期盼和想象力。

The protagonists of Expo 2011 Xi'an are various fascinating gardens. Significant international gardens named with participating countries are the epitome of Continents, Master Gardens designed by the international first-rate landscape architects compose a grand banquet in the Design Field; Creativity Gardens from famous universities around the world reveal contemporary scholars' and students' thoughts on urban landscape and environment nowadays. Moreover, traditonal landscape design is fully shown by Hongkong/Macau/Taipei Garden, Domestic Gardens, and Provincial Gardens, named with the names of provinces and cities, and also Feature Gardens, named with local flavors, which reveals Chinese wisdom on landscaping. Also, the parks of excellent enterprise representatives themed with new ecological materials are filled with imaginations and hopes for the future.

Functional Partition

园艺规划

从功能上主要分区为长安花谷内环广场、国际园、内地园、体验区、专类园、企业园、VIP区域、后勤服务区，所有展园均根据自己对本届世园会主题"天人长安·创意自然——城市与自然的和谐共生"的理解，以本地区、本行业或本企业（组织）的特有文化、先进技术、生活理念、建筑风格为指导思想，结合园区现有的环境特点，设计布局，便于各个相邻园区之间自然过渡，达到统一协调的设计效果。

The Expo Site is functionally divided into Chang'an Flower Valley Inner Square, International Gardens, Domestic Gardens, Experience Gardens, Feature Gardens, Enterprises Gardens, VIP Zone and Logistics Service Zone. Based on the understanding of the Expo's theme "Eternal Peace & Harmony between Nature & Mankind, Nurturing the Future Earth-a City for Nature, Co-existing in Peace", the designs of all the gardens and parks are guided by the cultural characters, advanced technology, life philosophy and architectural style of specific regions, industries or enterprises, and associated with existing environmental features of the Expo Site. Hence, adjacent parks and gardens achieve natural transition and coordination to show a perfect design effect of the entire Expo Site.

Landscape Design for Five Theme Gardens

五大主题园艺景观设计

五大主题园艺景观设计包含以下五个部分：长安花谷，五彩终南，丝路花雨，海外大观，灞上彩虹。

长安花谷是本届园艺博览会的入口主轴线和标志性景观，具有极强的艺术效果和标志性，用不同色彩的花卉描绘出"天上"景观，展示从古至今人们对"天"的认识和想象，在178天内将进行5次样式和花卉更换。

五彩终南是秦岭的缩影，地形丰富，地貌奇特，展会期间将布满鲜花，绚丽多彩。

丝路花雨则利用花卉、绿化雕塑、节点广场等景观元素，表现历史悠久的丝绸之路。

海外大观以欧洲园林为主，集锦其他国家和地区的园林艺术。

灞上彩虹则结合水面和滨水建筑，使游客远距离感受水与花交相辉映的美丽画卷。

The Five Theme Gardens are the Chang'an Flower Valley, Colorful Plants from Qinling Mountains, Flowers along the Silk Road, Overseas Collections and Flower Rainbow over the Ba River. Chang'an Flower Valley is the landmark landscape and main axis at the entrance of the Expo, delivering strong artistic effect and significance. It utilizes colorful flowers to depicture celestial landscape and display people's understanding and imagination of heaven. The style and flowers will be renewed five times during 178 days.

Colorful Plants from Qinling Mountains reveal the landscape of Qinling Mountains. With varied and peculiar terrain, it will be filled with bright and colorful flowers during the Expo.

The Flowers along the Silk Road embody the time-honored Silk Road through landscape elements such as flowers, green sculptures and squares.

Overseas Collections integrate European gardens as well as horticulture of other countries and regions.

Flower Rainbow over the Ba River combines the waterscape and waterfront architecture to bring the beautiful picture of water and flowers shining together for tourists in the distance.

Landscape Plazas

景观广场

2011西安世界园艺博览会景观广场共33个，分为景观广场和服务区广场，广场总面积31,376平方米。为了让游客享受游园过程，园区设置了26个休憩节点。

景观广场在设计上力求体现出世界园艺博览会主题核心"天人长安•创意自然"，以生态文明理念为指导，体现天人合一，城市与自然和谐共生的思想。景观广场和休憩节点的设计以其所处环境因地制宜，设计手法、材质元素多变，用色朴素，与周边景观浑然一体，更好地衬托出了各个主题展园，同时也形成了良好的服务环境和景观效果，其中更以2号、9号、10号、15号广场设计特色尤为显著。

Covering a total area of 31,376 m², the Expo 2011 builds 33 plazas, which are divided into landscape plazas and service plazas. To feast the tourists, 26 lounge spots are established.

Designers guide their works with ecological civilization concept, trying to reflect the idea of harmony between mankind and nature, that is, city life and nature, so as to be in correspondence with the theme of the Expo, "Eternal Peace & Harmony between Nature & Mankind, Nurturing the Future Earth-a City for Nature, Co-existing in Peace". The designers take different skills and elements according to the local circumstances in terms of landscape plazas and lounge spots. The colors are quite plain, just like nature itself. All of the above make the Expo gardens a place with nice service and landscape. Especially Plaza #2, #9, #10 and #15 are distinguishingly designed.

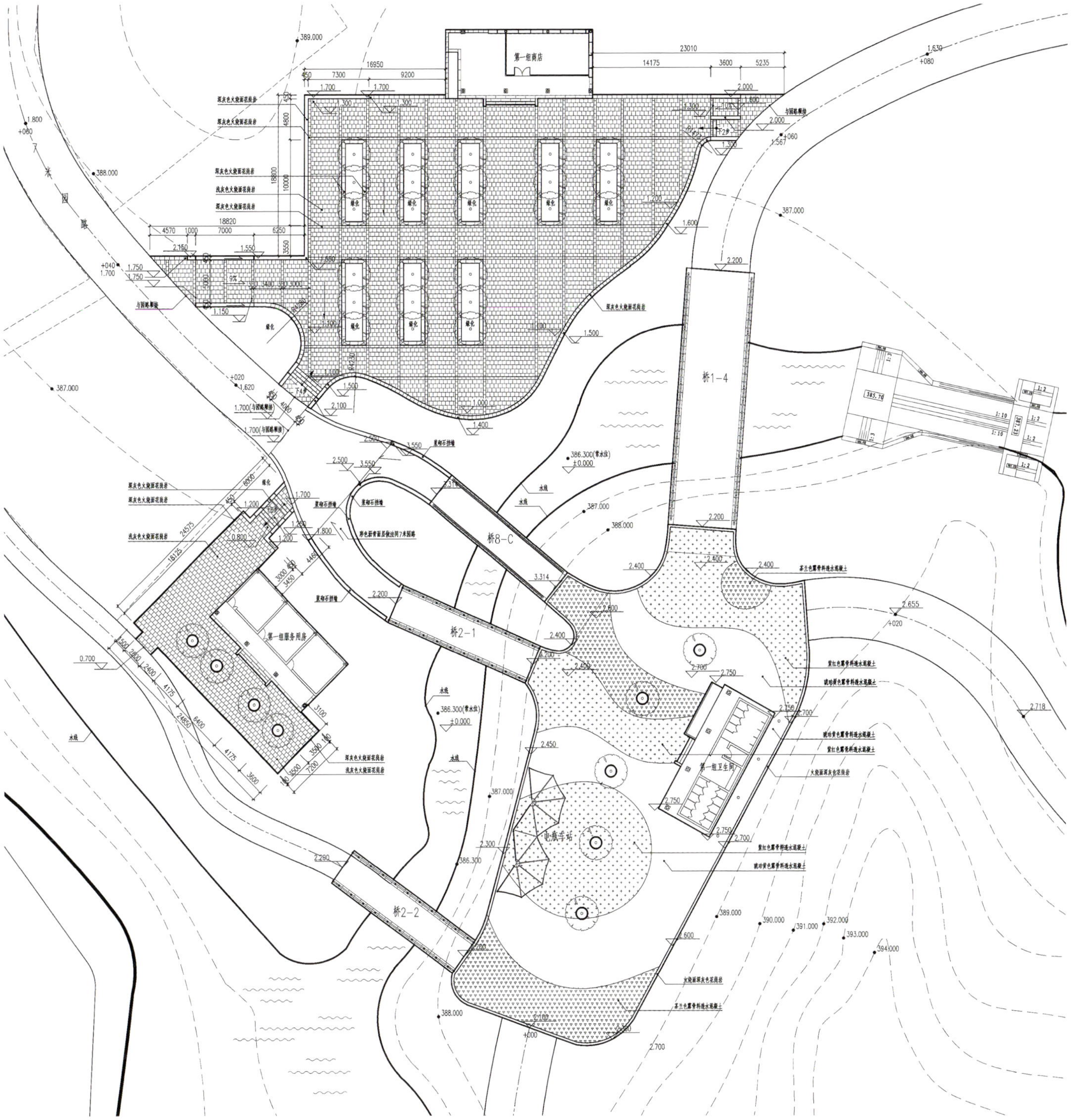

2号景观广场

2号景观广场为综合性服务区广场，位于国际园与长安园交界处，为客流较大的交通枢纽，面积3,255平方米。承担在室外游览、候车、休息、餐饮、购物多项公共空间的功能需求。设计充分利用现状地形，以服务功能为主，北侧为停留空间，运用石材与陶砖结合，树池座凳形式为主，南侧为多彩透水混凝土铺装的交通空间。

Landscape Plaza 2

The Landscape Plaza 2 is the Comprehensive Service Zone which is located at the junction of International Gardens Section and Chang'an Garden. It is one of the main traffic hubs in the Expo Site with an area of 3,255 m², which includes many public functional areas for outdoor visiting, waiting, resting, catering and shopping. Located on the north of the plaza is a wandering space for visitors paved by stones and earthenware bricks and mainly decorated with tree-grate seats. At the south, it is a traffic space with the colorful pervious concrete pavement.

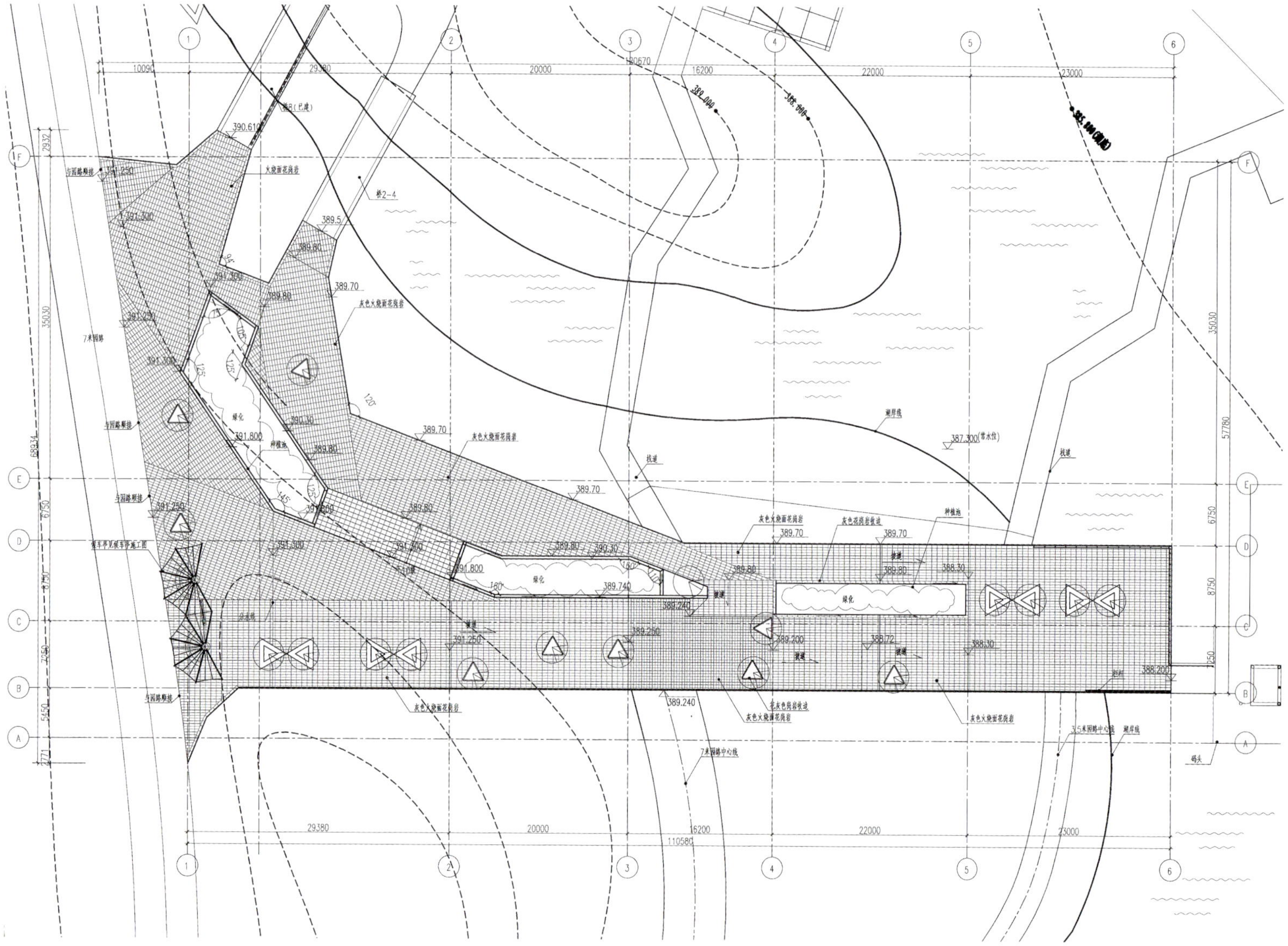

9号景观广场

9号景观广场,位于灞上人家北侧,为条形广场,延伸至湖畔码头,面积2,662平方米。连接多个景观桥梁栈道,集游览、候车、码头休息为一体,设计以现代折线感为主,与灞上人家设计风格统一,形成空间感觉上的延续。

Landscape Plaza 9

Located at the Section of Romance by the Ba River, the Landscape Plaza 9 is a strip-shaped plaza with an area of 2,662 m². It extends to the lakeside dock connecting several landscape bridges. The plaza is a comprehensive space with touring, bus-waiting, and docking functions. The design of this plaza is mainly applied by modern foldline style to echo with the Romance by the Ba River and to show the space-extending.

10号景观广场

10号景观广场，位于世界庭院西侧，为大型综合性服务景观广场，面积4,180平方米。依地形变化，创作出多层次的丰富空间，青砖灰瓦与防腐木配合景石的点缀，使广场带着鲜明的地域元素，突出表现本土企业园的特点。

Landscape Plaza 10

The Landscape Plaza 10 is located in the west of the World Gardens, covering an area of 4,180 m². It is a large-scale landscape plaza and provides various services for the visitors. Layered effect is added to the space according to its landform. Landscape stones are dotting among blue bricks, gray tiles and preservative wood, which highlights the local features, especially the characteristics of the local enterprises.

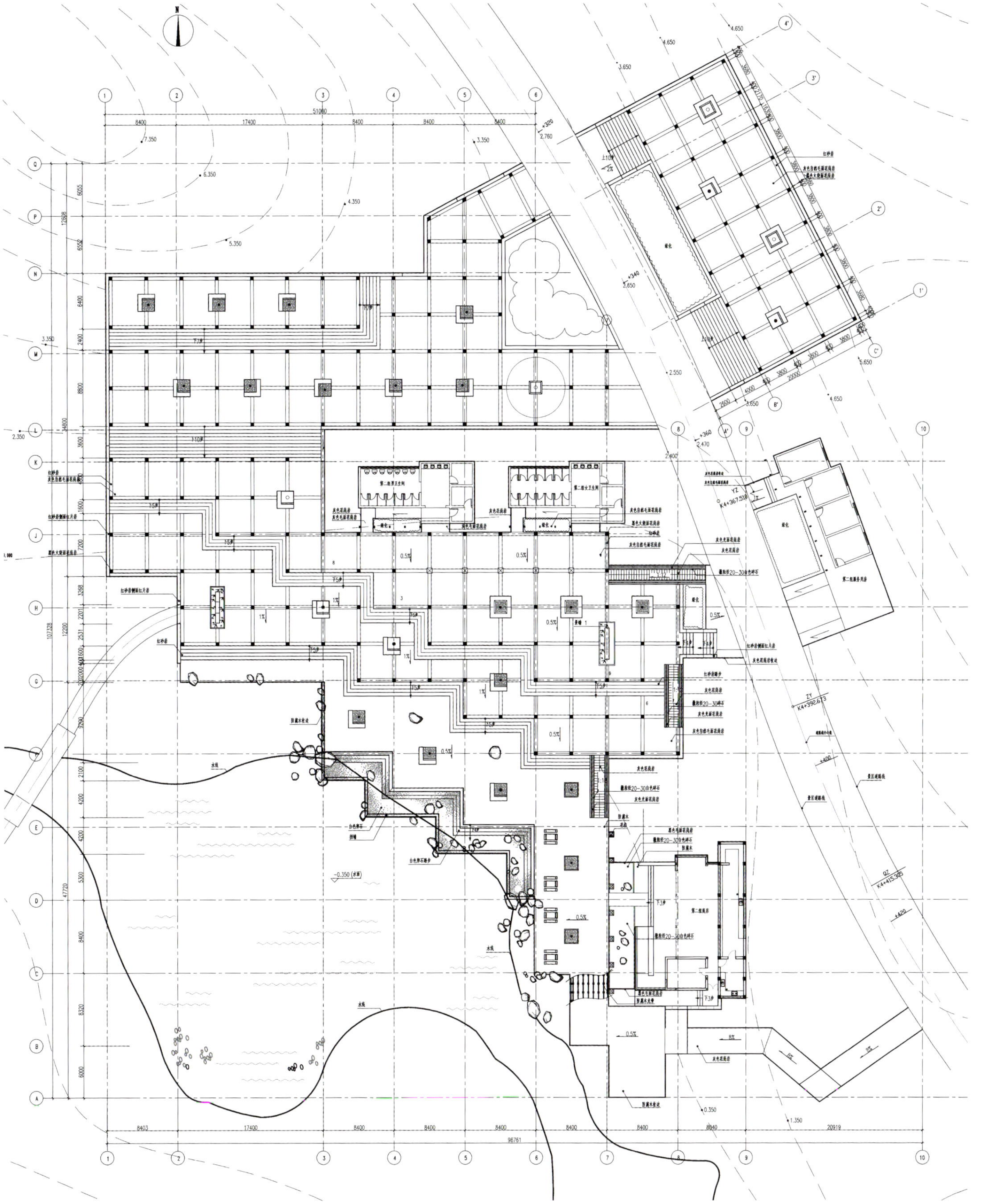

15号景观广场

15号景观广场,位于大师园中部,面积2,897平方米。广场设计通过极具现代感的折线对地形内保留大型乔木进行划分,使场地具有流动感,并体现出人工与自然的和谐共生。

Landscape Plaza 15

The Landscape Plaza 15 is located in the central part of the Master Gardens, covering an area of 2,897 m². The designers focus on preserving large arbor trees, shaping them by zigzag lines, and lending them a modern style, so that the flowing sense is the symbol of the symbiosis of the man-made and the nature.

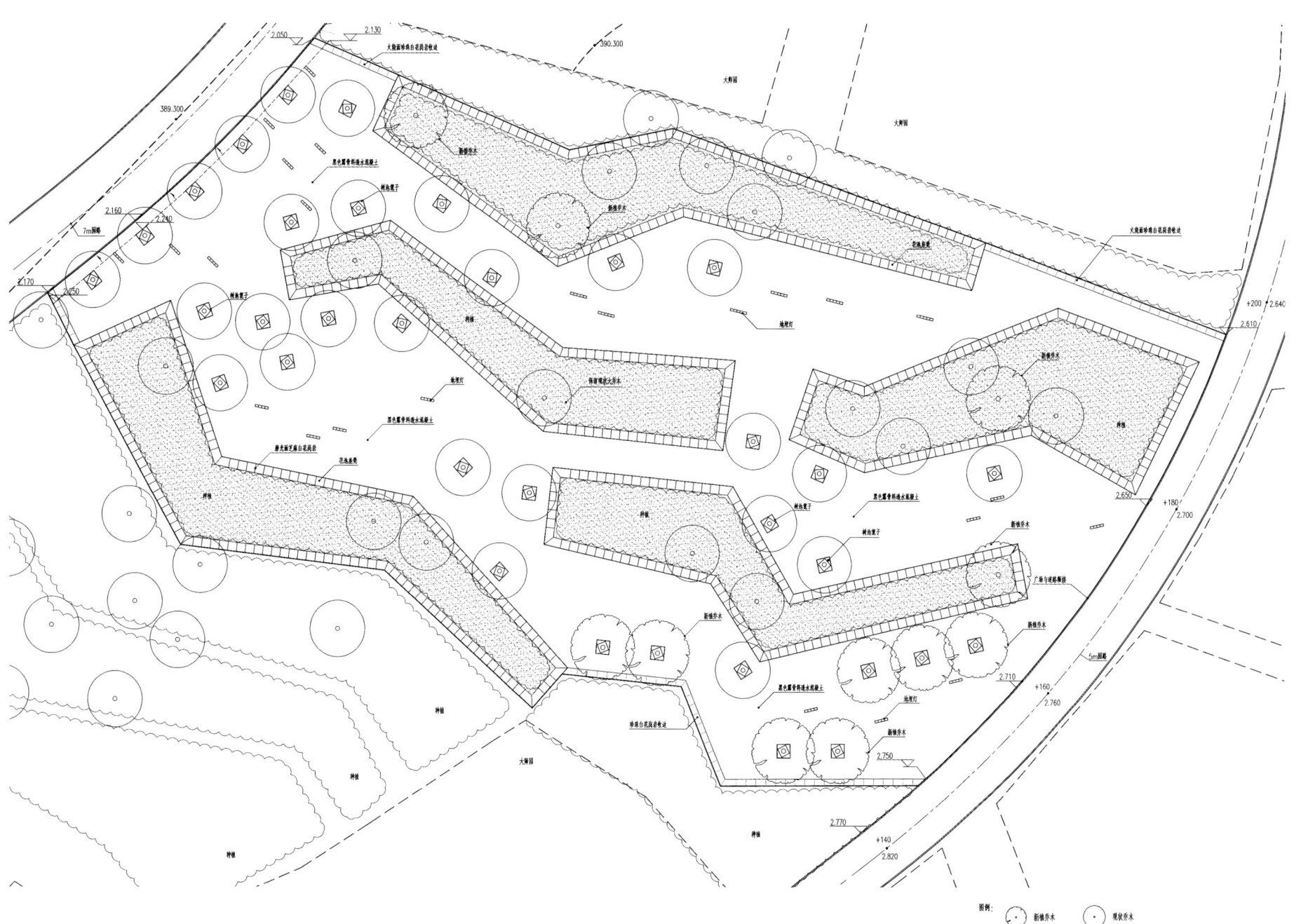

16号景观广场

16号景观广场，位于欧陆风情北侧，面积820平方米，为小型服务广场，彩色透水混凝土拼色配合灵动的曲线划分，使之更好地融入了周边环境。

Landscape Plaza 16

The Landscape Plaza 16 is located to the north of the European Avenue, covering an area of 820 m². It is a small service plaza with multi-colored pervious concrete pavement curved by vivid lines to perfectly melt into the surroundings.

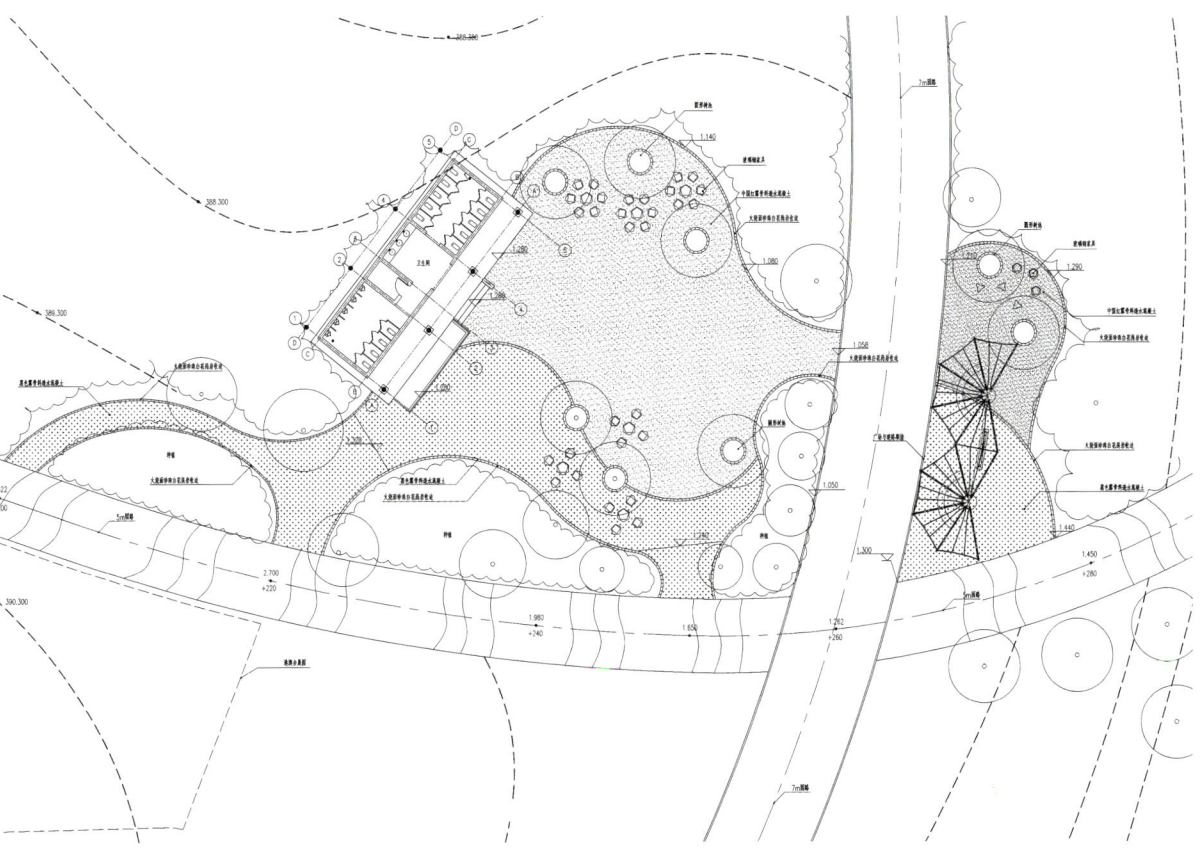

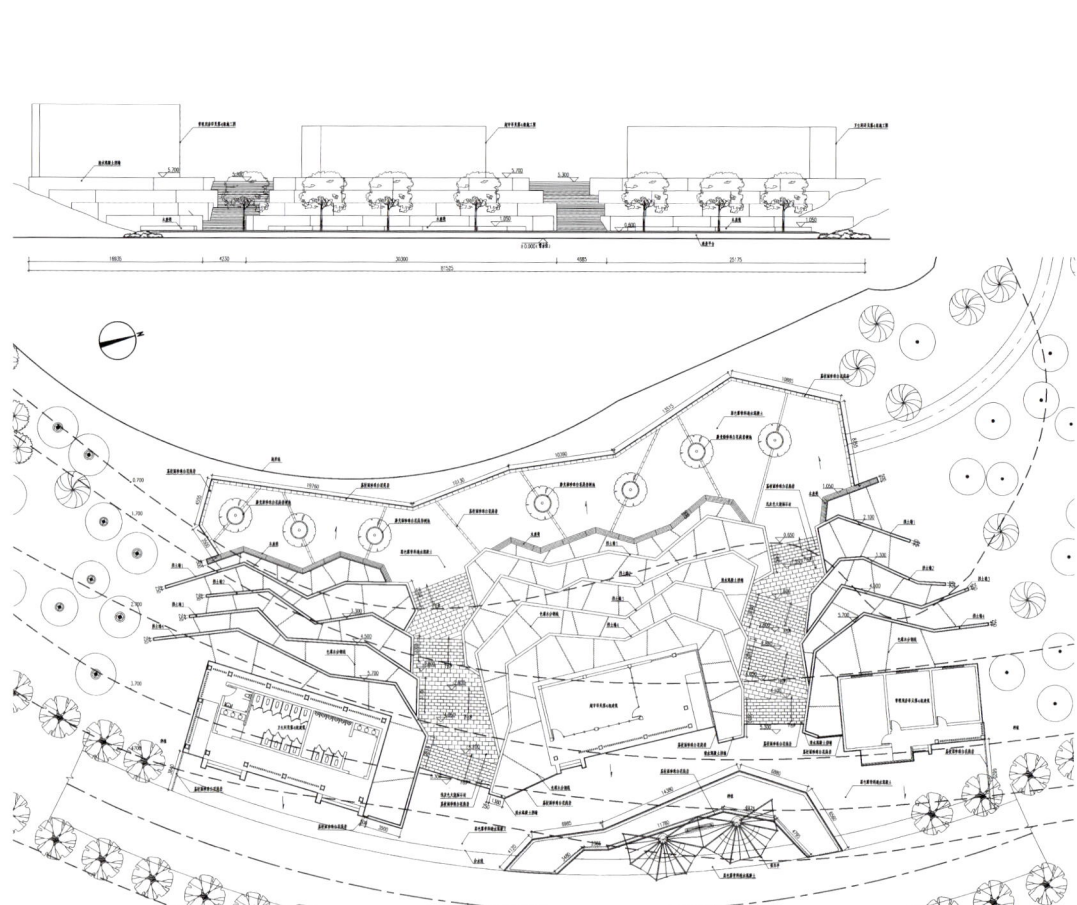

20号景观广场

20号景观广场,位于长安花谷北侧,广场面积约3,000平方米,包括了3座服务建筑与候车厅。广场整体呈梯田状,由现代的折线形挡墙进行划分,多彩的植物色块"嵌"入其中。

Landscape Plaza 20

Covering an area of about 3,000 m², Landscape Plaza 20 is located to the north of Chang'an Flower Valley. It consists of three service buildings and waiting rooms. The plaza is shaped as a terrace, which is divided by modern fold-line retaining walls. It looks great with different plants zones inlaid.

21号景观广场

21号景观广场，位于冒险岛东侧，面积740平方米。以红砂岩与青砖为主要材质，搭配木质景墙，平面布置上方圆结合，有张有弛。

Landscape Plaza 21

The Landscape Plaza 21 is located to the east of the Adventure Island, covering an area of 740 m². Red sandstones and blue bricks are the main materials, with wooden feature wall and the integration of square and round shapes on pavement.

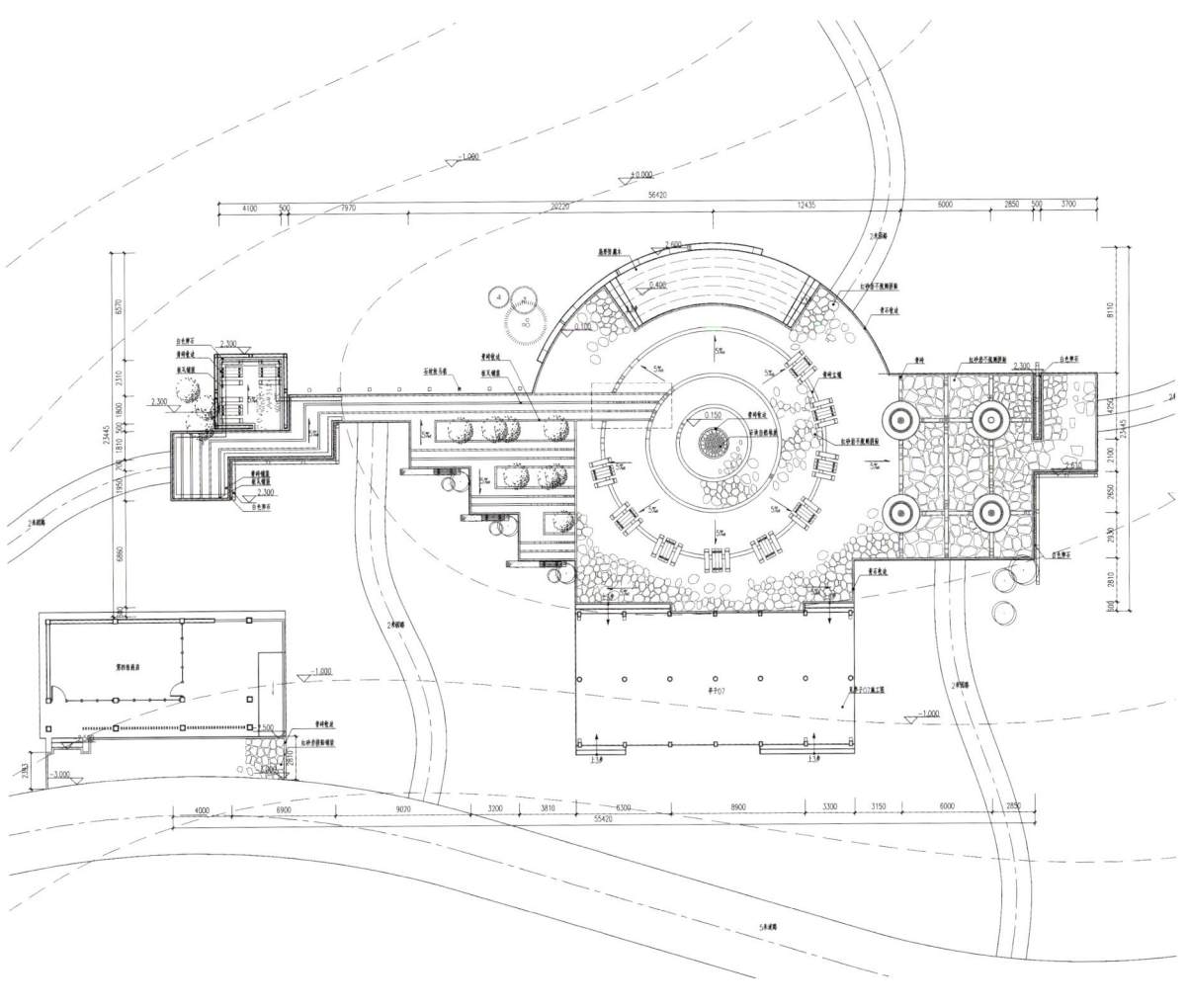

22号景观广场

22号景观广场,位于冒险岛西侧,面积1,470平方米。为综合服务景观广场。集服务购物、休闲观景于一体,材质以红砂岩与青砖为主。依地势而下,多层次空间延伸至湖边亲水平台。

Landscape Plaza 22

The Landscape Plaza 22 is located to the west of the Adventure Island, covering an area of 1,470 m². It is a multi-functional service landscape plaza unifying shopping, leisure, and sightseeing. Red sandstones and blue bricks are the main materials. The layered space slopes downwards along the landform and extends to the waterfront deck.

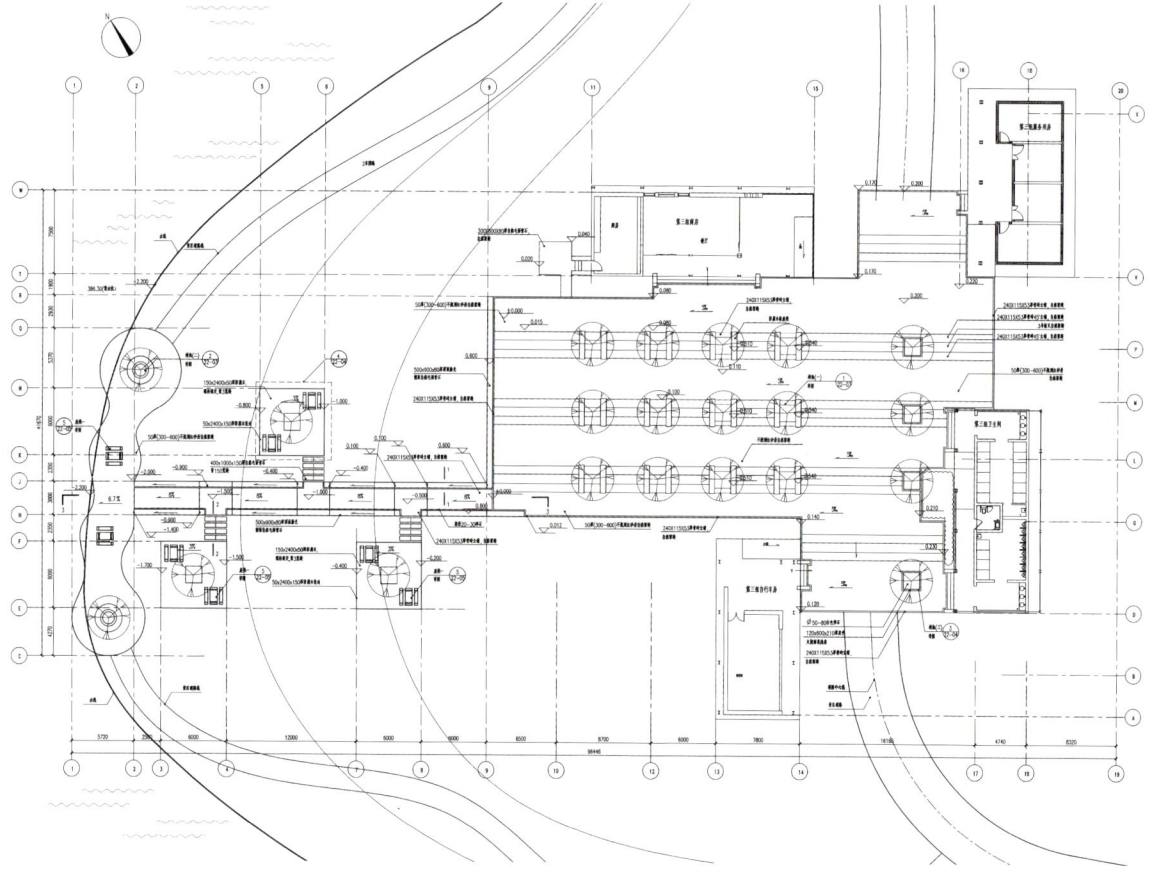

休憩节点
Rest Area

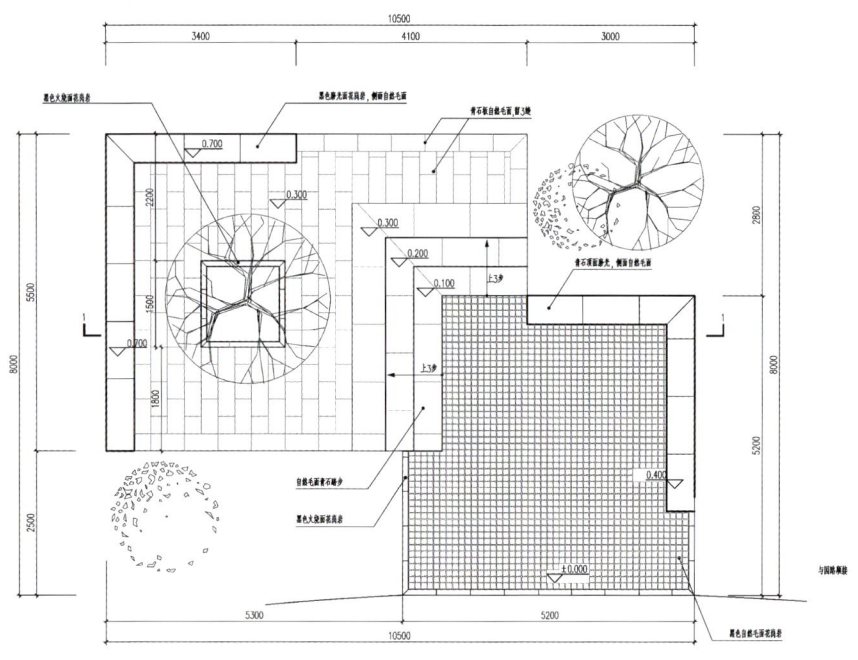

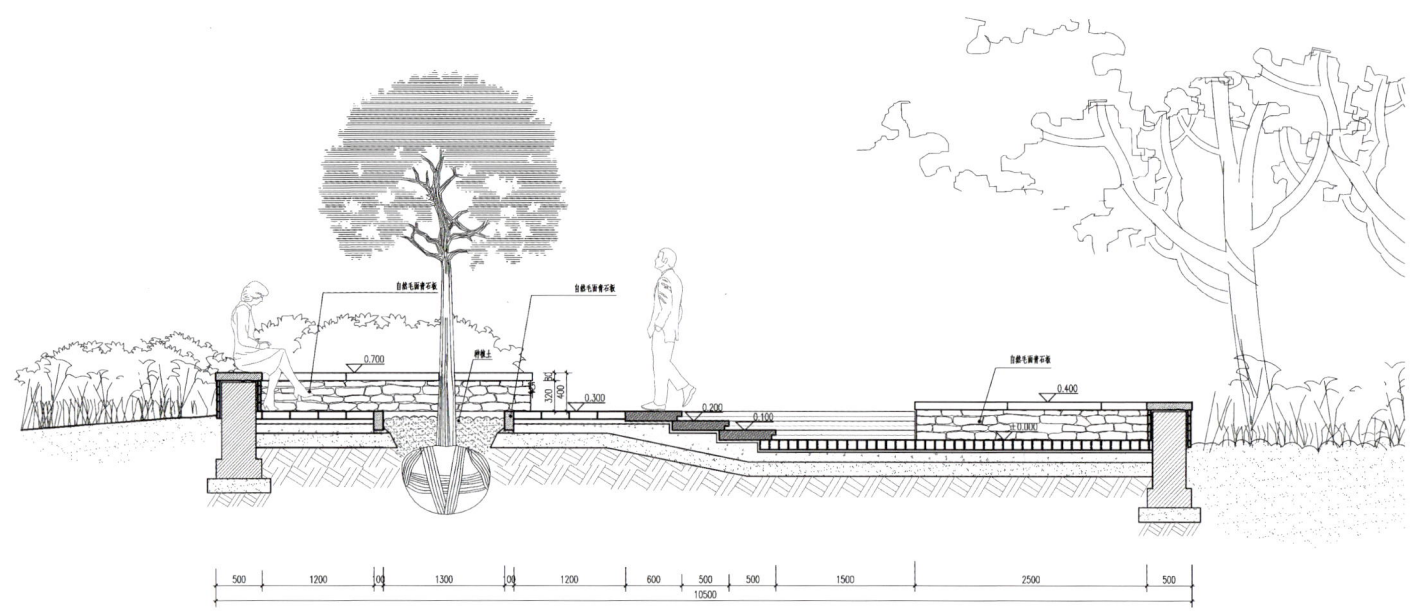

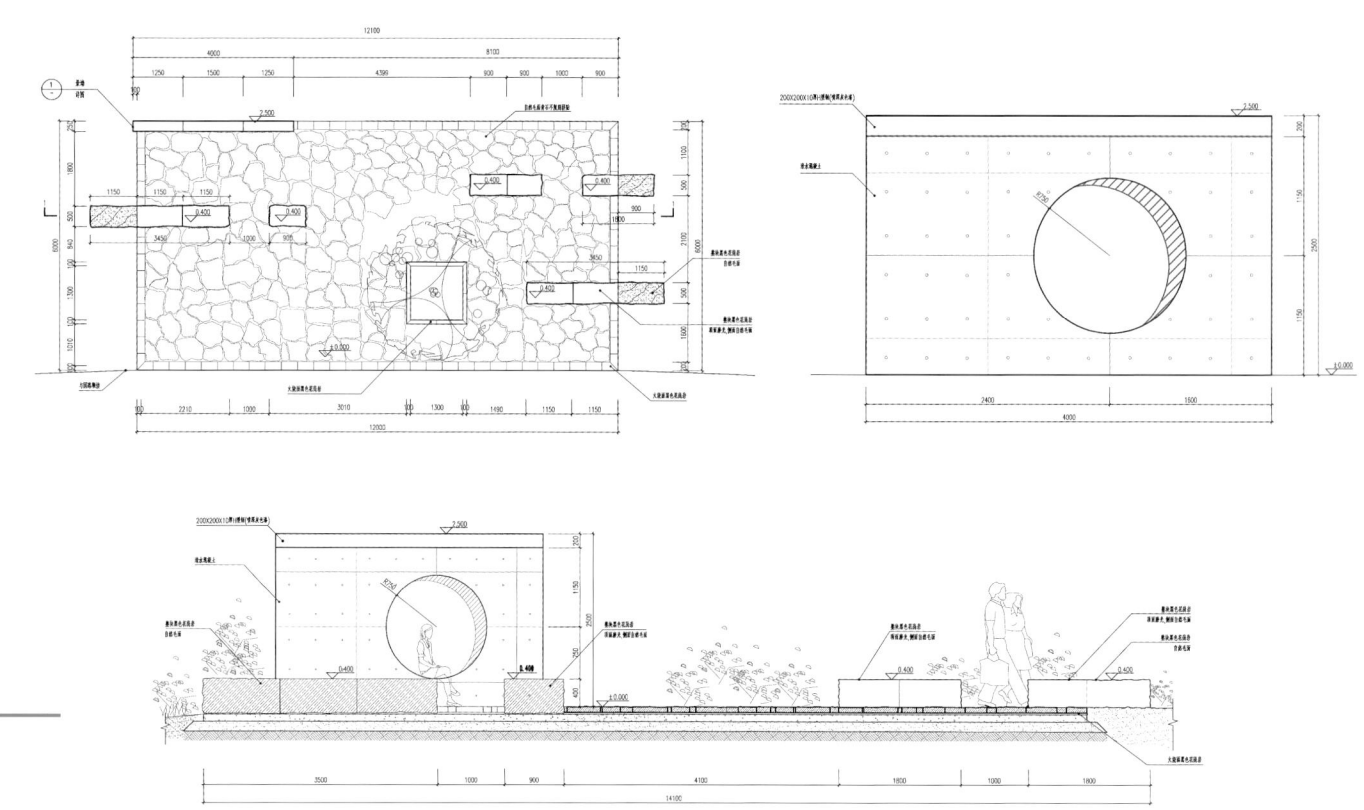
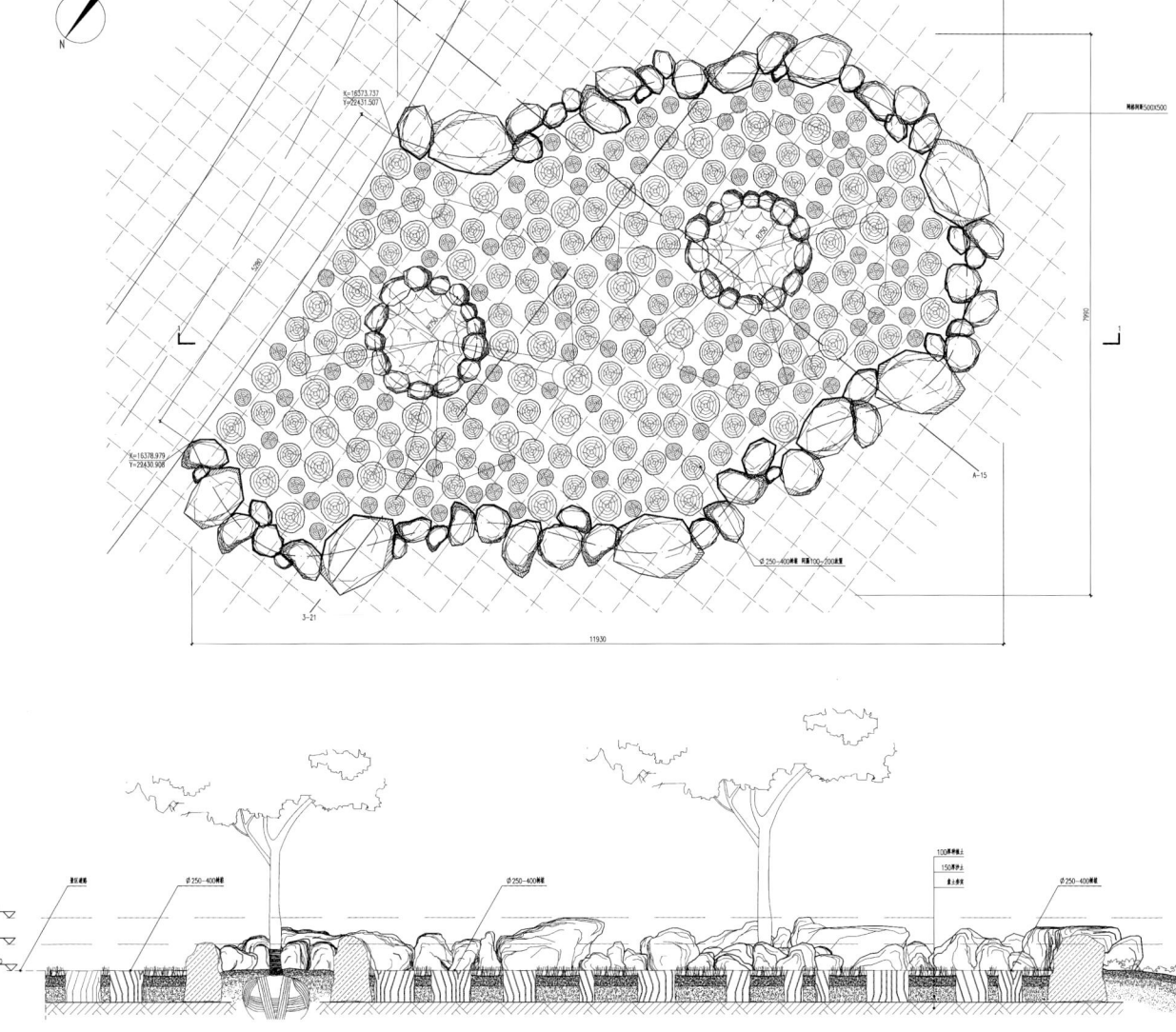

13000
6700
100
3700
100
2400
5900
2900
3000
1800
3600
5400
3000
3000
7100
13100

2800
800
1200
800
850
1100
850

1430
900
470
770
1100
770
800
1200
800
2800
2300
2500

2.500
2.300
0.400
2300
1400
3700
300
5400

500
2500
2100
400
300

N

12000
2500
1500
4000
1500
2500
5000
1800
3100
3800
3300

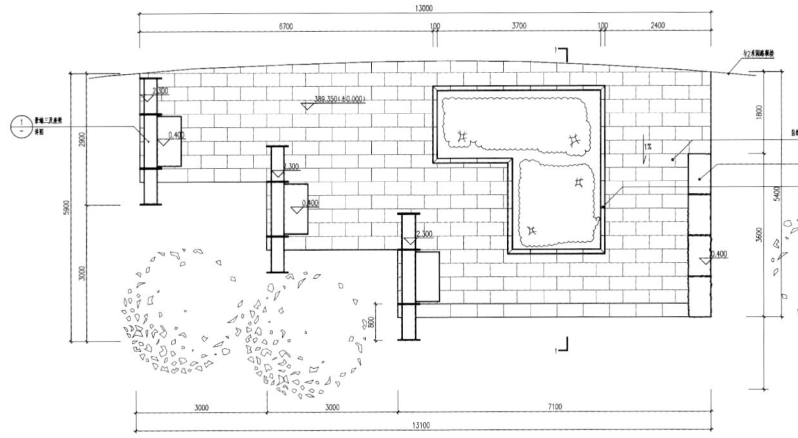

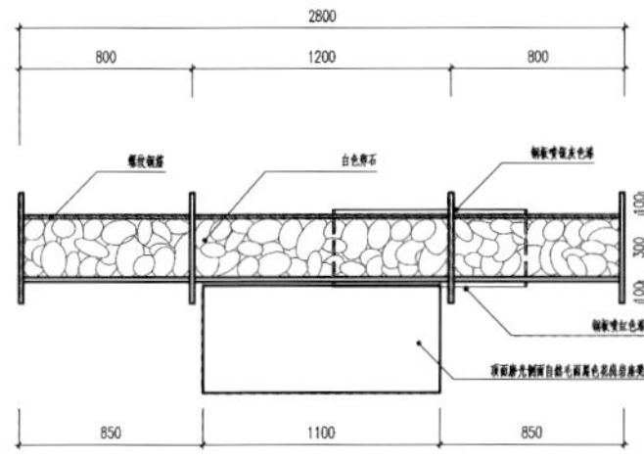

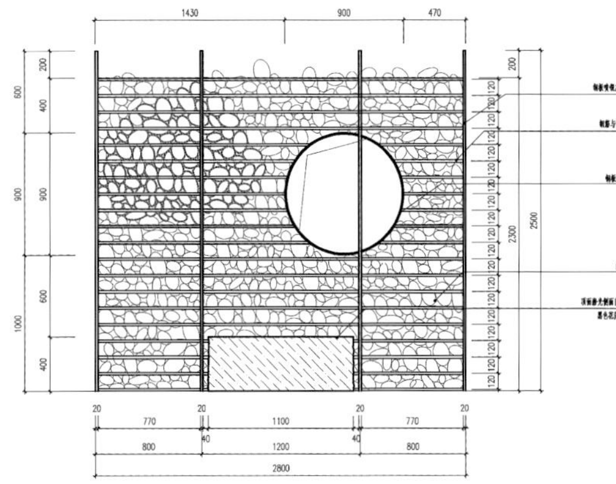

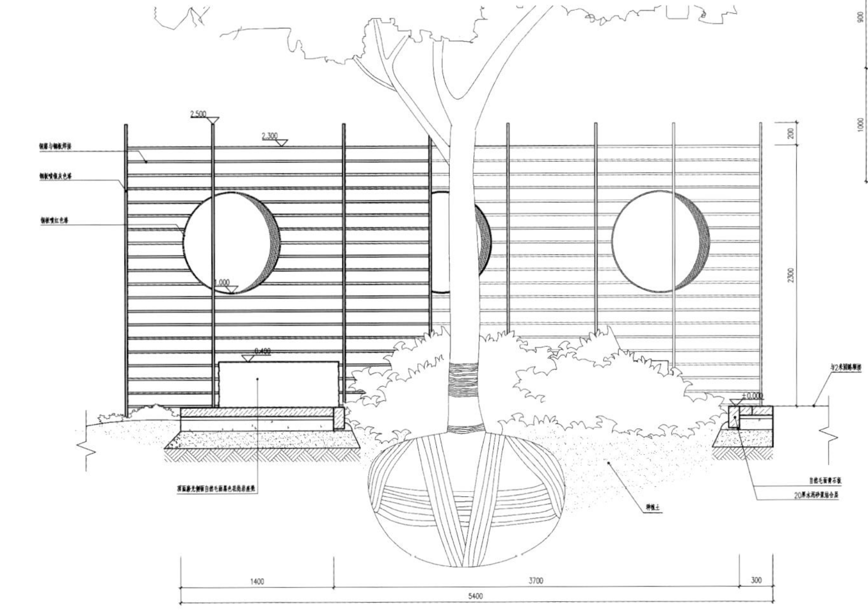

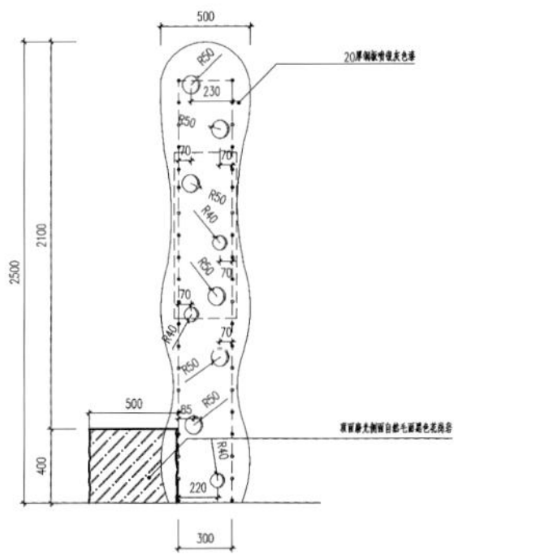

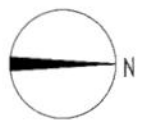

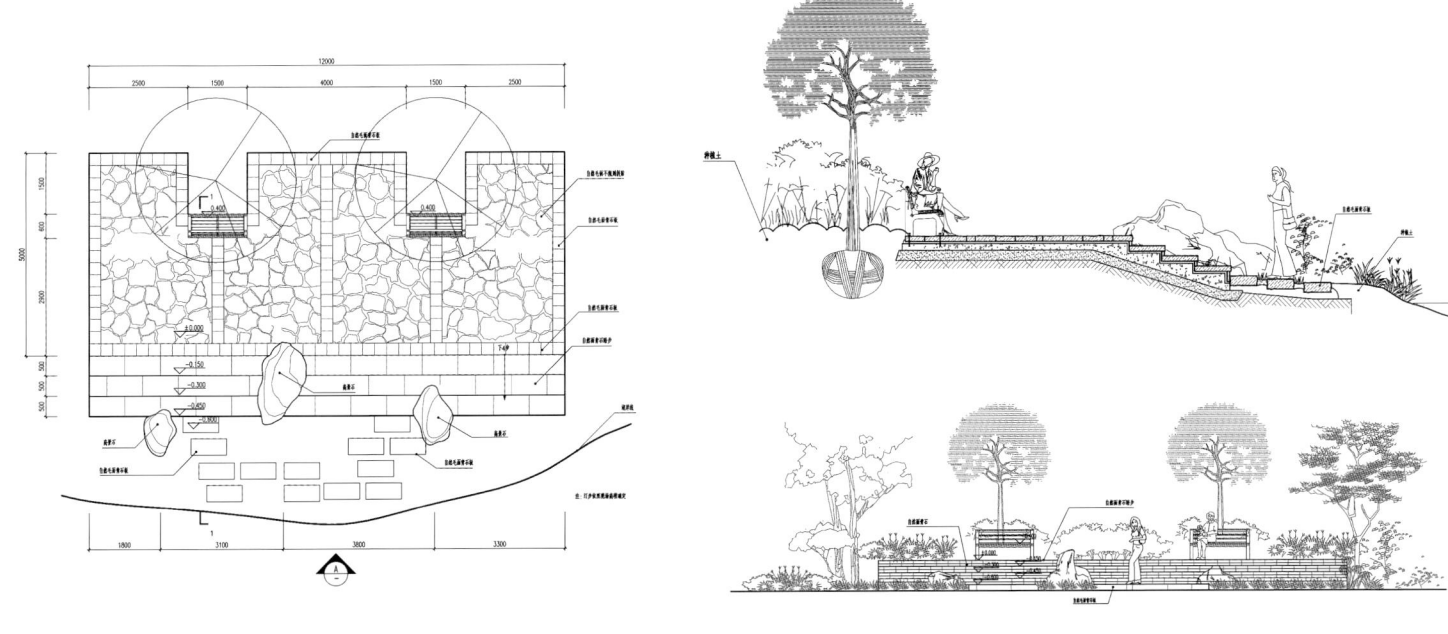

Landscape Pavilions

景观亭

园区内景观亭按造型分为8种，共建造11处。建造地点主要在环湖岸线附近，突出其亲水性，局部点缀于山地之间，充满神秘与趣味色彩，使游客们从多种视角感受世园风光。

According to the appearance, the total 11 landscape pavilions of Expo 2011 are divided into 8 types. They are mostly built along the lakeshore line, showing their water accessibility. While part of them spread themselves among the hills, adding some kind of mystery and interest. Thus tourists may feast themselves with beautiful scenery through various visions.

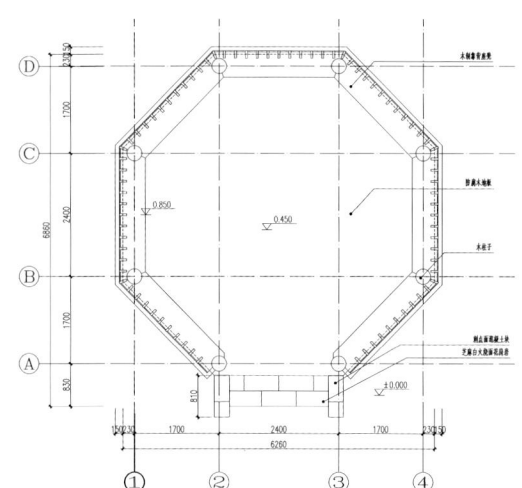

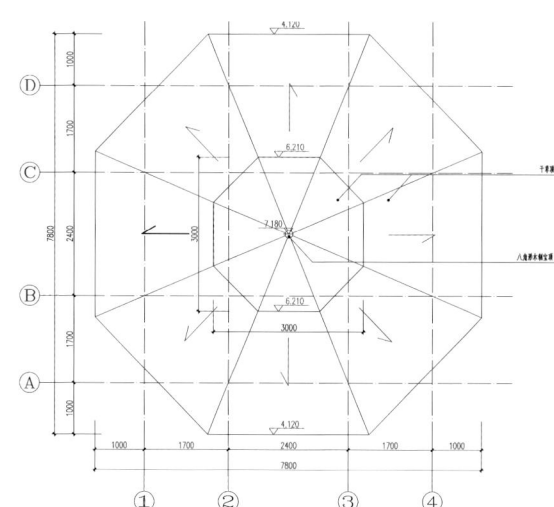

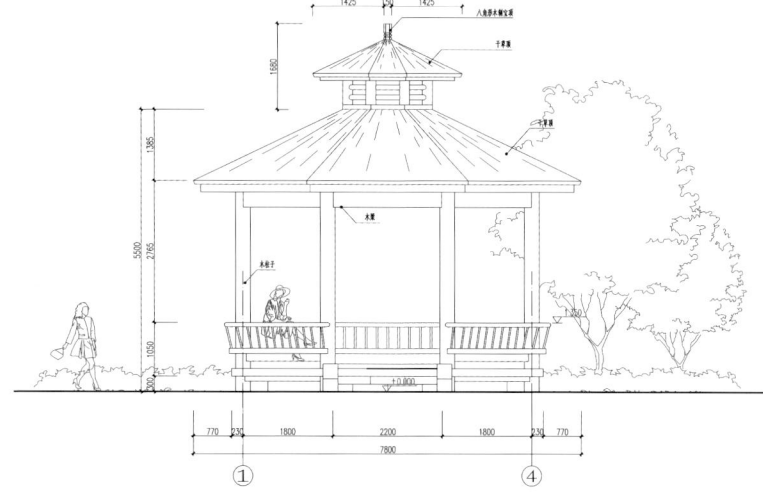

3号景观亭

3号景观亭为八角亭，占地面积约36.8平方米，高7.18米；亭子顶部挂设干草，端头配有八角形木质宝顶，整体造型别致、古朴。

Landscape Pavilion 3

The 7.18-metre-high Landscape Pavilion 3 is an octagonal pavilion, covering an area of about 36.8 m². The top of the pavilion is covered with hay. The stipple on the top is set with an octagonal wooden cap. The leading architectural form is simple, ancient and exquisite.

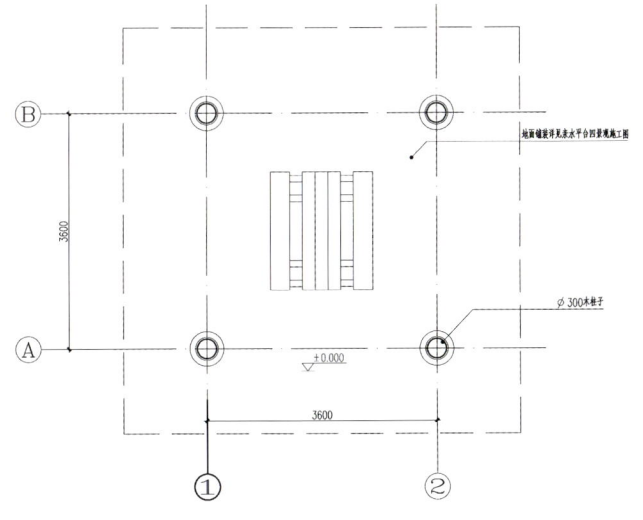
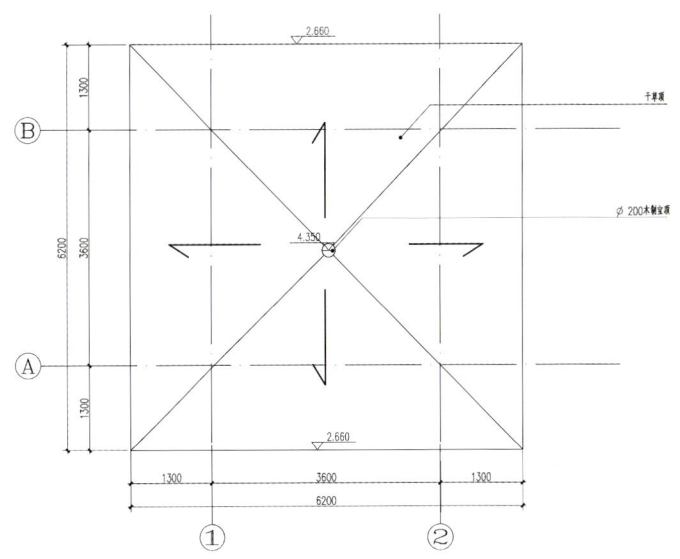
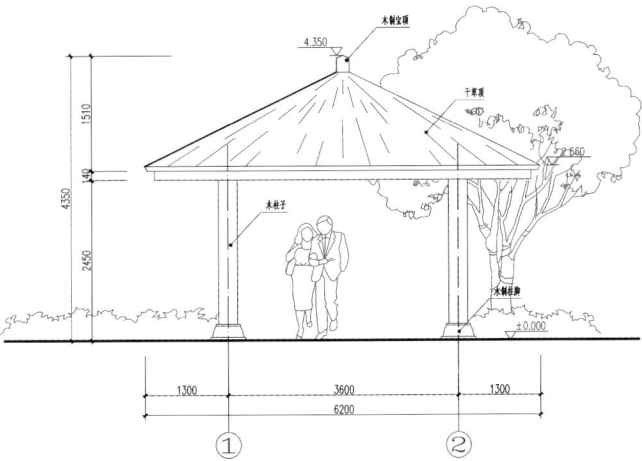

4号景观亭

4号景观亭为四角亭，坐落于4号亲水平台上，占地面积约38平方米，高4.35米，亭子顶面挂设干草，整体简洁、自然。

Landscape Pavilion 4

The 4.35-metre-high Landscape Pavilion 4 is a square pavilion, covering an area of about 38 m². It is located on the waterfront deck 4. The succinct and natural pavilion is covered with hay on the roof.

7号景观亭

7号景观亭为一个造型别致的廊亭,占地面积约157.25平方米,高6.64米,亭子顶部为防腐木屋面,内部设有木质桌凳,四周设有美人靠。

Landscape Pavilion 7

The 6.64-metre-high Landscape Pavilion 7 is an exquisite corridor pavilion which covers an area of about 157.25 m². The pavilion is decorated with preservative wood, with wooden tables and chairs in, and surrounded by armchairs.

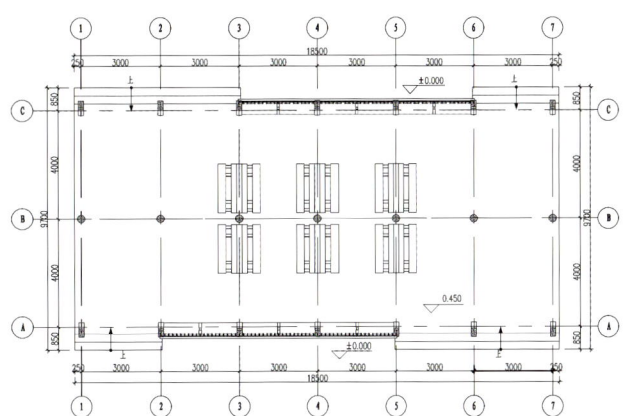
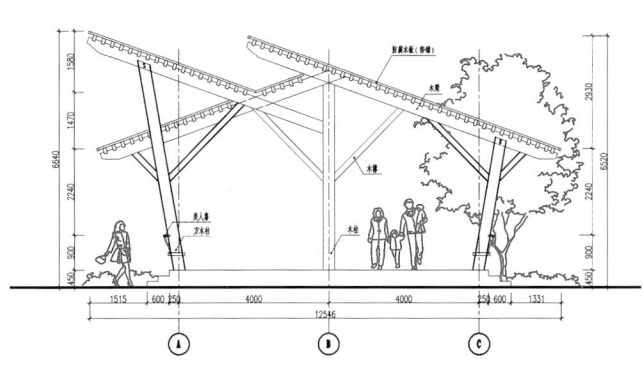

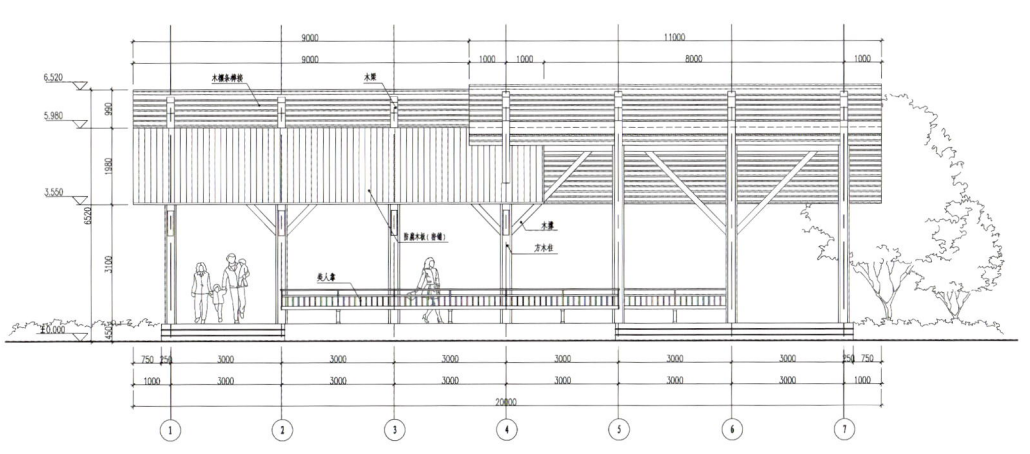

Garden
Planning
展园设计

展园设计共分为国际园、世界庭院、大师园、创意园、内地园、省内园、专类园、企业园,每个展园别具特色,风格迥异,让人们充分领略园林、园艺、建筑、艺术之美。

The Expo Site is divided into International Gardens, World Gardens, Master Gardens, Creativity Gardens, Domestic Gardens, Provincial Gardens, Feature Gardens and Enterprises Gardens. Every Garden has distinct feature and style, making people fully understand the beauty of garden, horticulture, architecture and art.

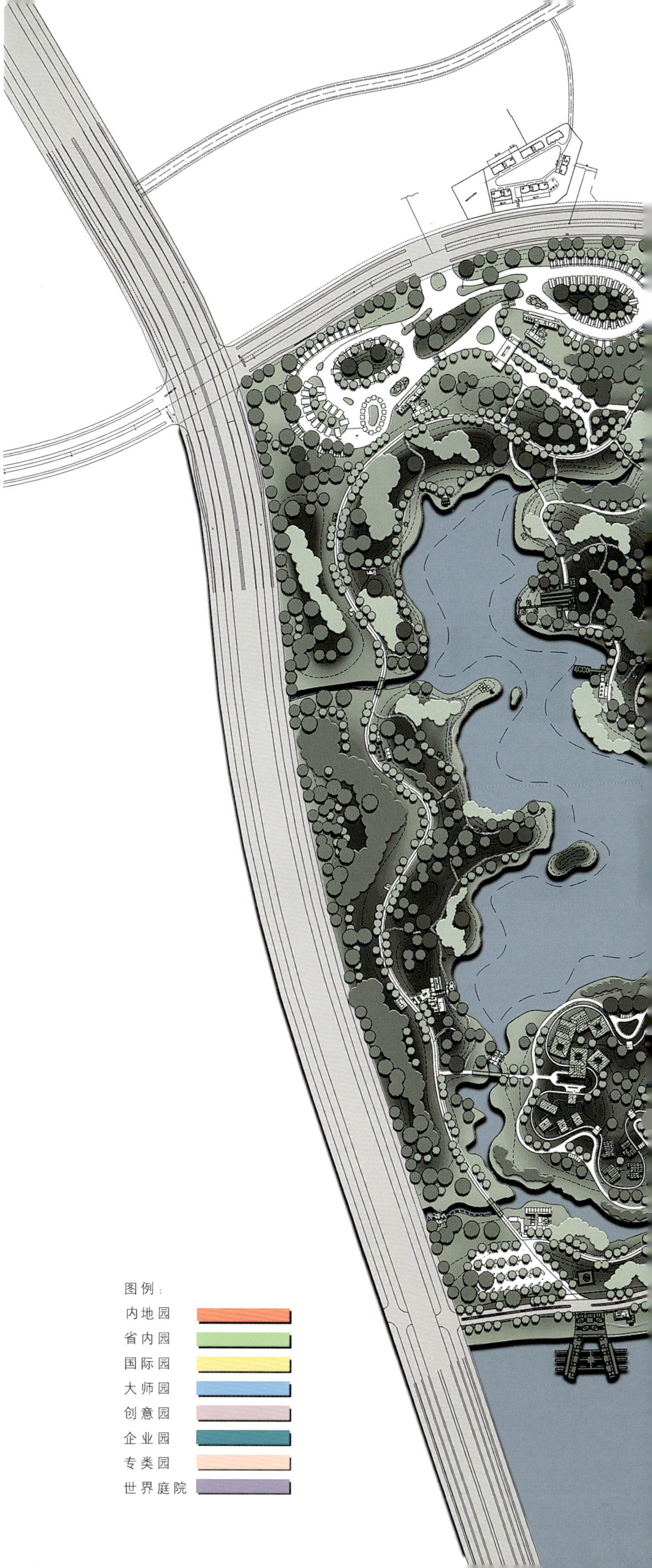

图例:
- 内地园
- 省内园
- 国际园
- 大师园
- 创意园
- 企业园
- 专类园
- 世界庭院

8号景观亭

8号景观亭为一个四角坡屋顶亭子，坐落于3号亲水平台上，占地面积约27平方米，高4.745米，整体以木结构为主，现代、明快。

Landscape Pavilion 8

The 4.745-metre-high Landscape Pavilion 8 is located on the waterfront deck 3, covering an area of about 27 m². It is a square wooden pavilion with a slope roof in modern and simple style.

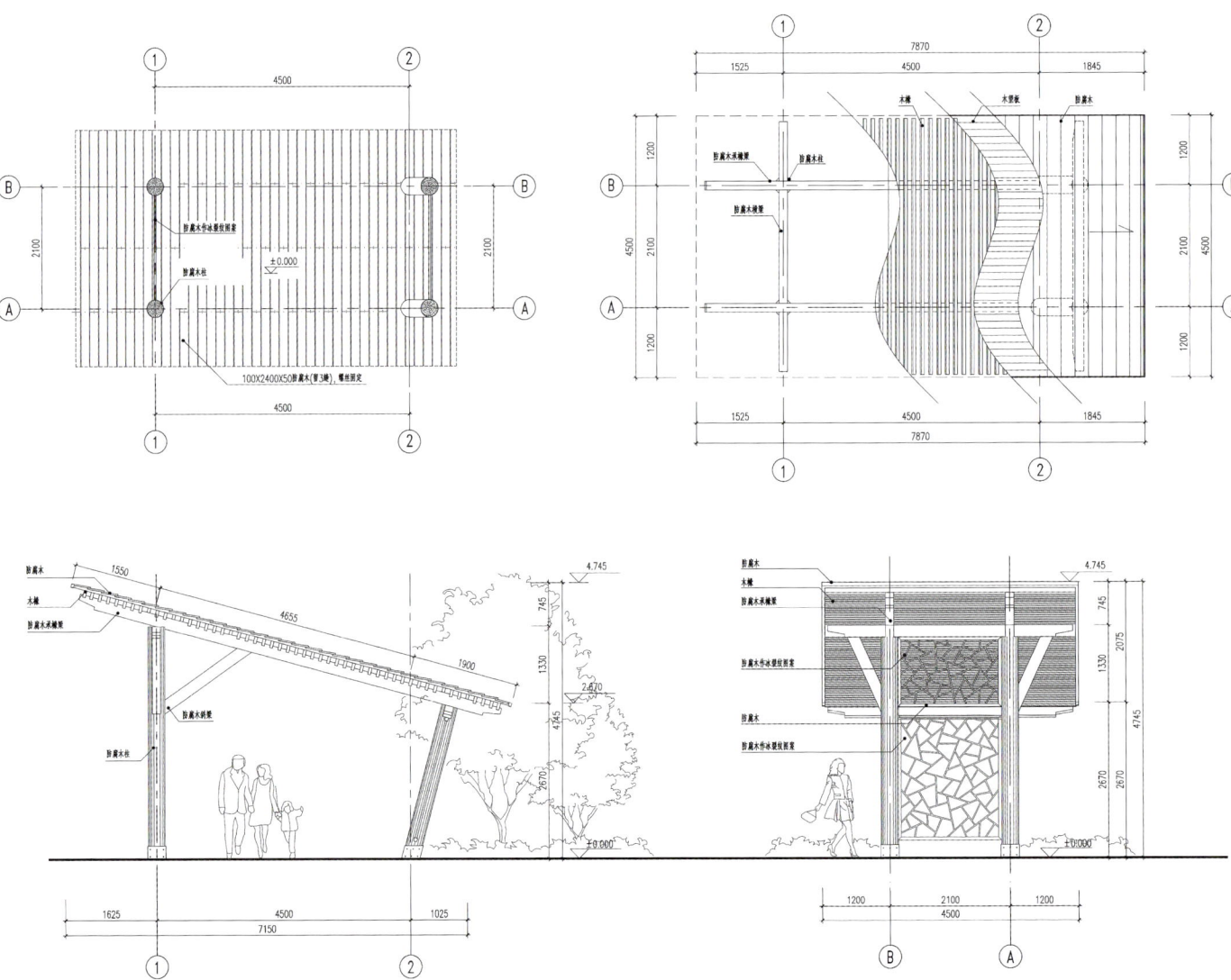

International Gardens
国际园

土耳其园 Turkey Garden

总面积约715平方米，旨在营造一种平民化的土耳其园林风格。设计将整个景观区域处理成两个院子，既相对独立又相互联系，分别以欧式喷泉，许愿树及拜占庭风格的亭子为中心，分配景观小品及流通路线，结合传统土耳其民居建筑，营造一种浓郁的土耳其式风格。

Turkey Garden covers an area of 715 m². It aims at creating a grass-root Turkish garden style. The whole garden is divided into two connected yards. A European fountain, a wishing tree and a Byzantine pavilion are set in the center of these two yards. Artistic landscape decorations, paths and traditional Turkish dwellings create a strong flavor of Turkish style.

希腊园 Greece Garden

总面积约500平方米，用现代科技表现希腊地中海地区的旅游观光文化和建筑特色。展园分为5个部分，每个分区都有一个中心元素。在每个区域都有一个画报横幅与屏幕，来表现各自不同的文化、旅游、建筑和美食特色。

The 500 m² Greece Garden shows the feature of tourist attractions, culture and architecture of Mediterranean region by modern technologies. The Garden is divided into five parts, each with a central unique landscape element. Also, the Garden applies a special pictorial banner and screen to create a visual feast of Greek culture, tourism, architecture and food.

国际竹藤园 Garden of INBAR

总面积约1,450平方米，国际竹藤园是以德国文化机构歌德学院赠送的艺术品——竹子展厅为主要建筑物的花园。梯形的园区内，3个布满各类竹子的开阔展厅等边分布，灯光透过玻璃幕墙，在静谧的夜晚更添大自然的神秘气息。

Occupying an area of 1,450 m², the Garden of International Network for Bamboo and Rattan (INBAR) features artistic bamboo exhibition halls, a gift from the German cultural organization Goethe Institute. In the trapezoidal Garden, three big exhibition halls full of bamboo are arranged as an equilateral triangle. In the evening, light comes out of glass wall, adding mysterious atmosphere to the nature.

阿富汗园 Afghanistan Garden

总面积约1,251平方米，以草药园、塔什库尔干园、帕格曼园和玫瑰园为骨架，配以凉亭、鹅卵石路、伊斯塔利夫陶器、雕刻木门等中亚典型庭院元素，充分表现了阿富汗古老的历史和丰富的民族特色。

Afghanistan Garden occupies an area of 1,251 m². It is designed with a framework consisting of the Herb Garden, Taxkorgan Garden, Paghman Garden, and Rose Garden. Within the framework, there are classical Central Asian elements such as pavilions, cobbled roads, Istalif potteries and carved wooden doors. The garden fully reveals Afghanistan's long history and distinctive national features.

缅甸园 Burma Garden

总面积约1,680平方米，缅甸展园采用东南亚建筑风格，从风格、色彩、理念方面充分体现东南亚的开放精神。佛教风格的寺庙建筑居于展园后方，开阔的空间旨在体现通透性、流动性与功能性的融合。

Occupying an area of 1,680 m², Burma Garden adopts architecture style of Southeast Asia, showing the openness of Southeast Asia in terms of style, color and concept. A Buddhist temple is built in the rear of the Garden, of which the open design integrates openness, flexibility and functions.

荷兰园 Netherlands Garden

总面积约2,714平方米，荷兰展园是由荷兰创意设计联盟主持设计的生态花园。园内建有最具特色的荷兰风车，五彩的郁金香如飘带一般将风车缠绕。在游览的小路上几双异地样式的木鞋更增加了北欧风情。

Netherlands Garden covers an area of 2,714 m². It is an ecological garden designed by the Netherlands Creativity Design Alliance. The most distinctive set is the Dutch Windmill which is twined by ribbon-like multicolored tulip. Several pairs of sabots randomly put on the sight-seeing path to create a unique romantic ambient from Northern Europe.

瑞典于默奥园 Umea Garden, Sweden

总面积约1,062平方米，以欧洲特色植物意大利铁线莲、忍冬及白桦树为主。该展园有一瑞典特色展品——发光的石头灯，周围的植物在夜晚灯光的照应下光彩夺目。树林中独特座椅让人忍不住坐下静静地享受欧洲庭院的惬意。

Umea Garden, Sweden covers an area of 1,062 m². The Garden features plantation coverage, especially the European plants such as Italian clematis, honeysuckle, and silver birch. The unique exhibit of the Garden is a luminous stone, which represents the characteristics of Sweden. It can emit light by itself and light up the nearby plants in the evening. The specially arranged seats in the small forest invite visitors to stay for a while and enjoy the peace in such a European garden.

孟加拉园 Bengal Garden

总面积约1,163平方米，展园体现简约、清新的设计理念，开阔的展园风格使内部景物一览无余，力求表现低碳环保主题。表演区域更使游客可以近距离接触极富传统孟加拉风情的文化活动。绿植主要有小叶女贞、梨树等。

The 1,163 m² Bengal Garden is endowed with a panoramic view, showing the design concept of simplicity and the theme of low-carbon lifestyle. The performance area allows the visitors to appreciate the traditional Bengal culture closely. The vegetation in the Garden consists mainly of Ligustrum quihoui and pear tree.

意大利园 Italy Garden

总面积约1,091平方米，意大利展园是由意大利都灵理工学院主持设计的托斯卡纳花园。展园设计与名称紧密呼应，先进的园林园艺技术与精致巧妙的布局结构相结合，别致的树墙和草坪中洒下点点阳光，让人仿佛置身于明媚的托斯卡纳。

Italy Garden, also named "Garden of Tuscany", occupies an area of 1,091 m². It is designed by Politecnico di Torino. The design is closely related to the name of the Garden. Advanced garden technology is applied to the elaborate layout in the Garden. Flowing light spots on the lawn cast by the sunshine and tree-wall bring you to the sunny Tuscany.

巴黎园 Paris Garden

总面积约1,508平方米，时尚之都，城堡花园，整洁优美，有着传统的法国花园的特色。园区主题是被花园环绕的城堡，突显了欧洲城堡大气、庄严的感觉。通往城堡的小径、精心修剪的树木、整洁美观的喷泉，处处彰显时尚之都巴黎的风采。

Paris Garden, occupying an area of 1,508 m², reveals the fashion of Paris and characteristics of traditional castle gardens in France. The theme of the Garden is a castle amidst garden, showing the magnificence and majesty of European castles. Paris, the capital of fashion, has its charm brought to the Expo with panes leading to the castle, carefully arranged trees, and neat fountain design.

巴基斯坦园 Pakistan Garden

总面积约747平方米,巴基斯坦园重在体现莫卧儿王朝的建筑园艺理念,由一栋伊斯兰风格的单体建筑与四座花坛组成。花坛中心的黑玛瑙喷泉象征水是万物生命之源,花坛中红、黄、白、粉四色玫瑰分别象征了构成和谐世界的不同基本要素。

Pakistan Garden occupies an area of 747 m². The Garden highlights the architectural and horticultural style of the Mughal Dynasty. The main structure of the garden consists of an Islamic monomer building and four flowerbeds. The black onyx fountain at the center of the flowerbeds symbolizes that water is the source of life; the red, yellow, white, and pink roses blossoming in the flowerbeds embodies the fundamental elements essential for a harmonious world.

印度园 India Garden

总面积约502平方米,印度展园依照古印度寺庙风格建造,整个花园布局体现出完美的几何对称性,两座大门分列两侧,主体建筑位于园区正后方。展园内放置有供游客休息的石质座椅,金钟柏和木槿围绕展台四周,营造出静谧的氛围。

India Garden occupies an area of 502 m². The Garden features temple style of ancient India. The overall layout adopts geometric symmetry with two gates on either side and the main building sitting in the middle of the backside. Stone seats are placed in the Garden for visitors to relax. Arborvitae and hibiscus surrounding the building creates a peaceful atmosphere.

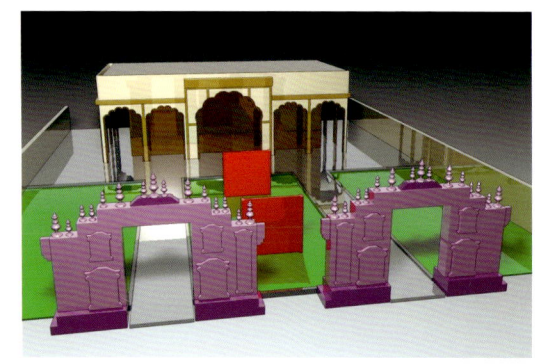

尼日利亚园 Nigeria Garden

总面积约2,149平方米,设计通过沙生植物、水生植物、热带雨林植物的变化,勾勒出"沙漠景观——稀树草原景观——热带雨林景观"的设计层级,用图腾雕塑体现了非洲的民风民俗和景观特色。植物配置多元,园路自然流畅,建筑风格独特,达到了层次分明、特点突出的景观效果。

Nigeria Garden occupies an area of 2,149 m². It outlines a "desert-savanna-tropical forest" landscape structure by rendering the different plants in desert, water and tropical forest. The totem sculptures give expression to the folkways and unique landscape of Africa. The well arranged Garden creates special landscape by various plants, curved path, and unique architecture.

尼泊尔园 Nepal Garden

总面积约1,087平方米,大量来自于尼泊尔国内的文化展品赋予尼泊尔展园浓郁的文化气息。

The Nepal Garden is 1,087 m². A large number of cultural exhibits from Nepal create a rich culture flavor for the Garden.

哈萨克斯坦园 Kazakhstan Garden

总面积约1,123平方米,主要有2个建筑物,分别是哈萨克斯坦展厅及东干协会展厅。建筑物以蒙古包为设计基础,寓意着崭新的哈萨克斯坦为古代文化赋予新的色彩。

The Kazakhstan Garden is 1,123 m². It mainly features two buildings: Kazakhstan Exhibition Hall and the Exhibition Hall of Donggan Association. Most of the architectures in Kazakhstan are yurt-style today, representing the ancient culture with modernism in the brand new Kazakhstan.

俄罗斯园 Russia Garden

总面积约2,288平方米,俄罗斯展园以传统恬静的乡村风格为主调,通过传统原木木屋、小取水井、露天浴场的安排,完美呈现了普斯科夫洲Izborsk小镇古老而悠久的农场原貌。被果园环绕的木屋使用整根木料加工,在俄罗斯是古老而传统的建筑工艺。

Russia Garden occupies an area of 2,288 m². The Garden features serenity of countryside life. The original appearance of ancient farms of Izborsk in Pskov is perfectly illustrated by the artfully arranged traditional wooden houses, small wells, and outdoor bathing place. The wooden house in the Garden surrounded by orchard is built with one-piece wood, which is a traditional building craft in Russia.

菲律宾园 Philippines Garden

总面积约574平方米，静谧小岛，雨林风光，椰风阵阵的温婉气息，仿佛使人置身于清新的海岸，热情的舞蹈，阳光的微笑，海水云天的景色，是让人流连忘返的好地方。

With the atmosphere of tropical rain forest, Philippines Garden occupies an area of 574 m². The Garden features a small peaceful island of Philippines, unique tropical landscape amidst coconut trees bringing to the visitors a fresh experience on seashore. It is an enjoyable relaxing place with blue water and white clouds for passionate dances and bright smiling faces.

玻利维亚园 Bolivia Garden

总面积约1,350平方米，玻利维亚展园重在突出安第斯山风貌，穿插以玻利维亚盐湖景观石树雕塑及印第安传统服饰和舞蹈，园区内部的石质广场、湖泊和亲水舞台的现场音乐演出，把玻利维亚历史文化及民风民俗以鲜活灵动的方式展现给所有游客。

Bolivia Garden occupies an area of 1,350 m². The Garden highlights the landscape of the Andes Mountain. The stone-tree sculptures of Bolivian saline lake lie around the garden, and live shows of Indian costume, dance, and music are presented on the stone-paved square. It will bring visitors to the history, culture, and folkways of Bolivia with liveliness and agility.

新西兰园 New Zealand Garden

总面积约1,167平方米，展园以高科技花园为主线，通过科学的设计方法和实用的设计理念，将甲板、植物、沼泽、湿地等元素完美融合，解决排水和植物根系对建筑物伤害的问题。园区内所有植物均从新西兰当地进口，有着非常高的观赏价值。

The New Zealand Garden occupies an area of 1,167 m². The landscape elements such as deck, plants, marsh, and wetland are perfectly integrated by scientific design method and practical design concept to solve the problem of damage caused by draining and plant roots. All plantations in the Garden are imported from New Zealand, providing high ornamental value for the visitors.

加蓬园 Gabon Garden

总面积约623平方米，方格造型的外装修，原生态方形景亭，碎石铺装与青石铺装体现中非风格，特色的建筑外观与世园会口号"绿色引领时尚"紧密结合，体现节能、低碳、环保的主题。

The Gabon Garden is 623 m². The exterior decoration of the cube-shaped ecological building features gridiron pattern. Central Africa style can be seen in the pavement made of gravel and bluestones. Special architectural style gives expression to the theme of ecology and echoes with the slogan of Expo 2011 Xi'an – "Green Leads the Trend".

德国奥尔登堡园 Oldenburg Garden, Germany

总面积约2,680平方米，展园内铺装特色道路，代表奥尔登堡历史踪迹，中央有一隐形花带，布置体现德国特色的花卉。靠近湖面的沙滩及亲水平台，给人一种亲临海边眺望大海的意境。

Oldenburg Garden Germany is 2,680 m². Special lanes are paved in the Garden representing the history of Oldenburg. An invisible flower belt in the center of the Garden grows special flowers from Germany. The beach and waterfront deck near the lake shows the artistic conception of seashore view.

日本北海道园 Hokkaido Garden, Japan

总面积约844平方米，园区完全是花的海洋，突显自然花卉的美丽。独一无二的紫色醇香的薰衣草，代表着北海道的美丽和浪漫。多种色彩的花朵交织，如同彩虹般绚烂。

Hokkaido Garden, Japan occupies an area of 844 m². The Garden is a sea of flowers, highlighting the beauty of flowers. Unique purple lavenders represent the beautiful and romantic Hokkaido. The interweaving flowerbeds in different colors look like a rainbow carpet.

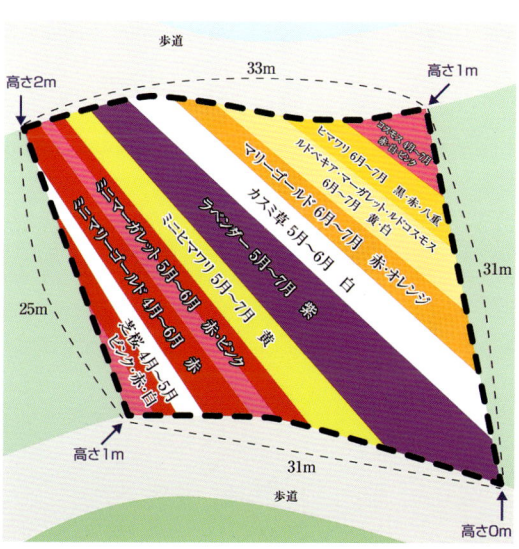

日本横滨风园 Yokohama Style Garden, Japan

总面积约804平方米，日本横滨风园采用"和谐的环境——新旧融合"的设计理念，主要采取江户时代的池泉漫步模式。在展厅上，采用日本最先进的迅速蒸发喷雾系统、绿化屋顶、太阳能风能等技术。

Yokohama Style Garden, Japan along the European Avenue occupies an area of 804 m². The Garden adopts the design concept of "Harmonious Environment – New in Old" with the promenading mode around pond and fountain originated from the Edo Time. The exhibition hall adopts the most advanced Japanese technologies such as spray system, green-roof, solar and wind power.

斯里兰卡园 Sri Lanka Garden

总面积约489平方米，可称作是佛教文化与园林园艺的完美结合。作为南亚的佛教国家斯里兰卡，其展园在紧扣佛教主题的同时也散发着南亚异域风情。

Sri Lanka Garden covers an area of 489 m². It is a perfect combination of Buddhism and garden art. Sri Lanka, a Buddhist country in South Asia, presents a garden with both Buddhist style and South Asian flavor.

日本奈良园 Nara Garden, Japan

总面积约501平方米，展示了日本园林园艺最精髓最核心的内容，其中的植物、石头以及布局等都体现了日本文化对园林艺术的诠释。

Nara Garden, Japan covers an area of 501 m². It presents the essence of Japanese garden art. The garden art in Japanese culture is fully illustrated by the arrangement of plants and stones.

韩国园 South Korea Garden

总面积约910平方米，主要展示韩国顺天特色亭子——爱莲亭，体现韩国特色文化。

South Korea Garden covers an area of 910 m². The main exhibit of the Garden is Aeryeonjeong, a Bower in distinct Suncheon style to show visitors the special culture of South Korea.

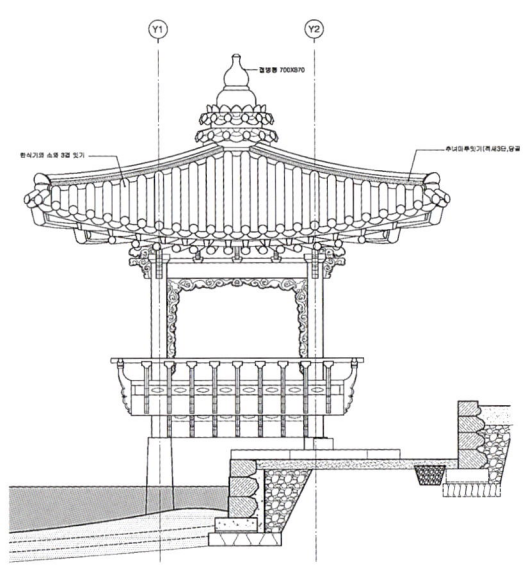

朝鲜园 DPRK Garden

总面积770平方米，展园特色以朝鲜历史悠久的主题公园为主，主要由3个部分构成，即"历史悠久之国"、"山清水秀之国"以及"自由和平之国"。青瓦白墙建筑、斗拱、云柱、绿色的瓦片、棕色的门，都充分展现了朝鲜的传统建筑风格。

DPRK Garden occupies an area of 770 m². The Garden is a theme park featuring the long history of Democratic People's Republic of Korea, featuring three themes including "country of long history","country of natural landscape" and "country of freedom and peace". The exhibition building with white walls, bucket arches, cloud columns, green tiles, and brown doors fully conveys the traditional architecture style of Democratic People's Republic of Korea.

泰国园 Thailand Garden

总面积约1,012平方米，建筑就像田园中升起的宫殿。一座极具古典风格的泰式建筑，将泰国多元化的艺术文化和现代化的建筑物浓缩在展馆的整体设计中，展现出浓郁的泰式庭院景观风格。

Thailand Garden covers an area of 1,012 m². The exhibition hall is a classical Thai-style building, which looks like a palace in the natural landscape. Rich elements of art, culture and modern architecture in Thailand are integrated into the overall design of the building, fully showing the Thai garden style.

World Gardens
世界庭院

世界庭院位于整个园区南部的一个约18,000平方米的小岛上,四面环水,是一个以展示西方园林的发展体系、造园艺术、园林形式为主的独立展园。世界庭院景区是集中展示世界各国各地区庭院特色的景区,集合了不同体系园林的空间形式、造园要素。世界庭院根据西方园林发展演变的时间顺序,设计游览路线和空间布局,让游人在短短的游程里全面地领略到人类历史长河中西方园林发展历史及园林风格特点,感受西方园林的不同魅力。根据各时期不同风格园林的空间形式、造园要素等特征,以某个经典且能代表当时园林特点的作品的片段为骨架,融合所在时代和国家的经典造园元素进行展园的设计。

古希腊园
古希腊园位于世界庭院的北侧,大约2,000平方米。希腊园分3个部分,从西向东依次为索罗斯广场、圣林、狮子门。希腊园以希腊德尔斐遗址广场(索罗斯广场)为骨架,以展示对西方建筑影响深远的希腊柱式以及完美比例的运用。

意大利台地园
占地约4,000平方米的意大利台地园设计截取法尔奈斯庄园的部分为主要骨架,采用中轴对称的构图形式,中间轴线是一条从平台到观景廊的宽大缓坡。

法国勒诺特尔园林
在勒诺特尔式园林里,采用主从分明,秩序严谨的几何网格。府邸位于最高处,起着统率的作用。同时,在贯穿全园的中轴线上,用花坛、雕像、泉池等加以重点装饰,形成全园的视觉中心。

英国自然式风景园
英国自然式风景园将假山、草地、亭子、花境、岩石等元素融合在一起,与周边的水面形成一个自然意境,一派田园风光显露于眼前。

西班牙伊斯兰园
西班牙伊斯兰园林又被称为摩尔园林,展园的设计原型来自西班牙著名的阿尔罕布拉宫狮子院,面积约1,000平方米。

The World Gardens Section is located on the 18,000 m² island in the south of the Expo Site. This Section is surrounded by water, showing the landscape design development of the Western gardens. In this Section, it exhibits various unique and classic gardens all over the world, which is a huge collection of many space forms and other design elements for garden landscape. The tour route and space layout in this Section is designed by the evolution of Western garden landscape to show visitors the difference of garden design between the West and China, fully presenting the fascination of Western gardens. The gardens are designed with typical landscaping elements of the specific time and country.

Ancient Greek Garden

Ancient Greek Garden is located in the north of the World Gardens Section, covering an area of about 2,000 m². The Garden consists of three parts: Soros Square, Sacred Wood, and Lion Gate from west to east. It takes the square of Greek Delphi site (Soros Square) as its main body to show the application of perfect proportion and Greek column style with a far-reaching influence on Western architecture.

Italian Terrace Garden

The Garden occupies an area of 4,000 m² in the World Gardens Section of Expo 2011 Xi'an. The design takes part of Villa Palazzina Farnese as the main framework and adopts the central axis symmetric layout. A wide gentle slope from the platform to the viewing corridor constitutes the garden's central axis.

French Le Notre Garden

Le Notre was a master of French classical garden design. The Garden adopts the geometric grids with clear subordination and rigorous order. The mansion is located at the highest point of the Garden, standing like a command. The central axis through the whole garden heavily decorated with flowerbeds, statues, fountains and pools is the visual center of the Garden.

British Natural Landscape Garden

British Natural Style Landscape Garden integrates elements such as rockeries, grass, pavilions, flower borders, and rocks against the surrounding water in the Garden to create a beautiful idyllic scene.

Spanish Islamic Garden

Spanish Islamic Garden is also known as Moorish garden with an area of 1,000 m², which mimics the famous Court of Lions in Alhambra Palace, Spain.

Garden with a Labyrinth in the Mountain

大师园·山之迷径花园

展园位于世界园艺博览会大师园1号地块,由西班牙新锐建筑事务所EMBT创立人之一的Benedetta Tagliabue女士担任主创设计师。展园的整体设计独辟蹊径,EMBT将中国传统的美学思想化入展园的平面布局创作。他们从中国山水画的肌理中提取灵感,将国画中的气韵脉络凸显出来,形成平面空间的划分区隔。EMBT将主入口设置在地块的东南角,而在西北向安排自由线形作为引导游客参观的主体流线。这种绚丽的处理手法夺人眼球,让观者充分领略到了EMBT在设计中对于复杂形式的控制能力。园中主道以竹篱为墙、藤片为顶,组合形成了作为展园核心的廊架结构,并以卵石铺地,攀缘植物盖满藤顶,让观者在游走的过程中,亲密接触自然,将人造景观与自然景观之间的界限进一步模糊。对于地块原本具有的标高落差,EMBT更视之为一种自然的"恩赐"。景观的高低起伏增强了园景的生命力,同时造成了展园内微环境与小气候的变化。EMBT根据不同植物的特性,将松柏类植物种植在展园中地势较高的地方,而花灌木和草本植物则被种植在地势较低处,通过逻辑性的安排,使展园具有清晰的层次。

自2010上海世博会中西班牙馆项目获得各界的肯定,EMBT在国内声名鹊起。该事务所以重视城市文脉著称,并且以精细繁复充满理性光辉的平面图让许多同仁都为之咋舌。在受到本届世园会的邀请时,设计师Benedetta Tagliabue坦言,她并不清楚中国人能否接受EMBT对于中国文化的解读,但是希望能够借此机会表达西方人眼中的东方文化。EMBT从西方人的立场,将传统的东方文化进行了富有现代精神的转译,从观者的体验出发,以多变的手法进行了空间的处理以及设计元素的整合;从一名建筑师的眼光出发,突破了基地的限制,平衡了人造之美与天然之美、空间体验与景观分布,将展园塑造成为连接东西方文化的纽带。

The Garden is located at the No. 1 Plot of Master Gardens Section in the Expo Site. It is mainly designed by Benedetta Tagliabue, one of the founders of Miralles Tagliabue EMBT, Spain. A new design style is developed for the overall design of the Garden. EMBT integrated Chinese traditional aesthetical ideas into the layout of the Garden. In the light of the venation of Chinese paintings, the layout and space division of the Garden looks like the painting. Its main entrance is set at the southeast end of the Plot, and the wandering alley in the northwest is the main line for visitors. This brilliant design attracts visitors' attention, from which they can fully appreciate the control capability of EMBT on complex forms in design. Bamboo fences act as walls of the main passages in the Garden, and vines as the roof. Screes are paved on the ground, and climbing plants cover the vine roofs, which blurs the boundary between artificial and natural landscapes, making people get close to the nature during the wandering. The elevation discrepancy of the Plot is fully taken advantage and considered by EMBT as a boon from the nature. The up and down of the landscape terrain added animation of the Garden, and caused the change of separated environments and climates. Plants like pines and cypresses are planted in the area at a high elevation, while plants like shrubs and herbs are planted in the area at a low elevation, according to their different characters. This logical setting endowed the garden with clear arrangement.

EMBT shot to fame in China since its design of Spain Pavilion won wide recognition in Expo 2010 Shanghai. EMBT is renowned for its emphasis on urban context. Counterparts are often astonished by its fine and elaborate layout design which is full of rational brilliance. When invited by this Expo, Benedetta Tagliabue confessed that she was not sure whether Chinese visitors would accept EMBT's interpretation of Chinese culture. However, she wanted to take this chance to show the eastern culture in the perspective of westerners. From the point of westerners, EMBT gave explanation with modern spirits to traditional eastern culture. They arrange the whole space and combine design elements with different forms to give visitors better experience. As an architecture firm, EMBT broke through the limit of the terrain and balanced the artificial beauty and natural beauty, as well as the space definition and landscape arrangement, thus making the Expo Site a ligament between eastern culture and western culture.

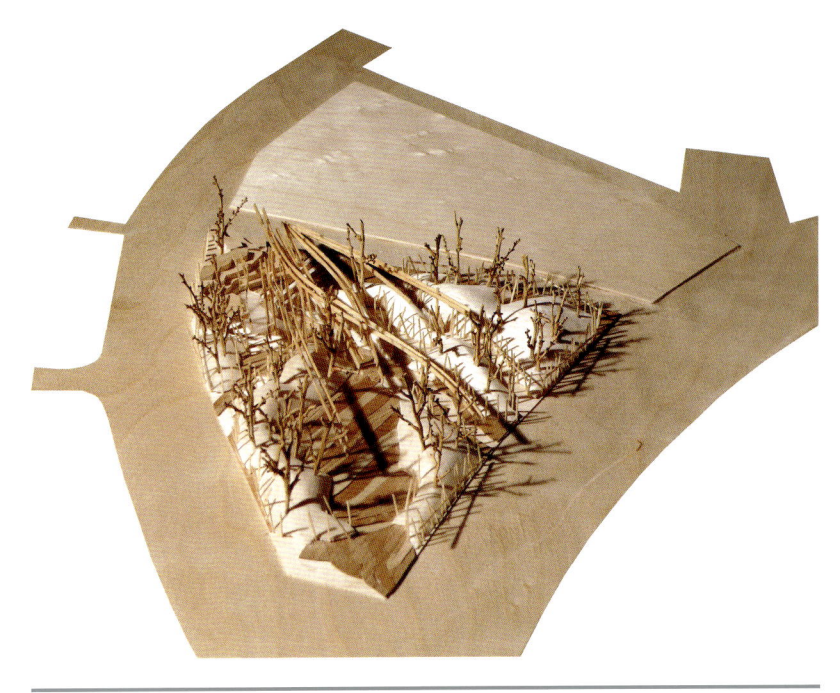

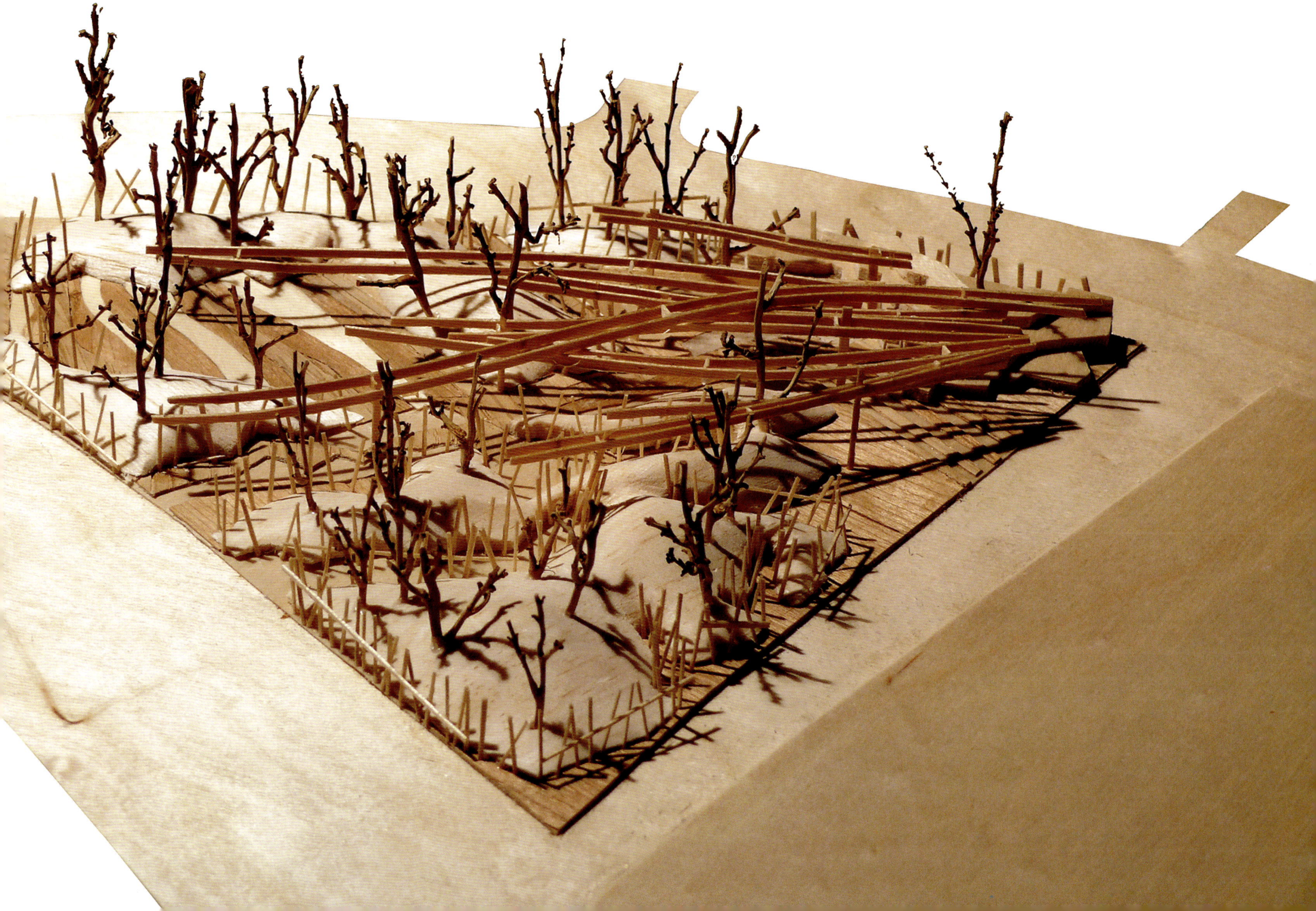

BAMBOO PERGOLA FOR FLYING VEGETATION

PARTHEROCISUS TRICUSPIDATA
WISTERIA SINENSIS
CAMPSIS GRANDIFLORA

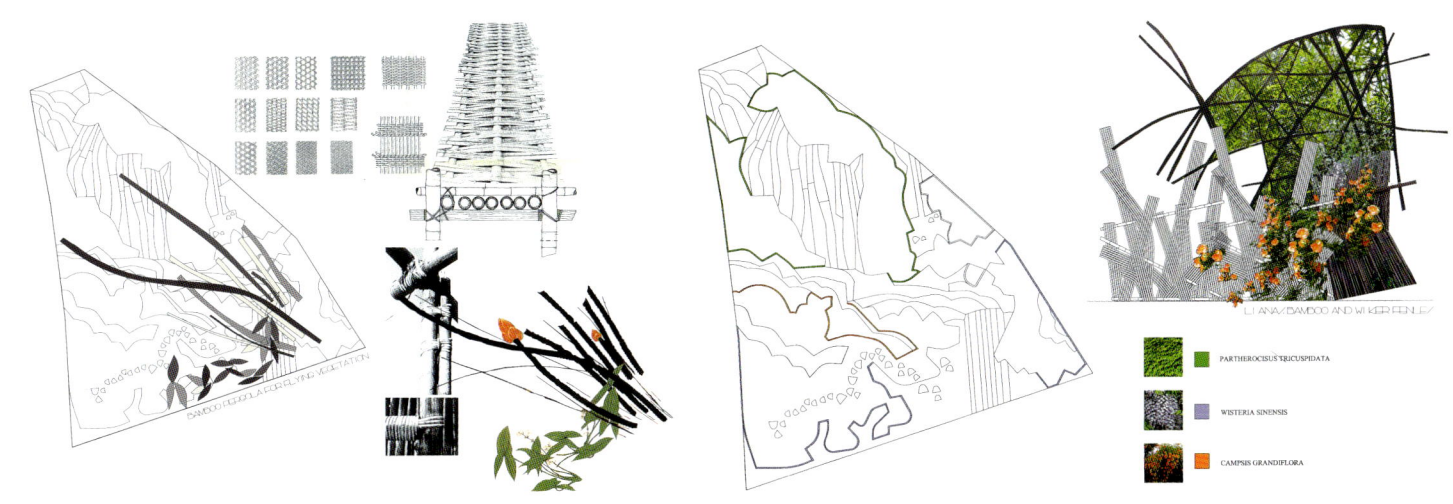

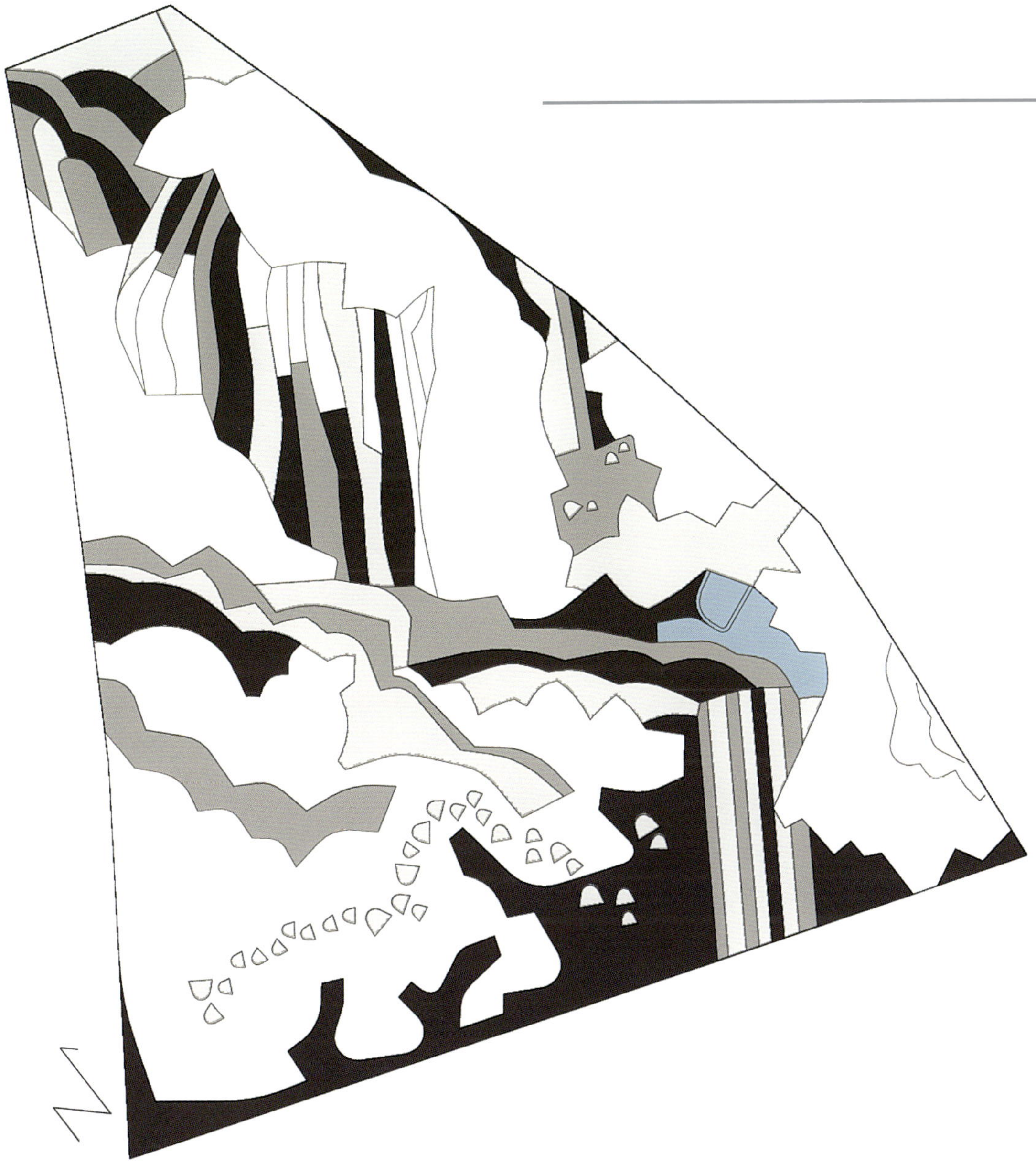

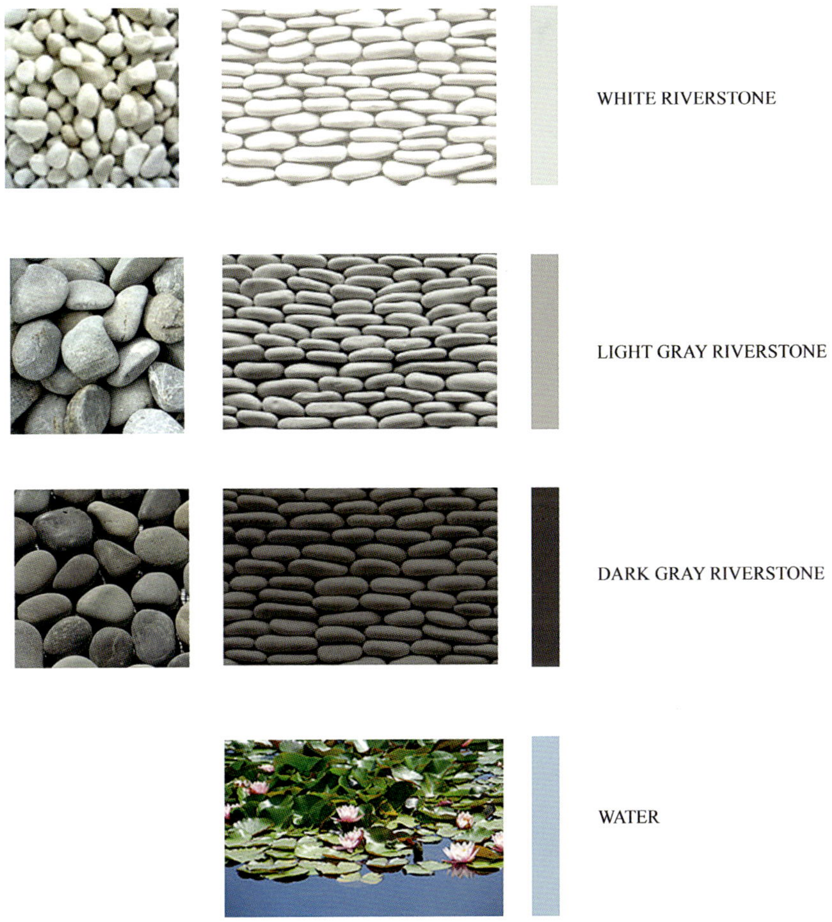

WHITE RIVERSTONE
LIGHT GRAY RIVERSTONE
DARK GRAY RIVERSTONE
WATER

Botanist Garden

大师园·植物学家花园

展园位于世界园艺博览会大师园2号地块，由英国Gross. Max.景观设计事务所主创设计师Eelco Hooftman先生领衔设计。展园由前庭、中心庭院和内室三部分组成，平面布局清晰简练，并且运用朴素的材质进行有机的组合，表达了不同空间的气质。前庭作为展园内外的过渡空间，选用简单的材质和适当的留白成为本案的设计策略。前庭以条形青砖为墙，波浪形陶砖铺地，在有限的面积中既体现了关中平原所具有的气度与情怀，又通过不同的纹理使得空间具有活力。在这块露天空间中，设计师使用电线将数个装有蝈蝈的圆形竹笼悬挂起来，让展园在起始处便具有了独特的魅力。绕过青砖照壁所进入的，就是中心庭院。一条同样是条状青砖铺就的折线小径穿过中心庭院，成为引导游客的线索，与前庭的墙壁进行呼应，并且让两个不同空间之间有了流畅的过渡。小径四周遍植水杉，在有着"活化石"之称的水杉树脚旁栽种着低矮的开花的植物和蕨类植物，与高大的水杉形成观感上的错落。在小径的尽头，有一个被园墙围合的空间，空间外墙是瓦片堆叠而成，瓦片之间形成的波浪形纹理，与前庭的地砖相似，从而将前庭古朴的气质延续了下来。正对小径处，有一个被绿植掩映的狭窄裂口，被充当为圆形空间的入口。进入内院，满眼都是白色开花植物，豁然开朗。这是一处向英国著名植物引种家E·H·威尔逊（1876-1930）致礼的花园，所有使用的植物都是威尔逊在中国各地搜集而来的品种。当游客们沉浸于花海中，情绪被调动到了最高点时，展园在此处点明了主题，而整体设计也随着最强音利落地收尾。

Gross. Max.景观设计事务所作为英国著名的景观设计机构，向来以"创意"作为设计的主导思想。在本案中，设计师通过前庭悬垂的蝈蝈笼，作为其对于创造"超现实体验"的强力手段，与他选用的代表了西安特色的元素以及丰富的植被一同，形成了立体的环境氛围。

Botanist Garden is located at the No. 2 Plot of Master Gardens Section. It is mainly designed by Eelco Hooftman, a chief designer of Gross Max Landscape Architects, Britain. The Garden is composed of three parts: front yard, central yard and inner yard. Its clear and simple layout applies the combination of simple architectural materials to express different atmospheres of different spaces. The front yard acts as a transition space between the outside and inside of the Garden. Simple construction materials and proper blank are used for the front yard. In the limited space, the walls built of banded grey bricks, and the ground paved with waved vitrified bricks display the special temper and affection of people in Kuan-chung plain. The different textures also add animation to the limited space. In this open-air area, several cages with crickets inside hung by electrical wires give special charm to the garden at the very beginning point. Along the grey brick made wall, visitors enter into the central yard. A trail paved with the same grey bricks crosses the whole central yard, serving as the guiding trail for visitors. The trail with its design coherent with that of the wall in the front yard serves as a natural transition between different spaces. Metasequoias with the title of "Living Fossil" are planted around the two sides of the trail, while short flowering plants and pteridophytes are planted at the foot of high Metasequoias, which gives a vision of different layers. At the end of the trail, there is a space surrounded by round walls. The outer wall is stacked with tiles. The wave texture formed between the stacked tiles assembles that of the bricks paved on the ground of front yard, thus extends the atmosphere of simplicity and sublimity from the front yard to the central yard. Against the trail, there is a narrow crack covered by green plants, serving as the entrance of the round space. When entering into the inner yard, visitors will see white flowering plants everywhere and feel refreshed instantly. This inner yard is built in honor of a famous plant introducing botanist E·H·Wilson (1876-1930). All species of the plants in the yard were collected by Wilson all over China. The garden points out its theme and also comes to its end when visitors are immersed in the sea of flowers with emotion being raised to the highest point.

As a famous landscape design institute, Gross Max Landscape Architects always takes "innovation" as its guiding idea for landscape design. In this project, the designer applies the hanging cricket cages in the front yard to create an experience of "super reality". Along with those elements full of Xi'an features and colorful plants, they form a vivid tridimensional environment.

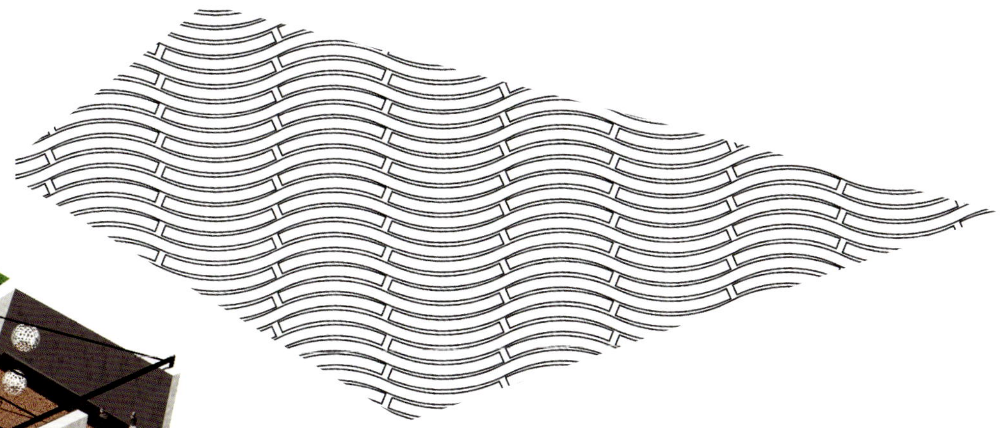

Passageway Garden
大师园·通道园

展园位于世界园艺博览会大师园3号地块，由澳大利亚Terragram事务所总裁Vladimir Sitta先生担任主创设计师。在本园中，明晰的平面表达和含蓄的景观呈现共同作用，形成了通道园丰富的内涵，同时代表了西方设计师对东方文明的又一种解读方法。展园的整体空间被南北向的四面墙体分隔为三大部分，墙体从西至东分别是冬墙、春墙、夏墙和秋墙，各自表面都根据指示季节的不同而被覆上了深浅不一的橙色系涂料。展园内的三个空间由一条水流纵贯相连，整体气脉首尾呼应，而不同空间带给游客的体验却各有不同。冬墙色淡，与颜色稍艳的春墙形成空间，与之后的夏墙一起形成了第一空间，主要还是起着接引作用。为了达到先声夺人的效果，设计师在此处设置了丰富的植被，高树与低矮的灌木错落而置，生机勃勃。而平坦光滑的步道旁，卧着的那一弯碧水在绵延向前时却并不安静，从水中氤氲而出的蒸汽仿佛昭示着一种力量正蓄势待发。红色的夏墙与春墙紧邻设置，由于地势和入口处相比有自然低陷，水流到了此处，随着力量喷薄而出过没了步道表面，明火与水流交织在一起，在有限的空间里表达了对伟大的自然力量的景仰；夏墙之后，便是金黄的秋墙，在两面墙之间便是展园的第二空间。水道在此处开始变窄，而步道也在此处戛然而止。与步道尽头相接的是柔软的沙地，游客到了这里心情可以放松下来，甚至连脚步都可以变得慢起来。当游客在前行的过程中看着自己留下的足迹，似乎内心也可以跟着节奏变得柔软；穿过秋墙的开口并不与之前墙体上的三个在一条线上的整齐切口相类似。为了给游客提供更多的游览体验，设计师将开口设置在秋墙的南端，并在开口的顶端悬挂大石，让沉浸内心的游客们猛然惊醒，注意到了位于第三空间内的月洞门。作为第三空间中最为醒目的标志，月洞门除了代表设计师从中国传统园林中得到的灵感以外，更是对生命的隐喻。这座并没有完全闭合的月洞门，在两端伸出的松树完全违背了自然规律，设计师通过这处人造景观，与表示"生生不息"的圆形一同，对"人定胜天"的观点提出了自己的疑问。

在小规模造景上有深厚造诣的Vladimir Sitta坚信时间将是考验设计的最好标准，他也将此作为自己的目标。在设计方案构思过程中，就四面墙的开口设计，Vladimir Sitta曾提出了25种设想，希望能以更为朴素的方法表达传统的审美情趣，而最后他只选用了3种。在本案中，Vladimir Sitta放弃了22种经过思考的设计手法，这是他站在西方人的立场上对于中国古典范例的解读，同时也表现了一个成功的设计师对于细节的一丝不苟。

The Garden is located at the No. 3 Plot of Master Gardens Section. It is mainly designed by Vladimir Sitta, president of Terragram, Australia. The combination of clear plane expression along with implicit landscape forms a rich connotation of the garden. Meanwhile, it represents another understanding of the eastern culture by a western designer. The entire space is divided into three parts by four walls facing north and south. From west to east, they are winter wall, spring wall, summer wall and autumn wall respectively. On the surface of the wall, different coatings of orange series are painted by characters of different seasons. The three parts of the Garden are linked together by a water flow. The end of the garden echoes with its beginning. Different spaces bring diverse experiences to visitors. The color of the winter wall is light and the spring wall is bright. The space between the winter wall and the spring wall, as well as the space between the spring wall and summer wall forms the first space of the Garden to mainly serve as a connection. In order to achieve an impressive effect, the designer arranges rich plants at this point. The high trees and low shrubs are well arranged and full of vigor. Along with the flat and slippy side walk, lying a blue water flow, wanderingly running forward and causing a little noise. The steam above the water seems to display a power which is poised for take-off. The red summer wall is closed to the spring wall. Due to the low terrain compared to the water entrance, when the flow arrives here, water submerges the surface of the side walk. The red flame and the water flow are interwoven together. It demonstrates the respect and aspiration to the great nature in the limited space. The golden autumn wall is set behind the summer wall. Between the two walls is constructed the second space. The waterway becomes narrow here, while the sidewalk ends abruptly. Connected to the end of the sidewalk is the soft sand. Vistors may ease their mood while stepping here and even slow down spontaneously. Looking at the footsteps left behind, the visitors seem to become relaxed. The opening on the autumn wall differs from the other three on the winter, summer and spring walls with tidy cuts. In order to provide visitors more adventure experiences, the designer sets the opening at the south end of the autumn wall with a big stone hung on the top. This special design will wake up the visitors immersed in their own world instantly and enable them to pay attention to the moon gate in the third space. As the most outstanding symbol in the third space,

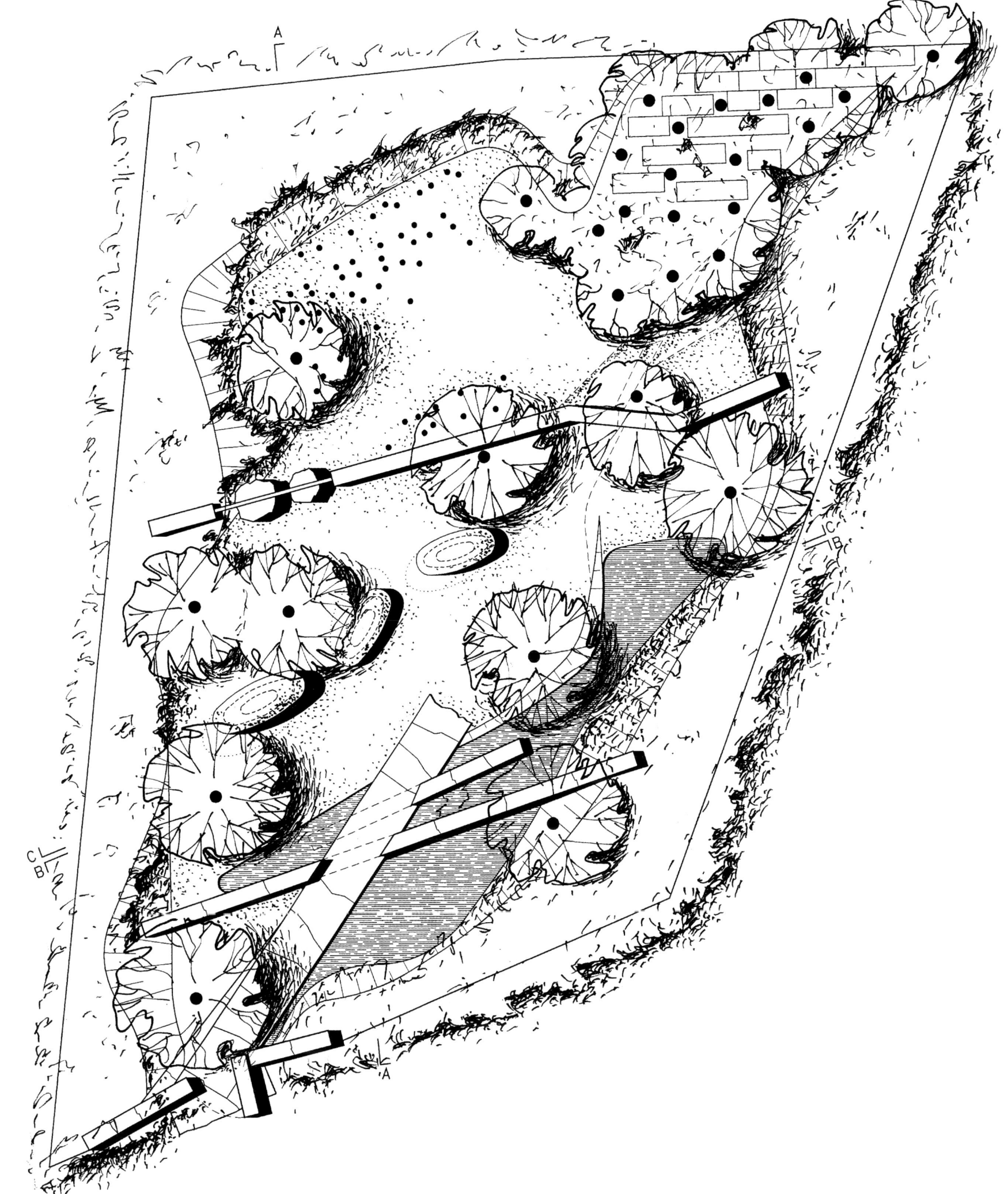

taking from the Chinese traditional garden, the moon gate is also a metaphor to life. This moon gate is not completely closed. The pine extended from the two sides completely violates the natural law. The designer presents his own doubt to the view that "man can conquer the nature" through this artificial landscape and the circle which represents endless reproduction.

Vlaimiir Sitta, with profound accomplishment on small-scale landscape architecture, firmly believes that time is the best standard to test the design and this is also his target. In the design process, Vladimir Sitta proposed 25 assumptions on the openings of the four walls. He wanted to show the traditional aesthetics through more plain ways. Finally, only three assumptions are applied. In this project, Vladimir Sitta gave up 22 thoughtful designs. It is his understanding of ancient Chinese classical cases at the point of a westerner, which also shows the meticulosity to the details of a successful designer.

Big Dig Garden
大师园·大挖掘园

展园位于世界园艺博览会大师园4号地块，由德国Topotek1设计公司主创设计师Martin Rein-Cano先生领衔设计。大挖掘园是一个完全开敞的展园，一个巨大的地洞是展园中绝对的主角。与毗邻的通道园和四盒院相比，它更像是一个过渡空间，一个填充了两园之间空白地带的装置，"然而它确实就是一个园林。"主创设计师Martin Rein-Cano强调。在大挖掘园中，游客可以在这个深不见底的地洞周围，聆听从洞内音频设备中发出的来自世界另一端的"声音"。设计师选择了阿根廷、美国、瑞典和德国作为外国站点，分别将代表这四个国家的不同声音作为东西方文化交流的媒介，吸引游客对那些未知国度进行探究和想象。

德国Topotek 1设计公司作为当今德国最有活力的新锐设计事务所之一，他们希望能从与众不同的角度发掘现代景观可以具有的更多可能性。在本案中，他们以最简单的空间布局，将主题进行凸显，从而使得主题的冲击力进一步加强，也让他们的设计在整个世界园艺博览会的展园中带上了实验性的色彩。然而，在这个看似简单的设计中，也包含了当代德国设计对于高科技的敏感和追求，展园中的地洞不仅是简单地挖了一个口径10米的洞，设计师考虑到世界是圆的，因此场所不再是二维平面，而是三维小球体。故而他们创造了一个10米宽混凝土壳层结构，类似于一个开口向上扩音器的外形，明确了设计师希望通过"大挖掘园"表达的东西方文化交流的内涵。

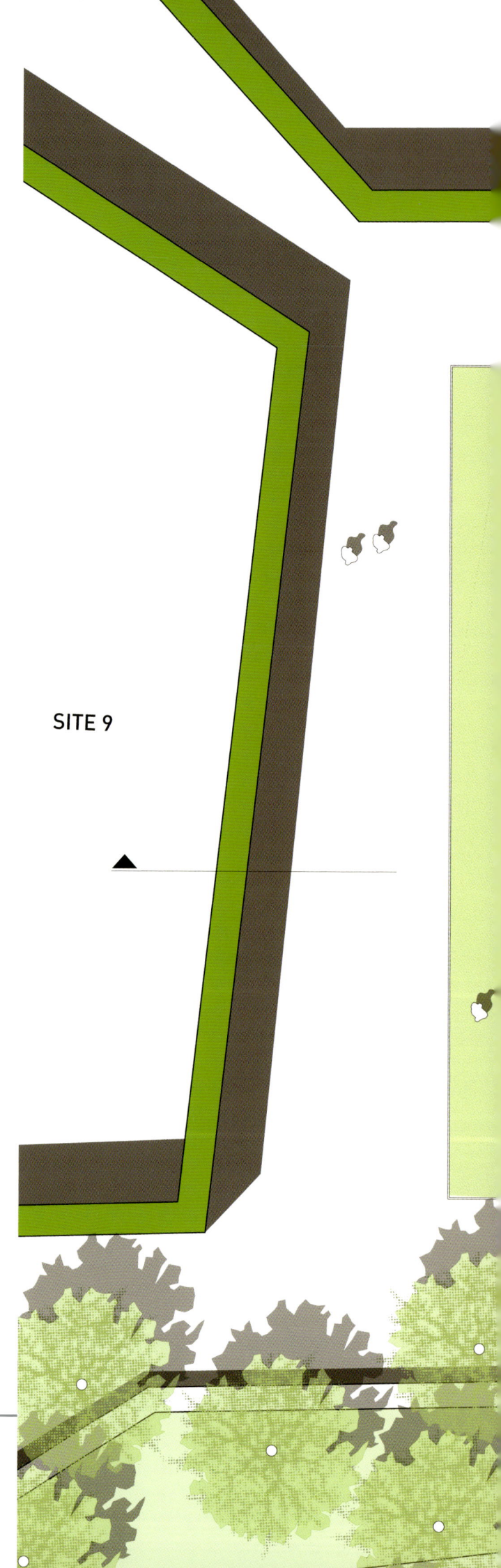

SITE 9

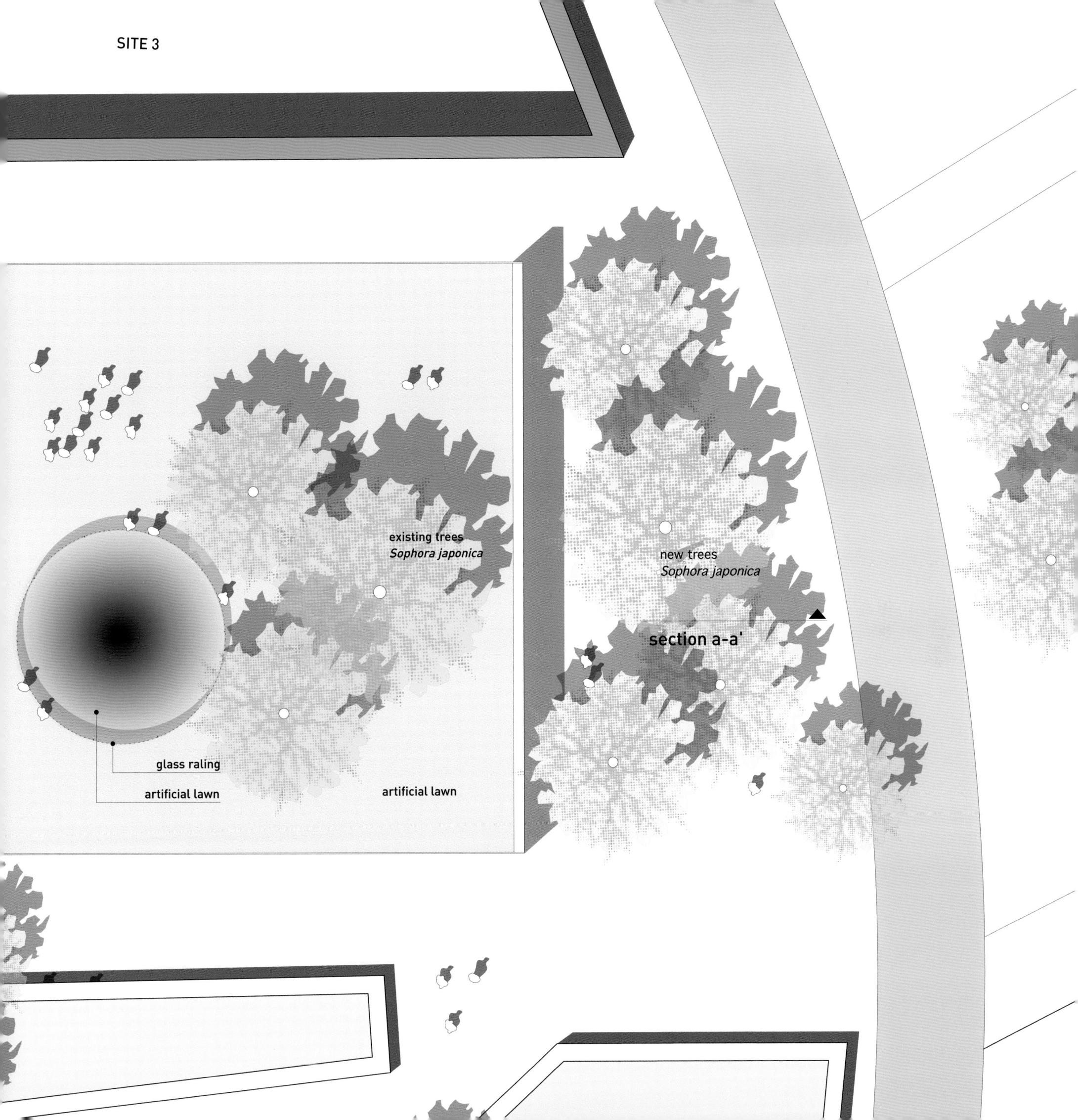

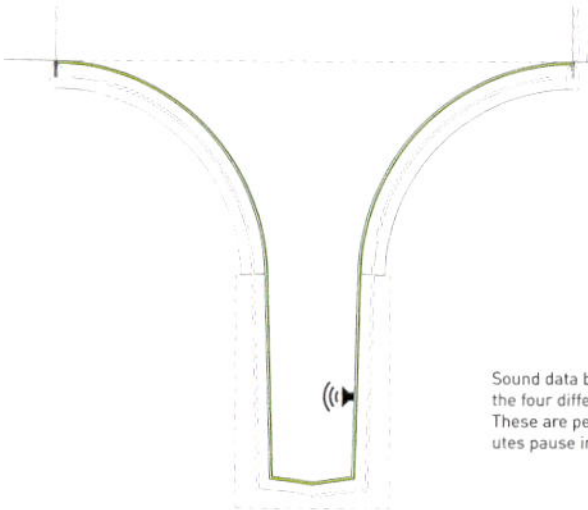

Sound data base consists of ca 200 recorded sounds from the four different sites on the „other side".
These are performed in a random interval with a 2-10 minutes pause in-between each soundscape performance.

Soundsystem

"And the grass is always greener on the other side!"

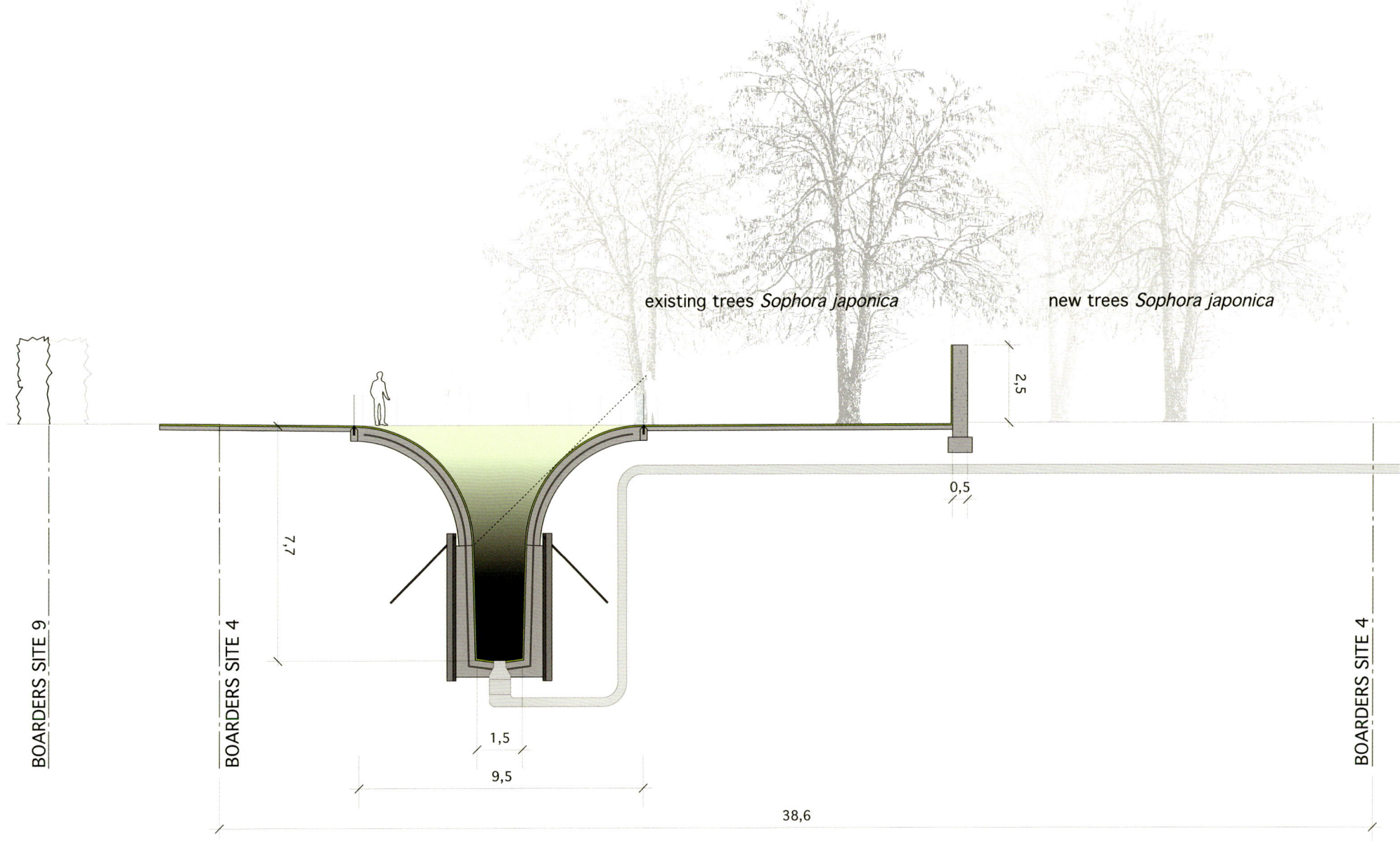

Elevation a - a'

The Garden is located at the No. 4 Plot of Master Gardens Section in the Expo Site. It is mainly designed by Martin Rein-Cano, a major designer of Topotek 1, Germany. The Garden is completely an open garden. A huge cavity is its main content. Compared with the Passageway Garden and the Four Boxes Garden, it looks more like a transitional space between the two. However, "it is indeed a garden", as emphasized by its major designer Martin Rein-Cano. In the Garden, visitors can listen to the "sounds" from the other side of the world in the deep cavity created by audio devices. The sounds from Argentina, America, Sweden and Germany are selected as the media for the communication between eastern and western cultures, to attract visitors' exploration and imagination on those unknown countries.

As one of the most progressive design corporations, Topotek 1 wants to explore more probability of forms for the design of modern landscape from unique point of view. In this project, they highlight the theme with a simple layout and further strengthen the impact of the theme. The design also brings some experimental characters to the Garden of Expo 2011. It seems quite simple, but contains sensitivity and pursuit of high-tech by German designers. The cavity in the Garden is not just a simple cavity with a diameter of 10 meters. Considering that the Earth has a spherical shape, the space in the cavity should not just be a two-dimensional plane, but a small three-dimensional sphere. Therefore, they have constructed a shell structure with a diameter of 10 meters, looking like an amplifier with its opening upward. This has defined the meaning of the Big Dig Garden as "Communication between Eastern Culture and Western Culture".

Landscape Garden of Map of China

大师园·山水·中国地图园

展园位于世界园艺博览会大师园5号地块,由法国Mosbach Paysagistes风景园林事务所主创设计师Catherine Mosbach女士领衔设计。展园设计平面创意从中国地图出发,结合西安的深厚历史与当地蓬勃进行的古城改造,以一组线性结构表达了四处尚未完工的工程面对西安地底密集的古迹所作出的退让。有着不同标高的线条互相之间形成的廊道,以不同季节着色的植物进行布置,除了将展园的空间进行划分,为植物的生长留出余地以外,更在竖向空间上形成错落交叠的视觉效果。

波尔多植物园的成功让Catherine Mosbach女士成为法国当代最具影响力的园林设计师之一。本案从设计师对中国画的定义出发,将"山水"进行了抽象的转译,最后为展园定下了"神似而非形似"的设计主旨。她借助当地的轮廓线设计了一个线性"符号"。为了使宏观背景下的"符号"能够融入展园这个微缩环境中,设计师将符号进行简化,并且以西方理性主义的目光,在线条上描摹出中国山水画中高低错落的意境和情致,同时和西安当地的现状结合起来,使得设计更富有时代性。产生于浪漫情怀的创意,实现于当代高科技设计手段的深化,使得这幅山水·中国地图将古典意象融入自然景观和立体的设计手法中,也为将来的园林设计指出了一条可行之路。

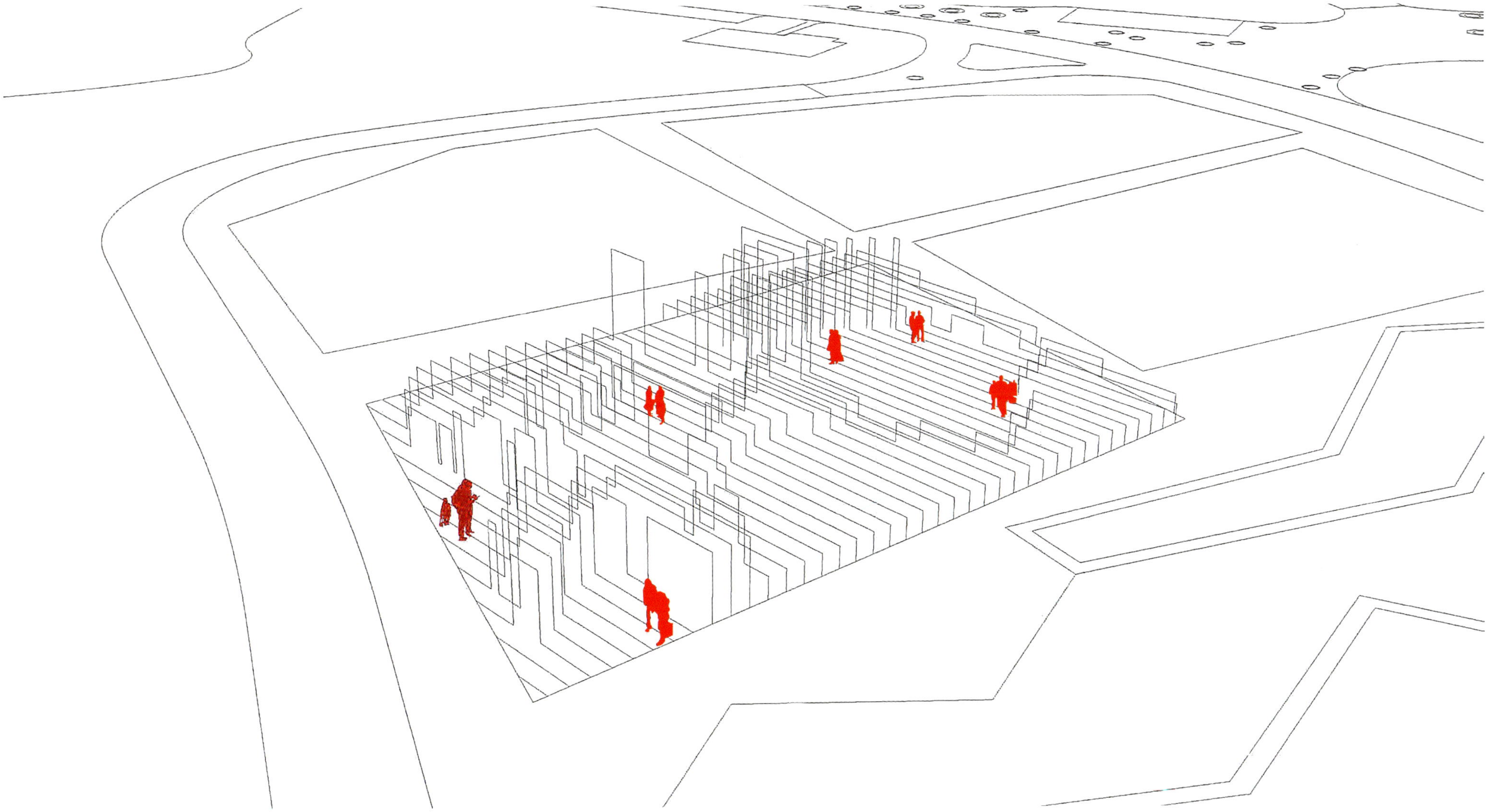

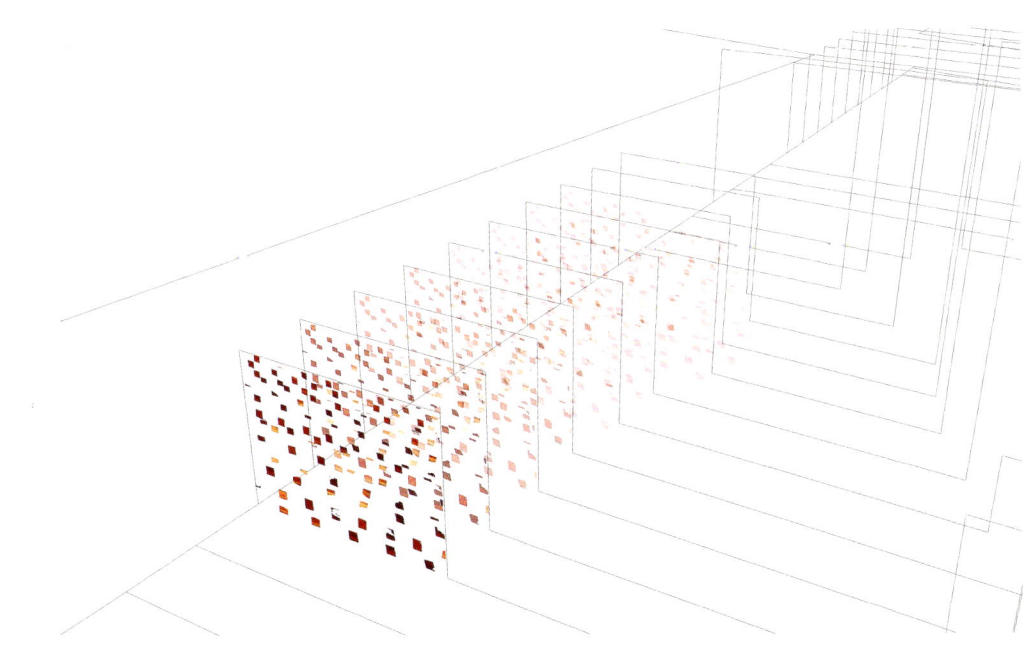

The Garden is located at the No. 5 Plot of Master Gardens Section. It is mainly designed by Catherine Mosbach, one of the chief designers of Mosbach Paysagistes, France. In the light of the map of China, the long history of Xi'an and the present flourishing reconstructions of this old city, a series of linear structures is used in the layout of the Garden to express the concession of four unfinished constructions to the abundant historical sites in the city. The corridors connected areas with different elevations are decorated with plants representing different seasons. These corridors not only divide the whole space into different parts, but also provide extra space for the growth of plants. Moreover, they brought in a random but overlapping vision in the upright space.

The success of Bordeaux Botanical Garden made Catherine Mosbach the most influential landscape designer in contemporary France. In this project, inspired by Chinese paintings, the designer set the "Similarity in Spirit but Form" theme for the whole design through abstract interpretation of "mountains and waters". She designed a linear symbol by virtue of local contour line. In order to integrate the "symbol" from macro-background into the micro-environment of the Garden, the "symbol" was simplified and endowed with conception and affection of Chinese paintings, from a view of western rationalism, combined with Xi'an's present situation, which makes the design full of time spirit. The innovation with romantic theme is fulfilled through present high-tech means, perfectly integrating ancient classic elements of Chinese paintings into the design of natural landscape. It opens up a possible way for the future landscape design.

Garden of Labyrinth

大师园·迷宫园

展园位于世界园艺博览会大师园6号地块,由Martha Schwartz事务所主创设计师Martha Schwartz女士领衔设计。作为各国风景园林设计师的共识,趣味性是园林设计中不可忽视的一点。在本案中,设计师摒弃了在设计平面上提取东方意象的方法,而是以更为简单直接的方法让游客可以碰触"东方",从而表达了自己对这古老文明的敬意。青砖作为关中建筑的代表材质,是本园最主要的材质。3米高的青砖墙构成了一个盒子,形成了本园的外观。园内空间被分成一系列没有封顶的狭长走廊。这些走廊宽度不同,每两堵墙之间留有空隙,以种植垂柳。每堵墙上都有数个拱形开口,游客穿过青砖拱门,进入不同的空间,并且于游览过程中在每个走廊末端的3米高的大镜子中看到视觉印象中被延长的走廊。不同走廊空间微妙地拓展和收缩,在尺度上不动声色地进行着改变,从而提供给游客不同的游走体验。花园后方的大三角镜面空间,仿佛是长时间探索后的馈赠,有着一片可变的、丰沛的绿色景观。游客并不能沿着原路返回,而是由展园侧面的两个封顶黑色走廊出园,在此处,游客会发现他们之前穿过的走廊,其实都是沿途设有单向镜面的单行道。在回程的时候,还能观察到人们身处迷宫时的举动。

Martha Schwartz女士的设计极其富有趣味性,在突破了基地限制,为游客创造更多感官体验后,她更以趣味十足的手法,安排了独特的参观路线,把人们处于迷宫窘境中的真实一面呈现出来。

The Garden is located at the No. 6 Plot of Master Gardens Section. It is mainly designed by Martha Schwartz, the chief designer in Martha Schwartz Partners. As commonly shared by landscape architects all over the world, being funny is always a point that cannot be omitted in landscape design. In this project, abstract eastern elements are abandoned by the designer, whereas, some simple objects are employed for visitors' contact with the "eastern" to express the designer's own regards to the old eastern culture. The blue bricks as representative construction material in Kuan-chung Plain are applied as the main materials in the Garden. The box built of blue brick walls at the height of 3 meters generates the outlook of the Garden. In the Garden, the space is divided into a series of narrow corridors without roofs, which are of varying widths. There is a space left between every two walls for the planting of weeping willows. Several arch openings are set on every wall to lead visitors to different spaces. At the end of each corridor, there is a big mirror at the height of 3 meters, which visually extends the corridor into the other side. The expansion and contraction of different spaces and changes in corridor width provide visitors with different experiences. The big triangular mirror space with a beautiful alterable green landscape at the back of the garden seems a boon for the visitors after long journey in the Garden. Visitors are not able to return by the original path. The two black corridors on either side of the Garden lead visitors to the exits, where they will realize original corridors they passed by are actually one-way passages with mirrors set at the ends. From the mirrors, visitors can watch the actions of other visitors trapped in the labyrinth. Martha Schwartz's design is full of fun, which breaks the limit of terrain and brings visitors more sensory experience. She also sets special visit line through colorful means to vividly display the embarrassment of visitors trapped in the labyrinth.

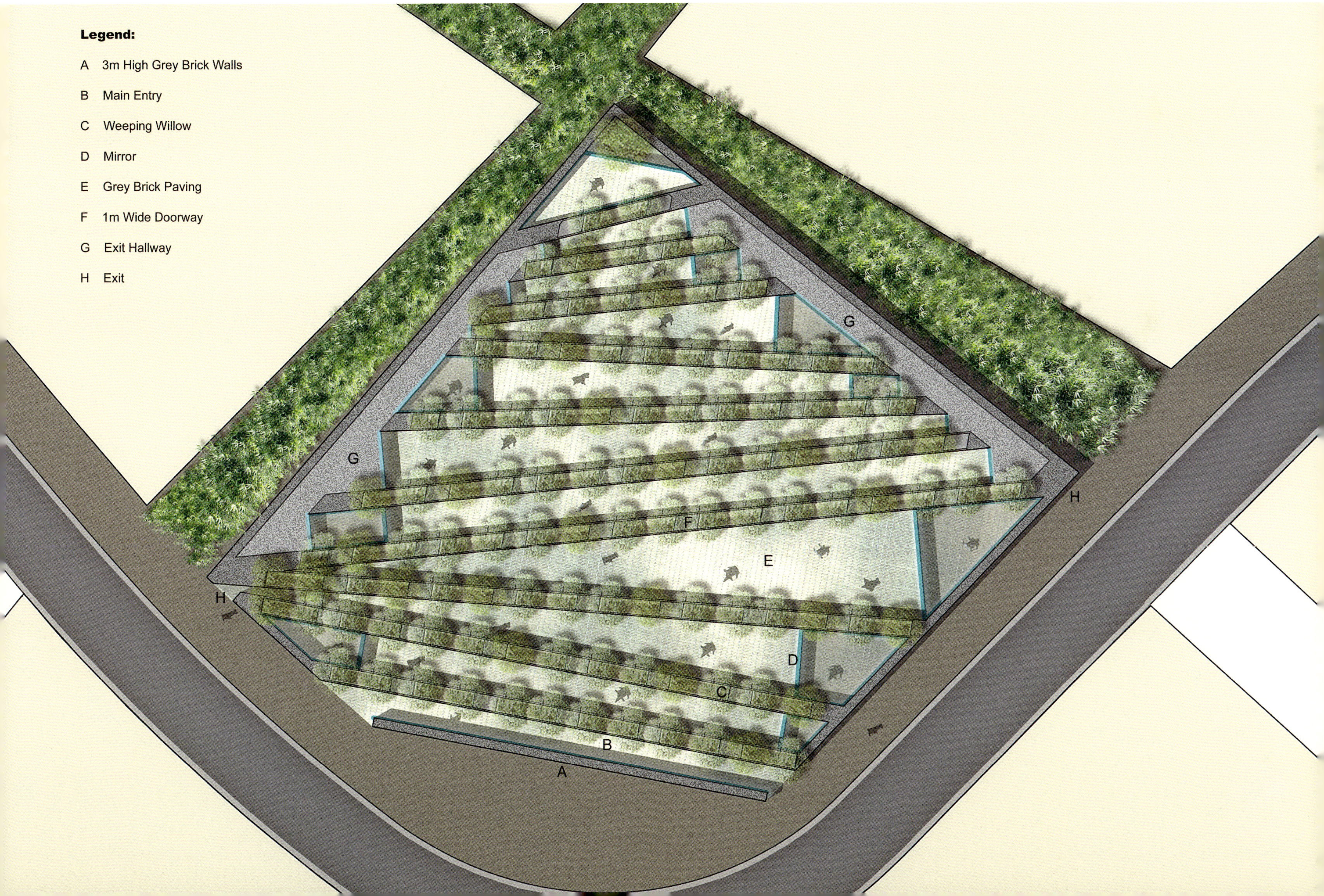

Legend:

A 3m High Grey Brick Walls

B Main Entry

C Weeping Willow

D Mirror

E Grey Brick Paving

F 1m Wide Doorway

G Exit Hallway

H Exit

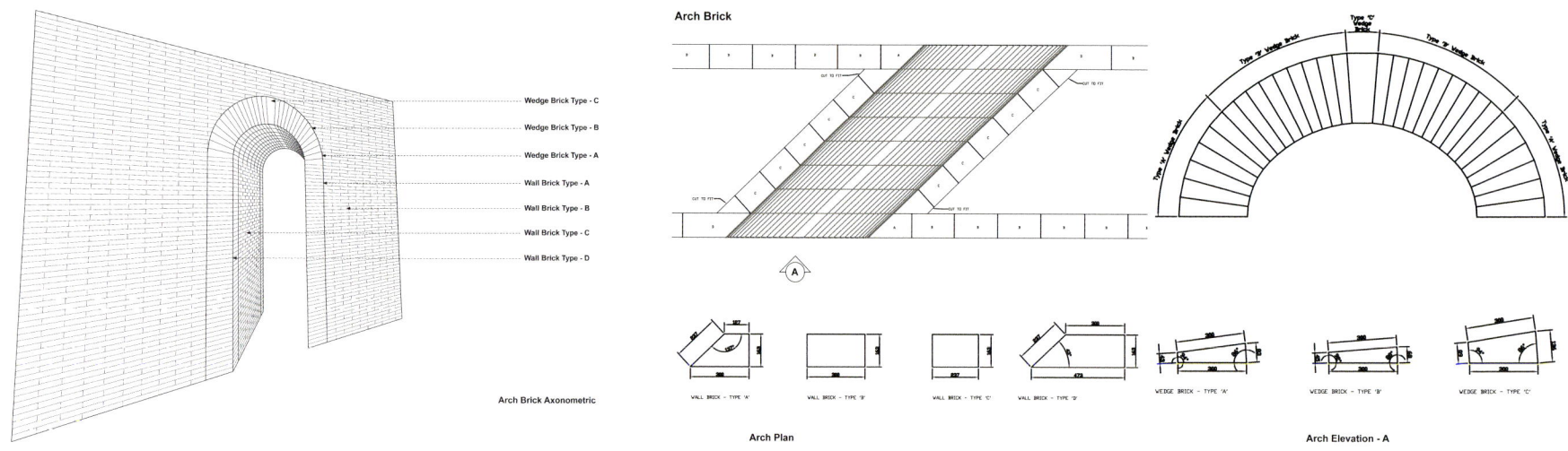

Arch Brick

Arch Brick Axonometric

Arch Plan

Arch Elevation - A

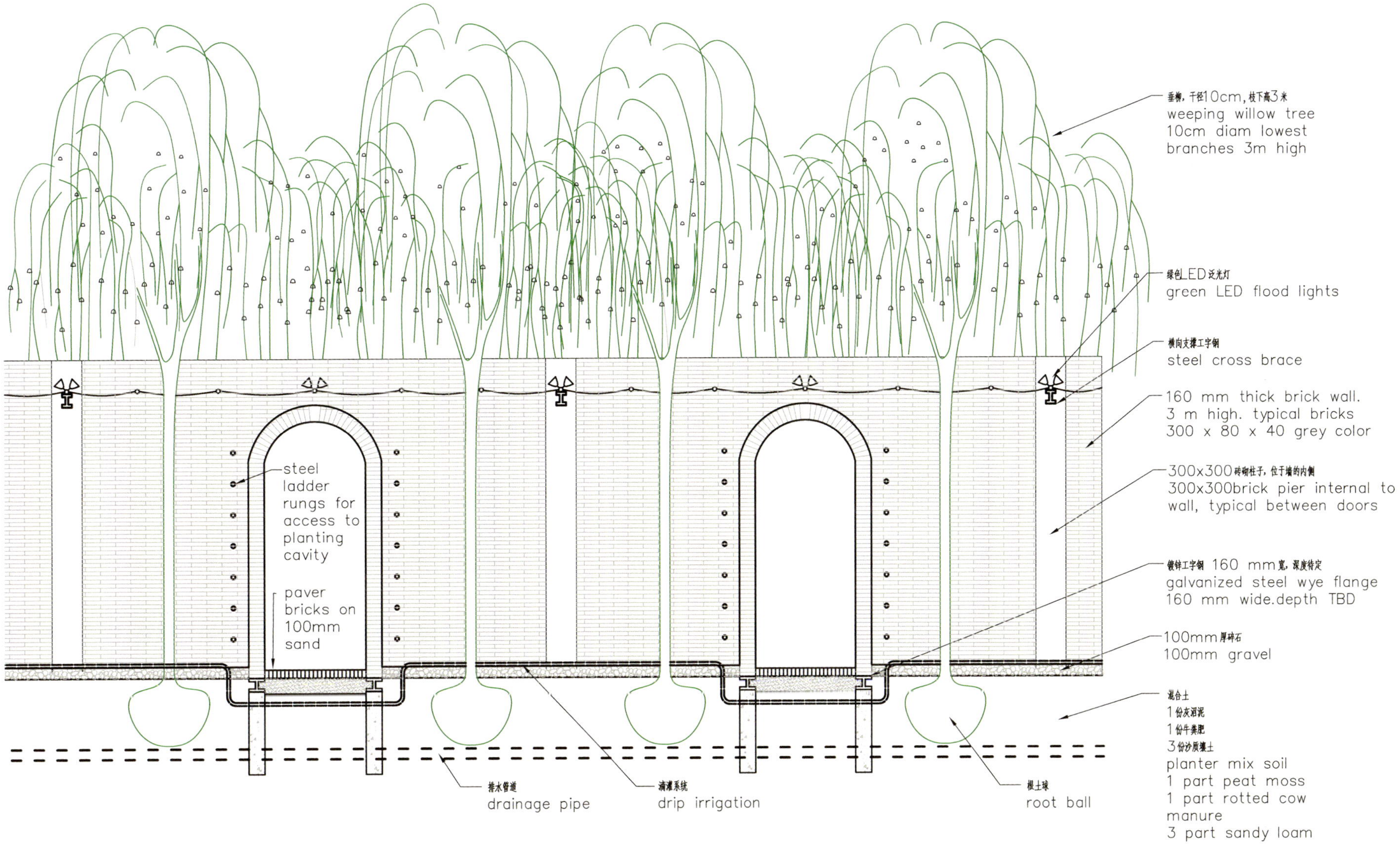

垂柳,干径10cm,枝下高3米
weeping willow tree
10cm diam lowest
branches 3m high

绿色LED泛光灯
green LED flood lights

横向支撑工字钢
steel cross brace

160 mm thick brick wall.
3 m high. typical bricks
300 x 80 x 40 grey color

300x300砖砌柱子,位于墙的内侧
300x300 brick pier internal to
wall, typical between doors

steel ladder rungs for access to planting cavity

paver bricks on 100mm sand

镀锌工字钢 160 mm宽,深度待定
galvanized steel wye flange
160 mm wide. depth TBD

100mm 厚砂石
100mm gravel

排水管道
drainage pipe

滴灌系统
drip irrigation

根土球
root ball

混合土
1 份灰泥泥
1 份牛粪肥
3 份沙质壤土
planter mix soil
1 part peat moss
1 part rotted cow
manure
3 part sandy loam

Garden of Bridges

大师园 · 万桥园

展园位于世界园艺博览会大师园7号地块，由荷兰West 8城市规划与景观设计事务所主创设计师Adriaan Geuze先生领衔设计。设计灵感取自中国山水画，设计师选取了"桥"作为塑造展园山水气质的手段，同时也希望能通过"桥"来表达对中国源远流长造桥史的敬意。本案借助地块临湖且位于整个世园会场地中较为中心的优势，利用人工造景与自然景观的协同，吸引游客进入这座在绿植掩映下的万桥园。本园主径由一条单向的狭窄羊肠小道与五座高低拱桥组成。朱红色拱桥将展园分为桥上与桥下两部分空间，行于桥下的游客在蜿蜒的砾石小道上被周围的茂密竹林所包围，仿佛进入了一座绿色迷宫，曲折徘徊却静谧深远，似乎可以在这里偷得片刻的清闲；而桥上的游客，则可以体会到何谓"登高望远"。在拱桥之上，游客视线开阔，尽揽世园胜景。

荷兰West 8城市规划与景观设计事务所丰富的从业经验与不同领域的合作经验，让Adriaan Geuze先生的设计很好地平衡了外观与设计内涵之间的关系。在本案中，设计师以精炼的设计语言突出了设计内涵，将"桥"和"径"作为整合全园的血脉，在视觉及游走体验上造成了高低错落的效果，同时吸收了东方传统造园中对于自然的理解以及东方文明中对于内省的重视。设计师选取了翠竹作为最主要的植物，既满足了本园内涵的需要，又符合了西安当地的气候，更体现了千百年来东方人具有的隽秀清雅，坚韧不拔的风骨气节。

The Garden is located at the No. 7 Plot of Master Gardens Section in the Horticultural Exposition. It is mainly designed by Adriaan Geuze, a chief designer of West 8 Urban Design & Landscape Architecture. Inspired by Chinese paintings, the designer takes bridges to create an atmosphere of "mountains and waters" for the Garden, also to express his regards to Chinese long history of bridge construction. In this project, fully taking advantage of the neighboring lake and its central location in the Expo Site, the designer combines artificial landscape and natural landscape to attract visitors to the Garden covered by green plants. The main road in the Garden is composed of a one-way narrow meandering footpath and five arch bridges with different heights. Those red arch bridges divide the Garden into two parts: on the bridge and under the bridge. The visitors walk on the footpath paved with screes, surrounded by luxuriant bamboo grove, which seems to enter into a green labyrinth. It is very devious and quiet, and visitors can take a rest here. The visitors on the bridges will get a bird's eye view of the whole Expo Site.

The West 8 Urban Design & Landscape Architecture, the Netherlands, is a company with rich career experiences and cooperation experiences in different areas, which enables Adriaan Geuze to perfectly balance the relationship between outlook and connotation of designs. In this project, the designer highlights the connotation of the design with concise style. "Bridges" and "footpaths" are taken as the veins of the Garden to give visitors up-and-down effect during the walking. He also takes the Chinese traditional gardens' understanding of nature and eastern culture's emphasis on self-examination into the design. The green bamboos selected by him as the main plant of the garden meet the connotation of this Garden, also fit the climate of Xi'an city. Moreover, bamboos incarnate the easterners' traditional temper of elegance and fortitude.

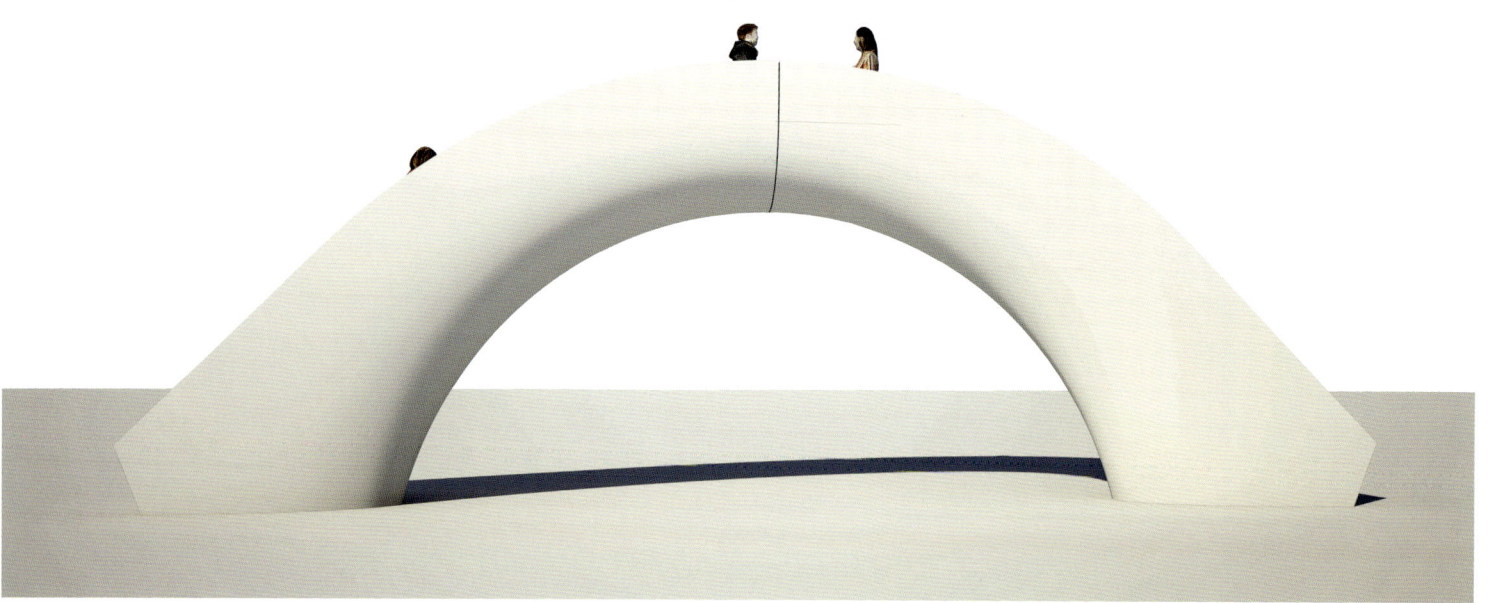

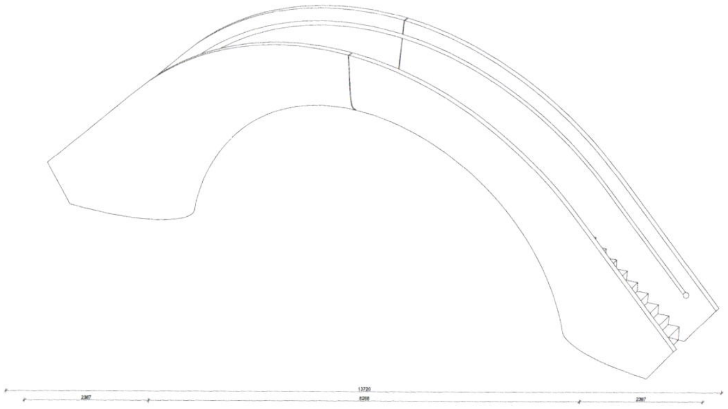

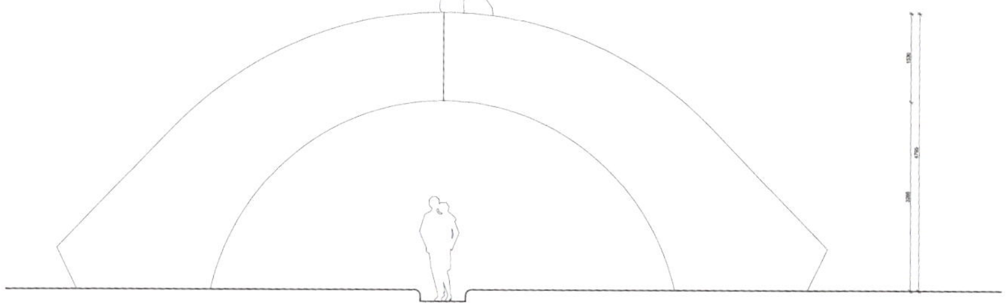

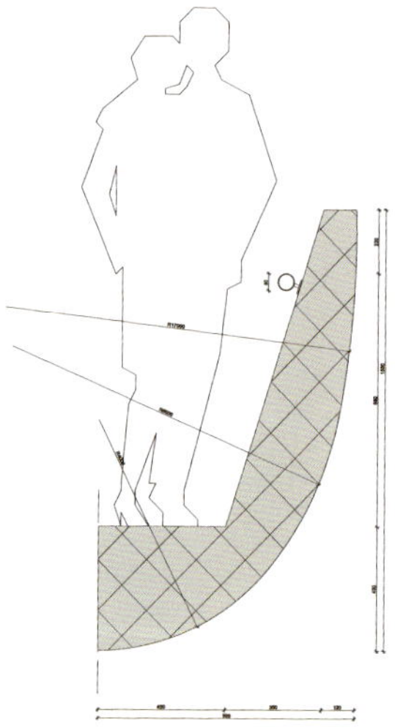

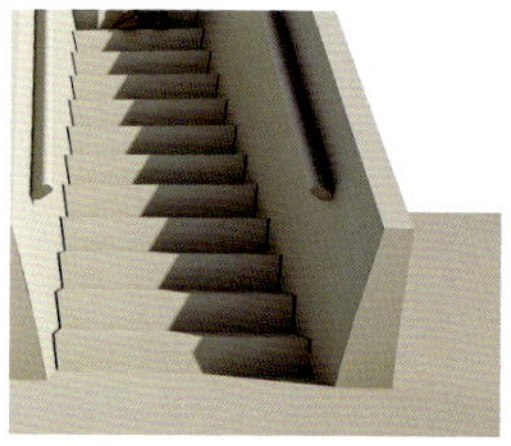

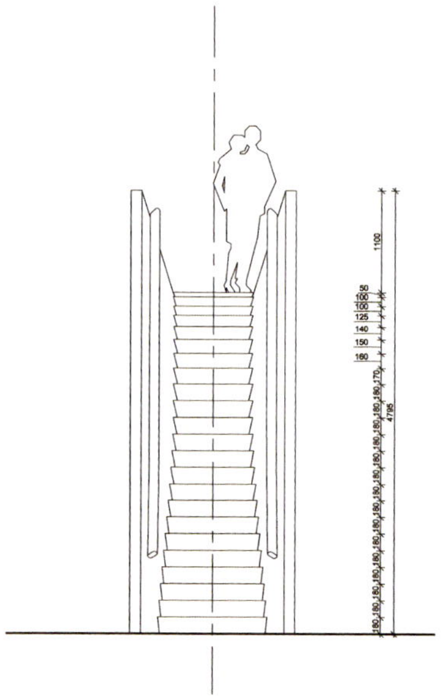

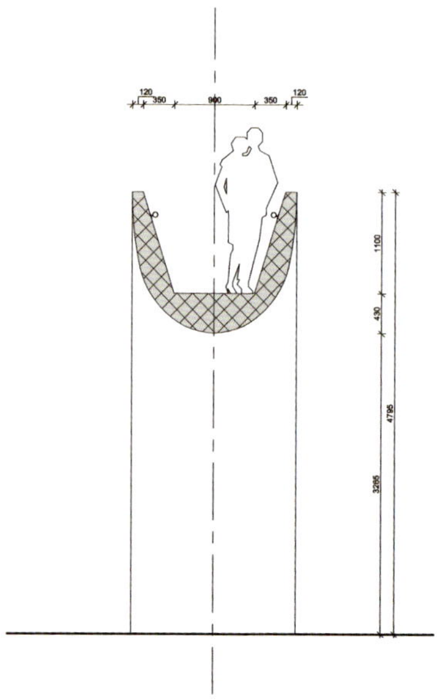
120
350
900
350
120
1100
430
4795
3265

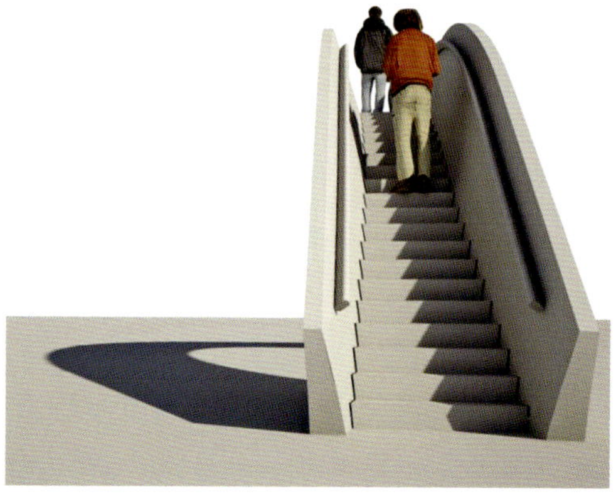

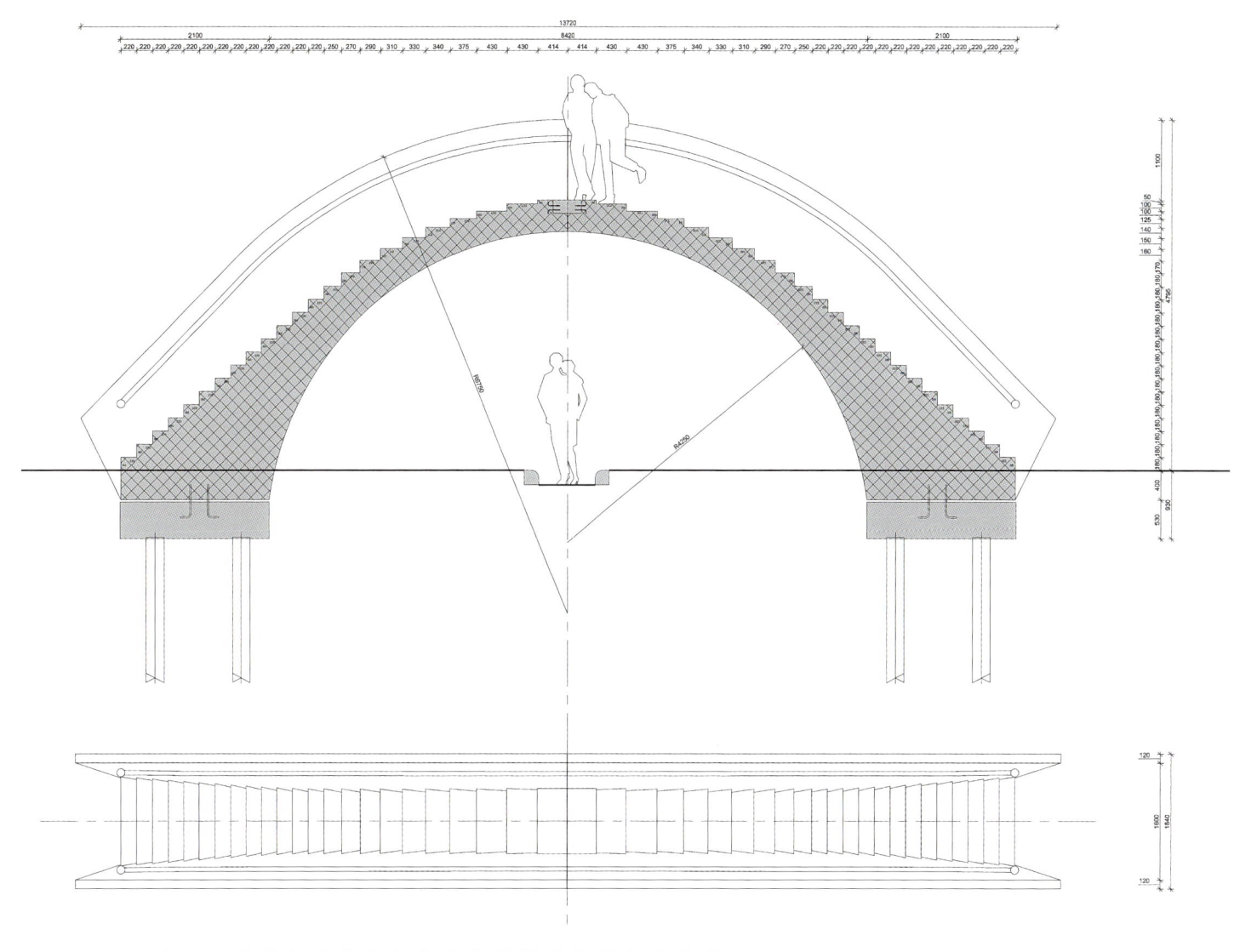

Loess Garden

大师园·黄土园

展园位于世界园艺博览会大师园8号地块，由丹麦SLA景观设计事务所创始人Stig L Andersson主创设计。设计将"土"与"水"作为代表中国文化的手段，由黄河携带的泥沙冲积形成的黄土高原，是一片肥沃的土地，孕育了无数的生命，同时也形成了中国引以为傲的黄河文明。设计师就在这个大背景下，以9个他创造的赤陶泥塑形象为线索，讲述了他眼中的中国文化如何绵延了千万年。40厘米深的不规则平底水池面积为315平方米，成为展园的中心地带，水池中置有喷泉，喷泉会根据收集的河水和雨水数量多少，对那片被黄土填埋的水池进行不定期喷洒，从而形成或干涸或湿润的黄土地面，对黄河流域的水资源进行了意象上的模拟。水池四周的砖路是展园最主要的步道，长140米。而横架在水池上的Z字形小桥，是由黄色铁炉条铺设而成。在水池之外与砖路交错种植的茂密植被，与水池中间稀疏的7棵大树形成对比，间或出现的赤陶雕塑，憨态可掬，为游客带来了惊喜与视觉的变化。

设计师Stig L Andersson先生以对于细微变化的敏感在行业内著称，他细腻的设计以及对自然纹理的追求赋予了本园平和的氛围，为游客提供了与自然最大程度的接触机会。同时设计师将雕塑作品与园艺进行有机的结合，仿佛通过项目的铺陈向游客讲述了一个意义深长的故事。

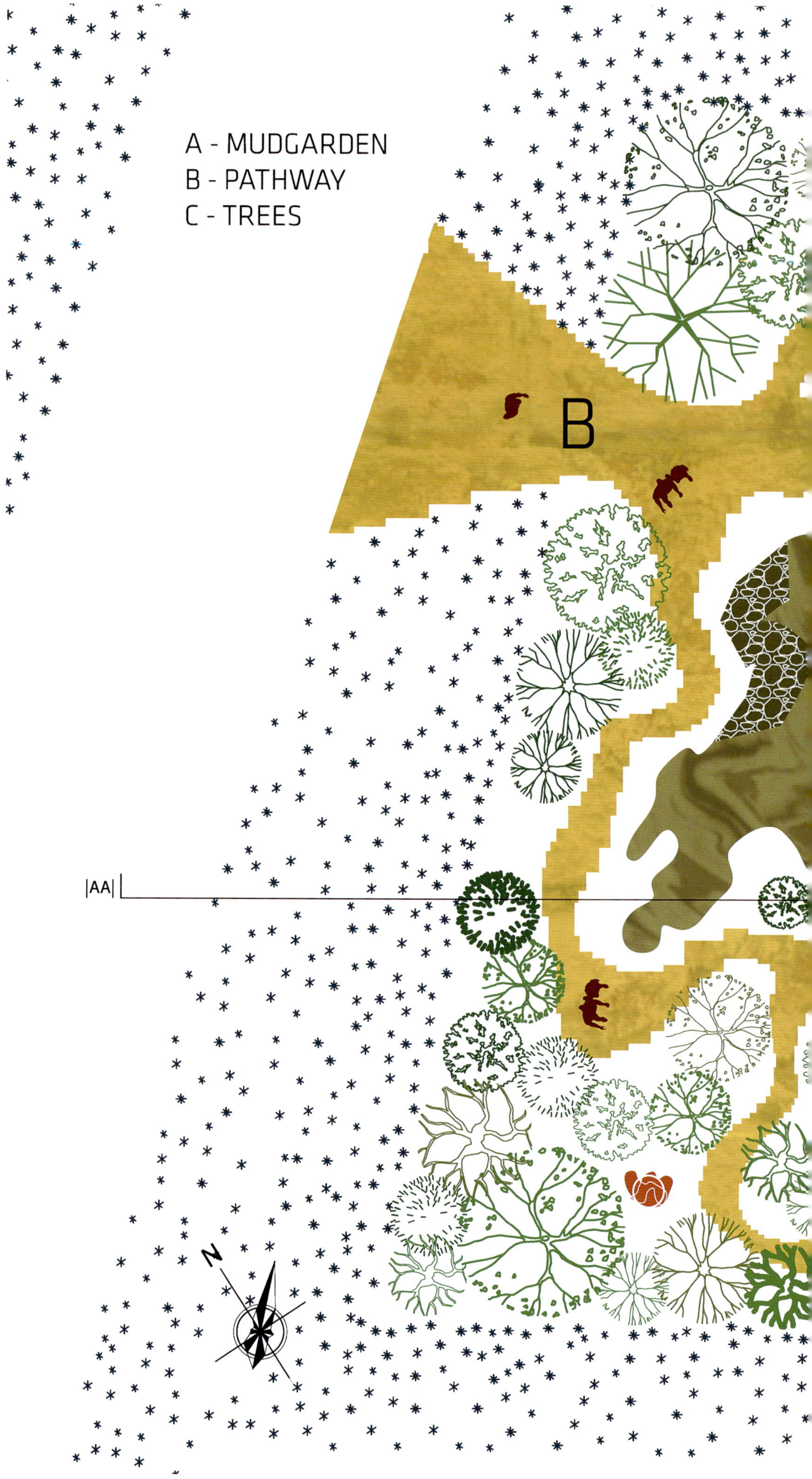

A - MUDGARDEN
B - PATHWAY
C - TREES

PUBLIC FLOW

GREEN CARPET

MUD BASSIN

STEEL EDGES AND ROCK SETTINGS

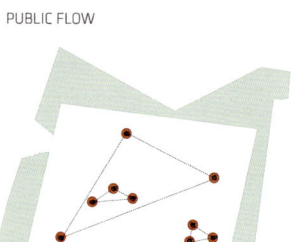
CLAY SCULPTURES

RESTING PLACES

PRESERVED TREES

NEW TREES

GOLDEN SPOTLIGHTS AT NIGHT

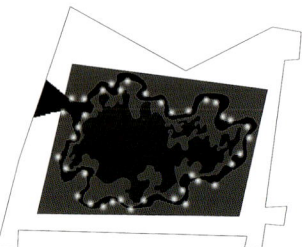
BOLLARDS WITH BLUISH LIGHT

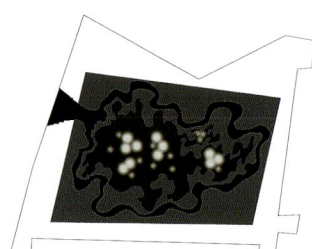
WATER JETS LIGHTED BY FIBER OPTICS

1. Earth is removed to make a void for the mud.

2. Galvanized steel edges are hammered into the ground.

3. Rocks are placed in some parts of the edge.

4. The void is filled with granulated mud.

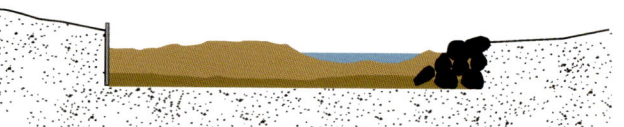
5. The bottom layer is kept moist to retain the water.

6. The surface varies from dry to wet.

The Garden is located at the No. 8 Plot of Master Gardens Section. It is mainly designed by Stig L Andersson, the founder of SLA, Denmark. The design uses water and soil to represent Chinese culture. The loess plateau formed by the sediments carried by the Yellow River is a rich land and breeds countless life. It also produced the great Yellow River Culture. Hence, the designer uses 9 terra-cotta clay sculpture created by himself as a clue for displaying Chinese culture in his own eyes, which continues for thousands of years. A 40 cm deep irregular pool with a flat base and an area of 315 m² becomes the central region of the garden. In the pool there is a fountain. The fountain could spray the pool covered with loess at an irregular frequency, which is decided by the amount of water collected by the fountain from the river and rains. This will generate a dry or wet loess ground, an abstract simulation of the water resource distribution of Yellow River Area. The brick way around the pool with a length of 140 meters is the main sidewalk in the garden. The Z-shaped bridge crossing the pool is built with the yellow iron sheets. There is a vivid comparison between the denseness of plants grown between the pool and the brick way and the sparseness of 9 trees in the middle of the pool. The lovely terra-cotta clay sculptures scattering in the garden bring a great surprise and visual change for the visitors.

Stig L Andersson, is famous for his sensitivity of subtle change in the industry. His exquisite design and pursuit of nature texture endow the garden a placid atmosphere, and provide a chance for the visitors' contact with the nature to the extreme. Meanwhile, the combination of the sculptures and the landscape of garden provide a way to tell the visitors a meaningful story through the proper arrangement.

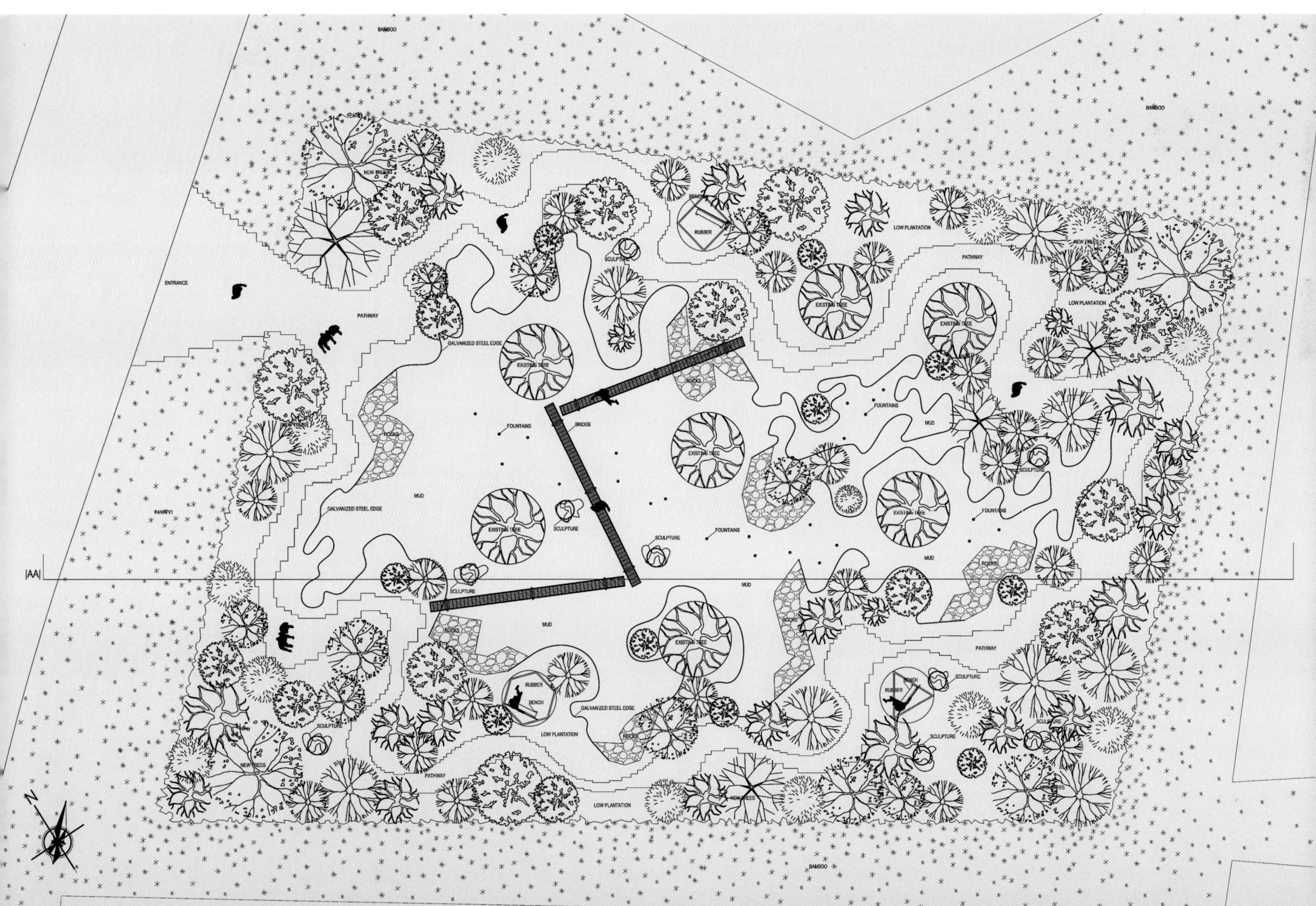

FOUNTAIN PRINCIPLE

STEEL BRIDGE PRINCIPLE

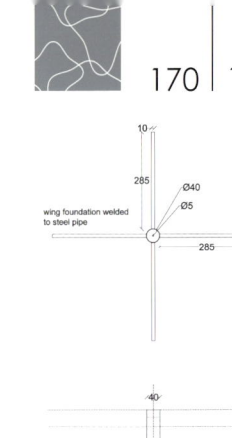

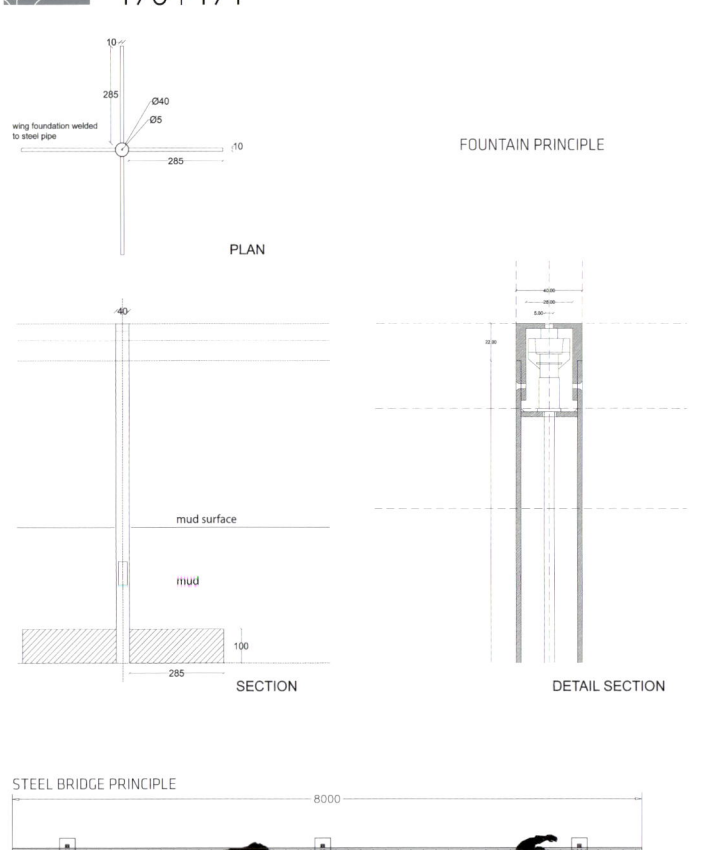

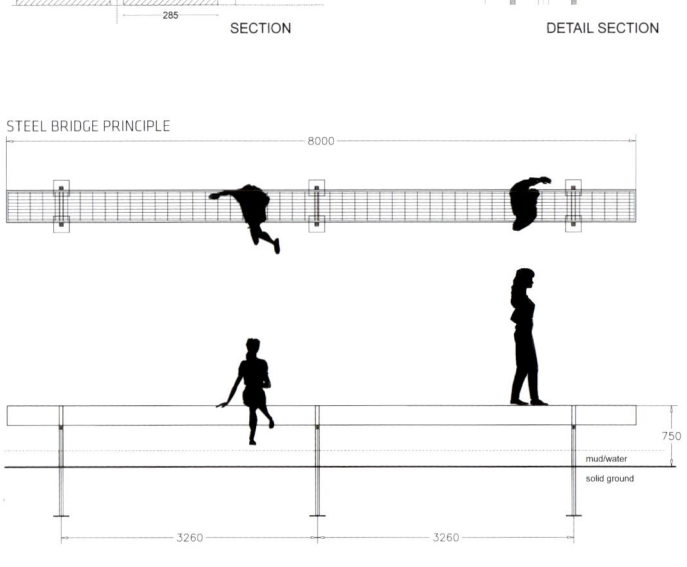

BRICK PATH PRINCIPLE

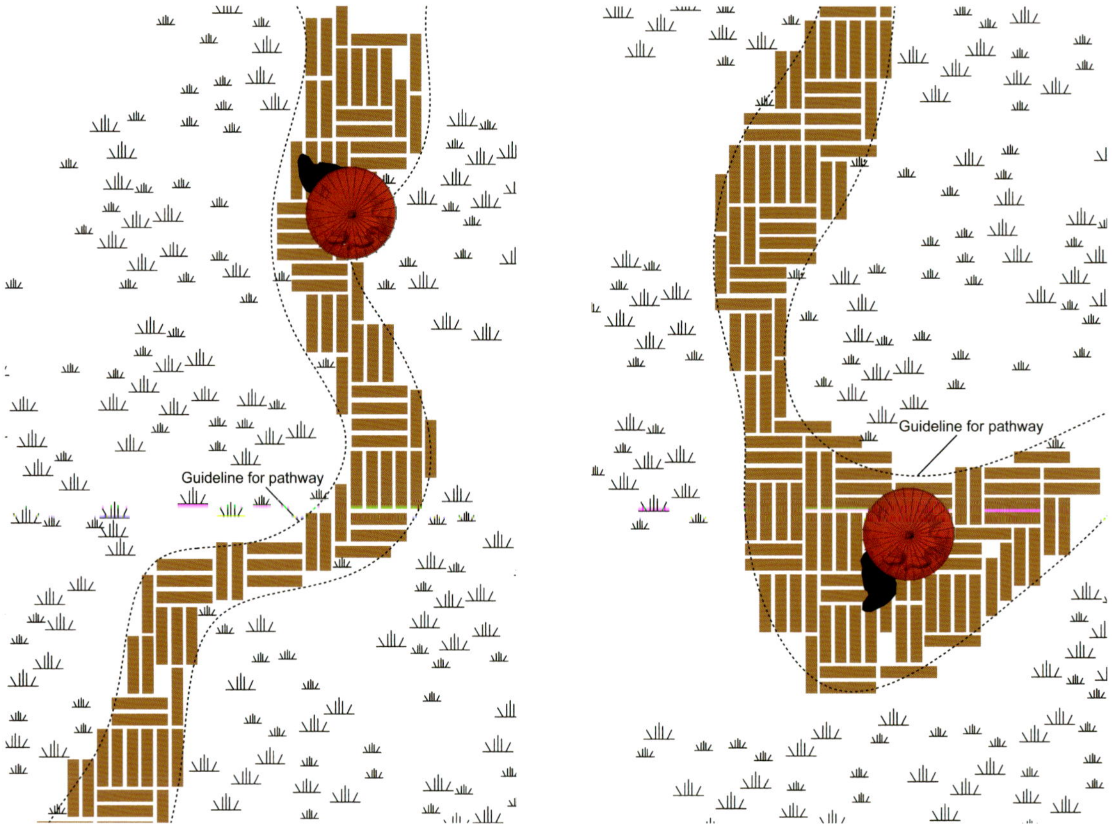

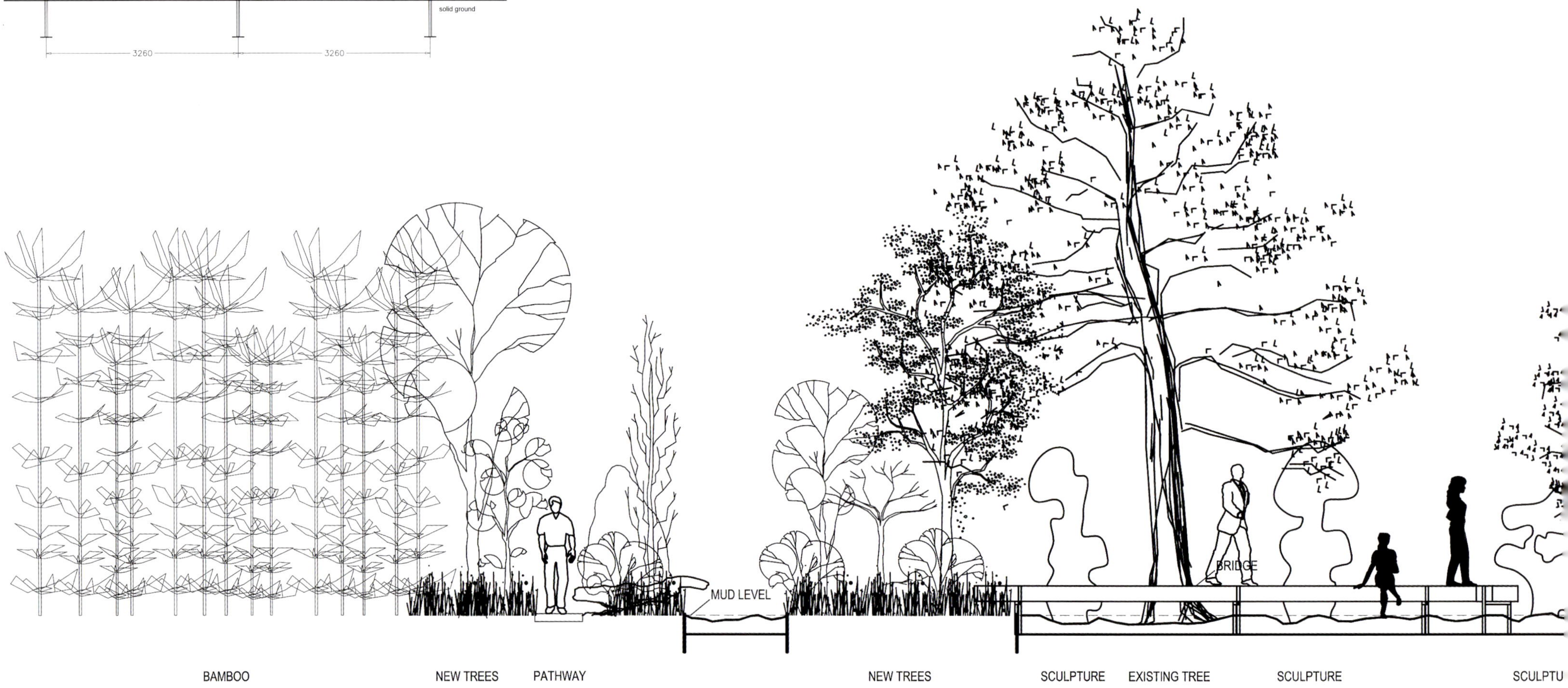

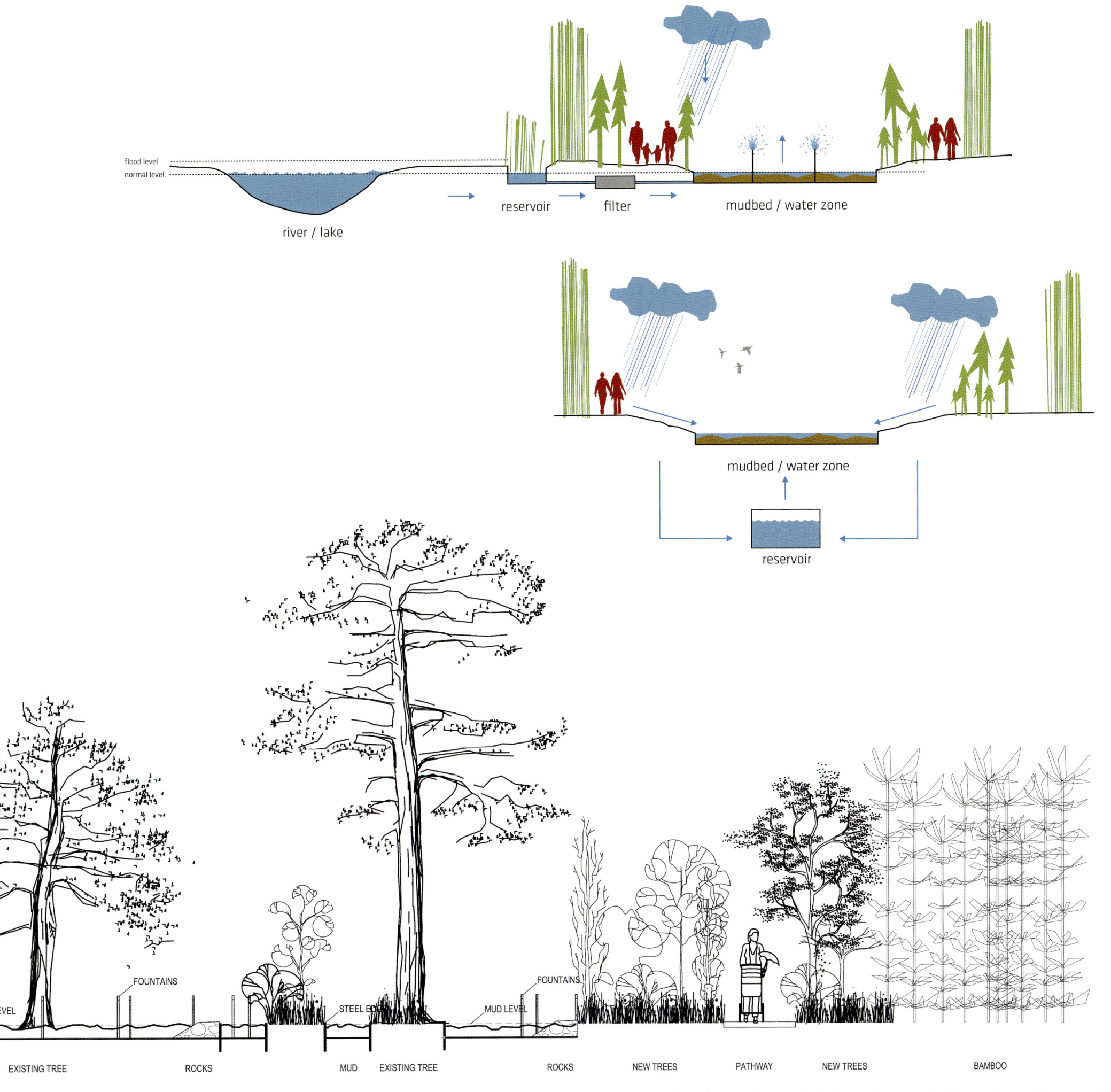

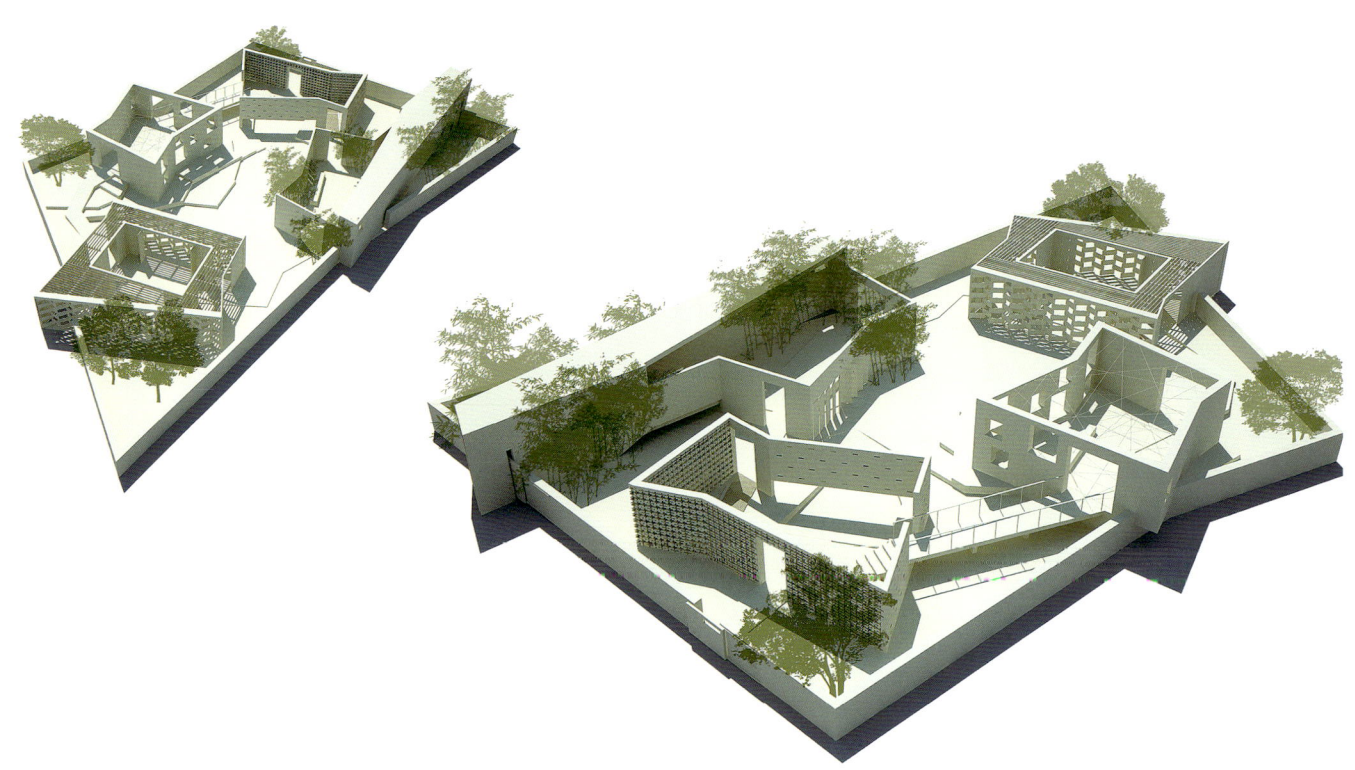

Quadrangle Garden

大师园·四盒园

展园位于世界园艺博览会大师园9号地块，由中国多义景观规划设计事务所主持设计师王向荣领衔设计。本园基地大致呈矩形，四周被1.6米高的夯土墙包围。在园内，设计师利用石、砖、木等自然材料建造了四个盒子，希望通过材料的变化和景观安排表现出四季轮回的主题。盒子与围墙一起，把本园分隔成1个主庭院，以及位于盒子后面和旁边约10个小庭院。花园的地面统一在整体流动的线性之中，并且在空白中点缀植物、白色和黑色沙石、瓦片，以及薄水等景观。从南墙的入口进入园中，在环形动线上依次安排了春、夏、秋、冬四个代表性庭院，以粉墙石材、木头、石头和金属、青砖来凸显四个庭院所代表的季节。在四个"盒子"的墙面上有不规则开口，进入一个"盒子"之后，可以透过这些开口观赏到其他庭院的景象，并且在中心庭院内安排其他的辅助路线，方便游客在四个"盒子"之间穿行。

主张"寻找中国现代景观的表现形式"的王向荣教授认为，设计的本质应该遵循自然的过程，将创作作为自然演进的一个部分，延续区域的地域特征。在本案中，设计师将传统园林中"步移景异"的审美情趣通过现代设计手法浓缩在一个微观环境中，表达了设计师作为一名东方人对自然景观和造园的理解与感悟，以及对于植根于自然的中国传统哲学所进行的深入思考。

The Garden is located at the No. 9 Plot of Master Gardens Section. It is mainly designed by Wang Xiangrong, the chief designer at Atelier DYJG, China. The Garden basically takes a rectangle shape, surrounded by loam walls with 1.6 meters height. The Garden is built with four boxes by some natural materials such as stone, brick and wood, to show seasonal changes. The boxes and the walls separate the Garden into several parts: a main courtyard and ten small courtyards behind or beside the boxes. Plants, white or black pebbles, tile and slim water decorate the void of the ground. Across the entrance on the south wall, visitors enter into four courtyards orderly arranged on a circle line: the Spring Courtyard, the Summer Courtyard, the Autumn Courtyard and the Winter Courtyard. Different construction materials of the walls like wood, stone, metal and blue brick are used to reveal different seasons of the yards. Some irregular openings on the walls of the boxes enable visitors to have a view of the sceneries of other courtyards. Some auxiliary paths are set in the central yard to facilitate visitors' crossing among the four "boxes". Holding the idea "to find the expression forms for Chinese modern landscapes", Wang Xiangrong insists that design in essence should follow a natural process and be put into one part of the nature evolution to continue to reveal the local features of the region. In this project, through modern design skills, Wang expresses the aesthetics of Chinese traditional gardens that "Scenery Changes with Every Step" in a micro-environment, which demonstrates the understanding of natural landscape and appreciation of gardens as an easterner. It also shows the designer's deep thought on Chinese traditional philosophy rooted by nature.

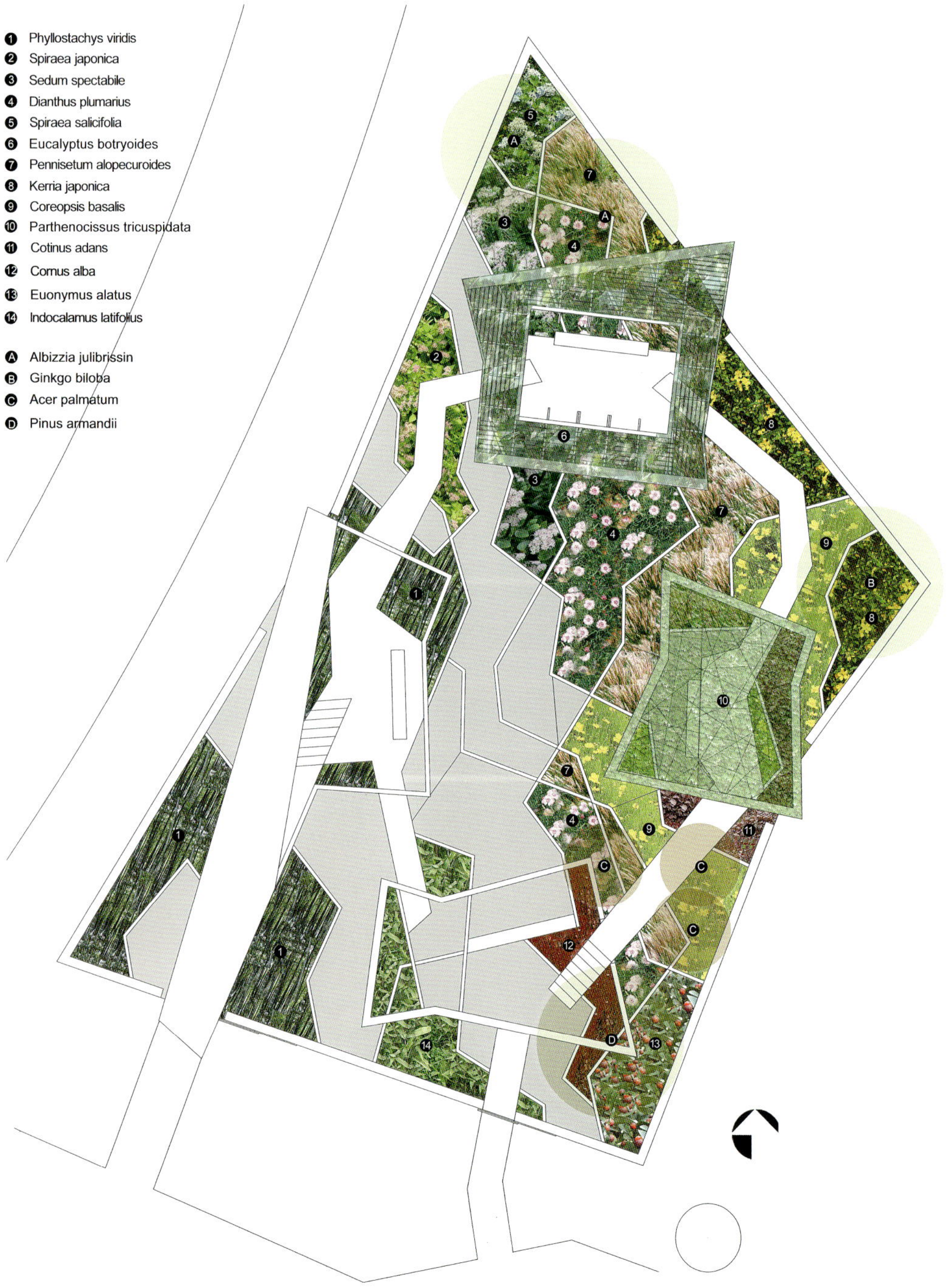

- ① Phyllostachys viridis
- ② Spiraea japonica
- ③ Sedum spectabile
- ④ Dianthus plumarius
- ⑤ Spiraea salicifolia
- ⑥ Eucalyptus botryoides
- ⑦ Pennisetum alopecuroides
- ⑧ Kerria japonica
- ⑨ Coreopsis basalis
- ⑩ Parthenocissus tricuspidata
- ⑪ Cotinus adans
- ⑫ Cornus alba
- ⑬ Euonymus alatus
- ⑭ Indocalamus latifolius

- Ⓐ Albizzia julibrissin
- Ⓑ Ginkgo biloba
- Ⓒ Acer palmatum
- Ⓓ Pinus armandii

- ① black sandstone
- ② white sandstone
- ③ tile
- ④ plant

四盒院
FOUR
BOXES
GARDEN

四盒院

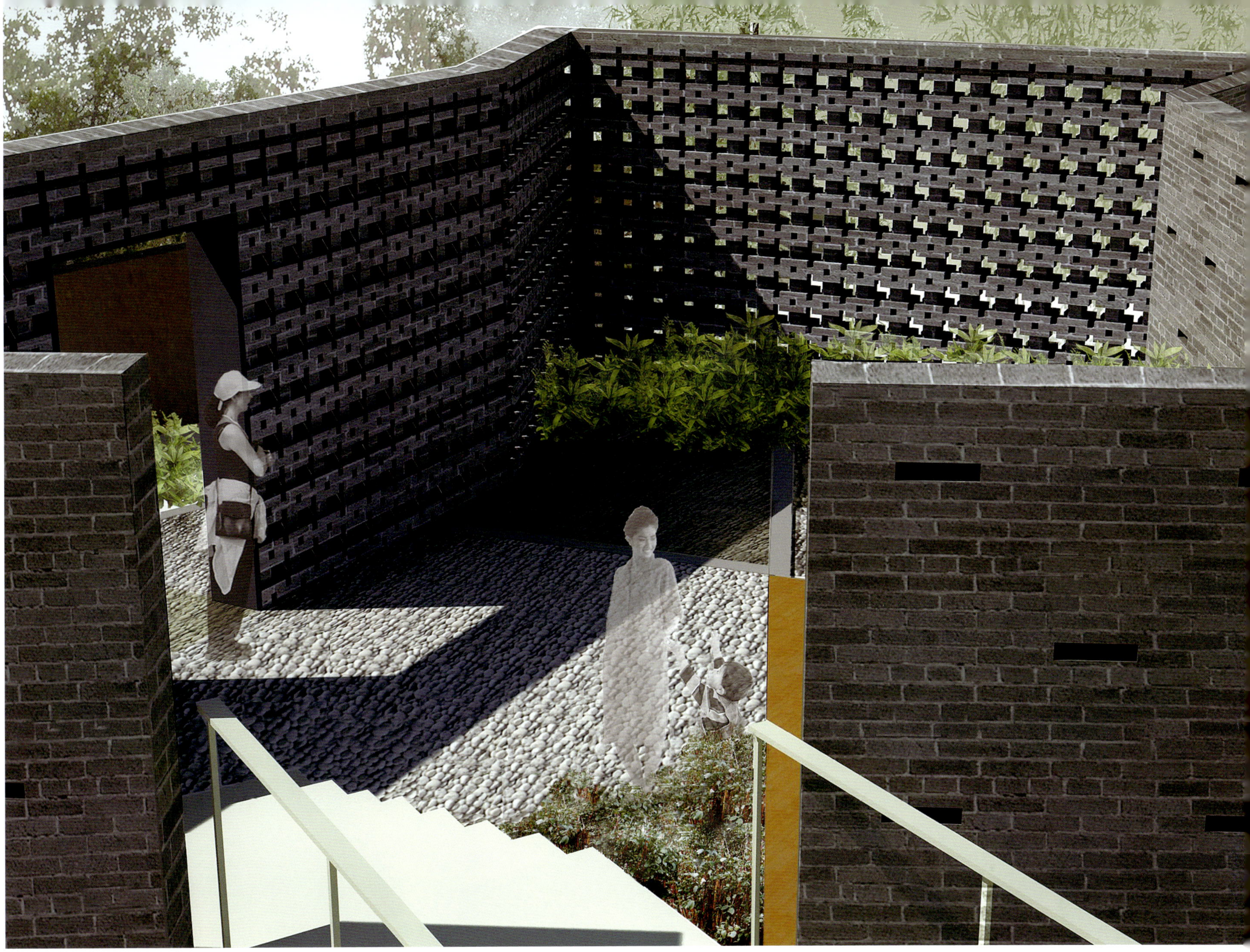

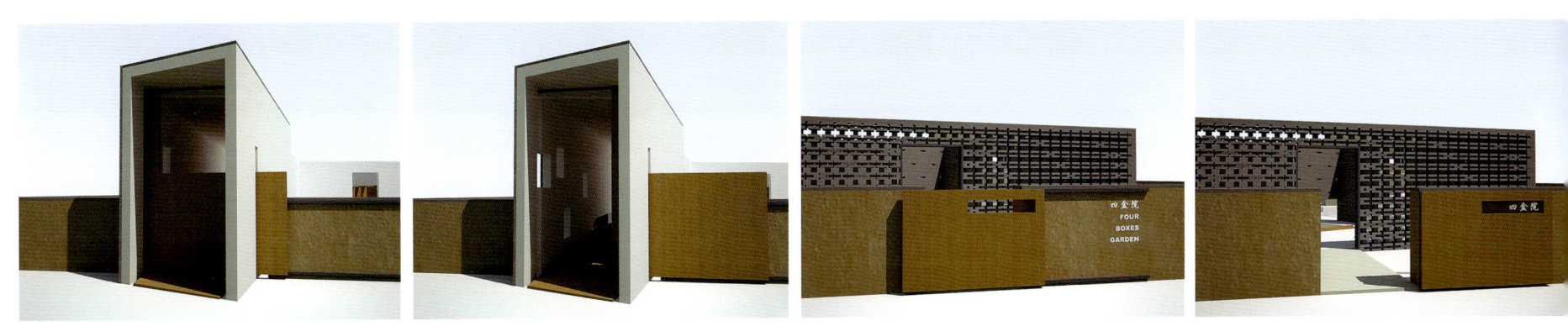

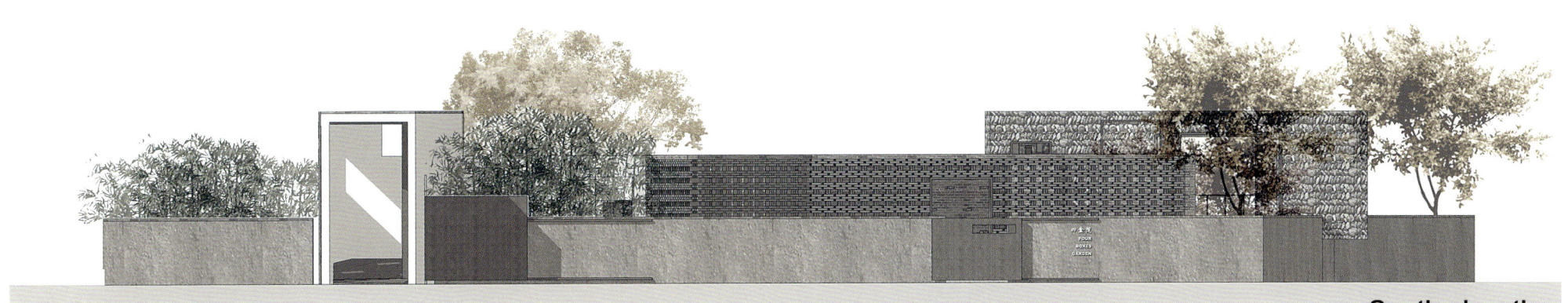

South elevation

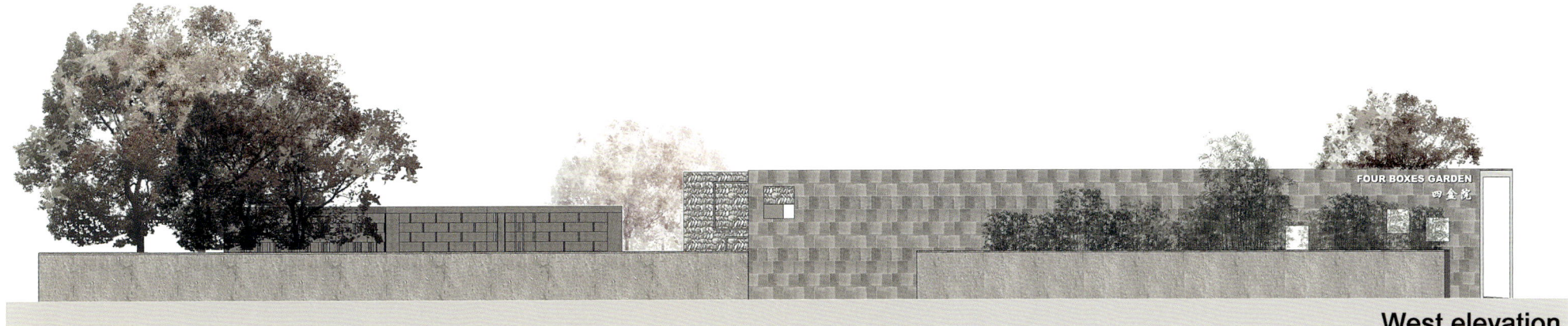

West elevation

Poem of Song
创意园·风的诗歌

展园位于世界园艺博览会创意园展区，由中国香港大学负责设计。"风的诗歌"所想要表达的是，如何在一座庭园中表达出身处城市环境的感受。城市中恒定的能量流，显现为风的不断流动，而设计中的风车装置将日夜转动，象征着香港充满着活力。"风"作为设计使用的主要元素，同时也为打造"绿色"展园带来了极大的可能。考虑到世界园艺博览会的展期为2011年4月28日至10月22日，如何以更节能的方式度过炎炎夏日，是本案对于新时代倡导的环保主题所做出的回应。设计考量了当地风的因素，在展园中搭建了风塔，利用现代科技手段，除了更大程度地利用自然风资源，也塑造了展园立体的景观层次。流动的风，恰如香港这座城市本身，流转不息、生气勃勃。

The Garden, designed by University of Hong Kong, is located at the Creativity Gardens Section in the Expo Site. It is named as "Poem of Song" to reflect agreeable life and vitality of Hong Kong. The design of the Garden applies the wind wheel that is running unceasingly day and night to symbolize the city of Hong Kong full of vitality. Meanwhile, as the main element of the Garden, the "wind" also creates a "green" landscape space for visitors. In consideration of the limited exhibition duration from April 28th to October 22nd 2011, one of the vital design points in this project is to set more energy-saving devices into the Garden to overcome the burning-hot summer in Xi'an, and to echo with the theme of environmental protection in the new era. The wind tower built in the Garden with high technologies shows that the designers make full use of the local wind resource and create a dimensional landscape here.

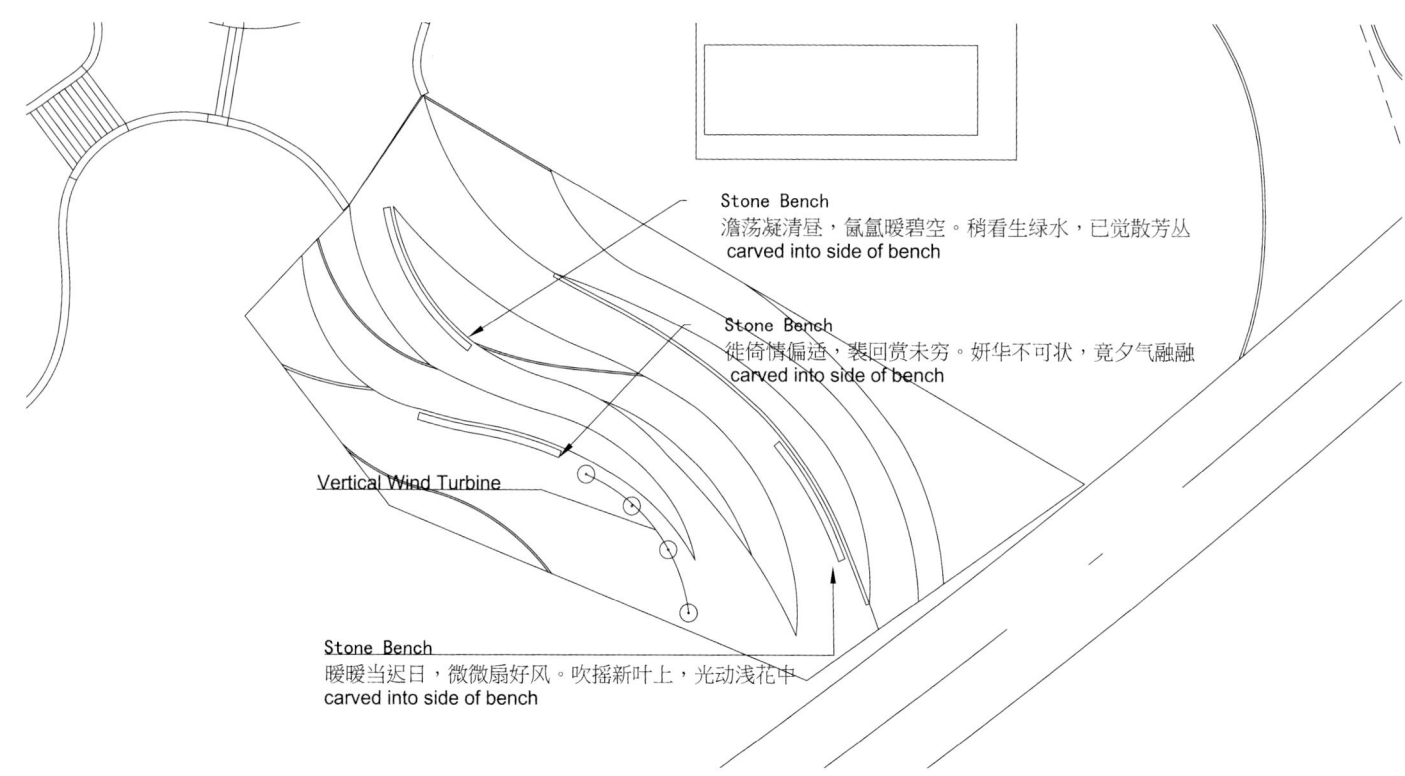

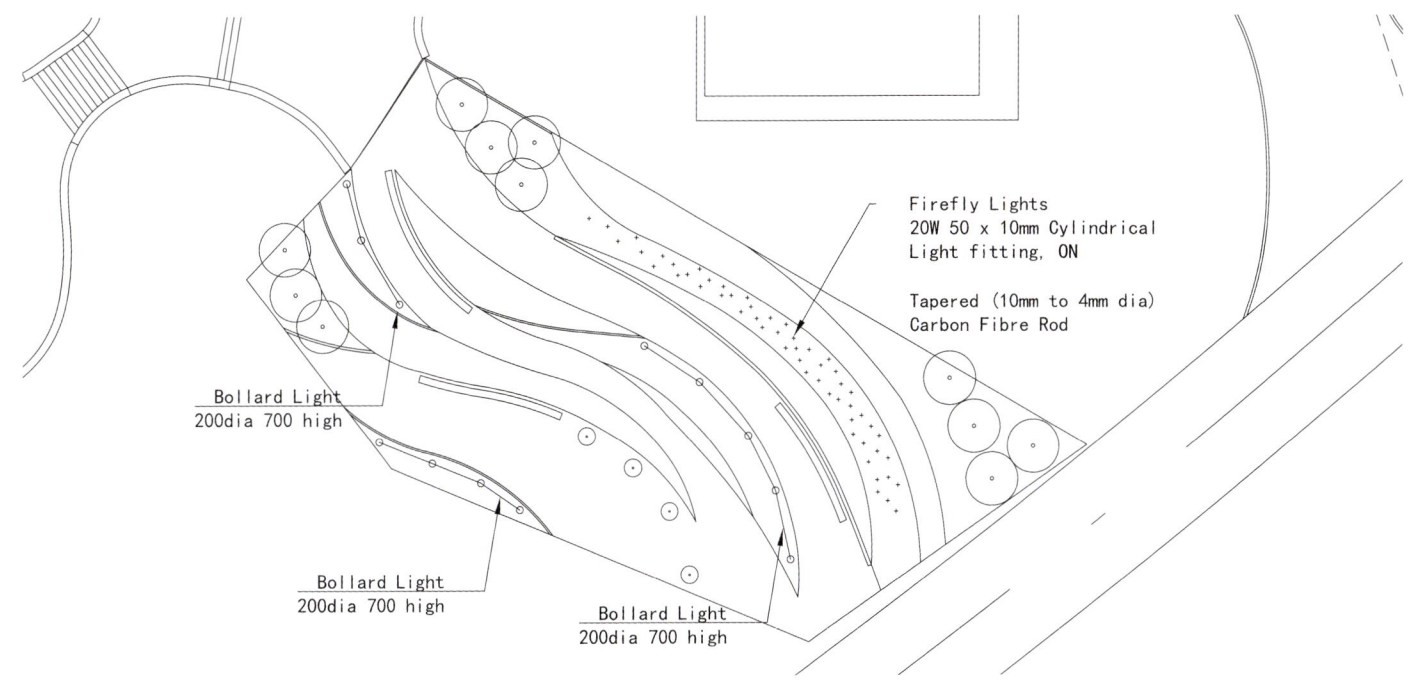

Firefly Lights
20W 50 x 10mm Cylindrical
Light fitting, ON
Tapered (10mm to 4mm dia)
Carbon Fibre Rod
Bollard Light
200dia 700 high
Bollard Light
200dia 700 high
Bollard Light
200dia 700 high

391
390
391
390.8
391.3
391.6
390
390.6
391.6
391.6
391.3
391

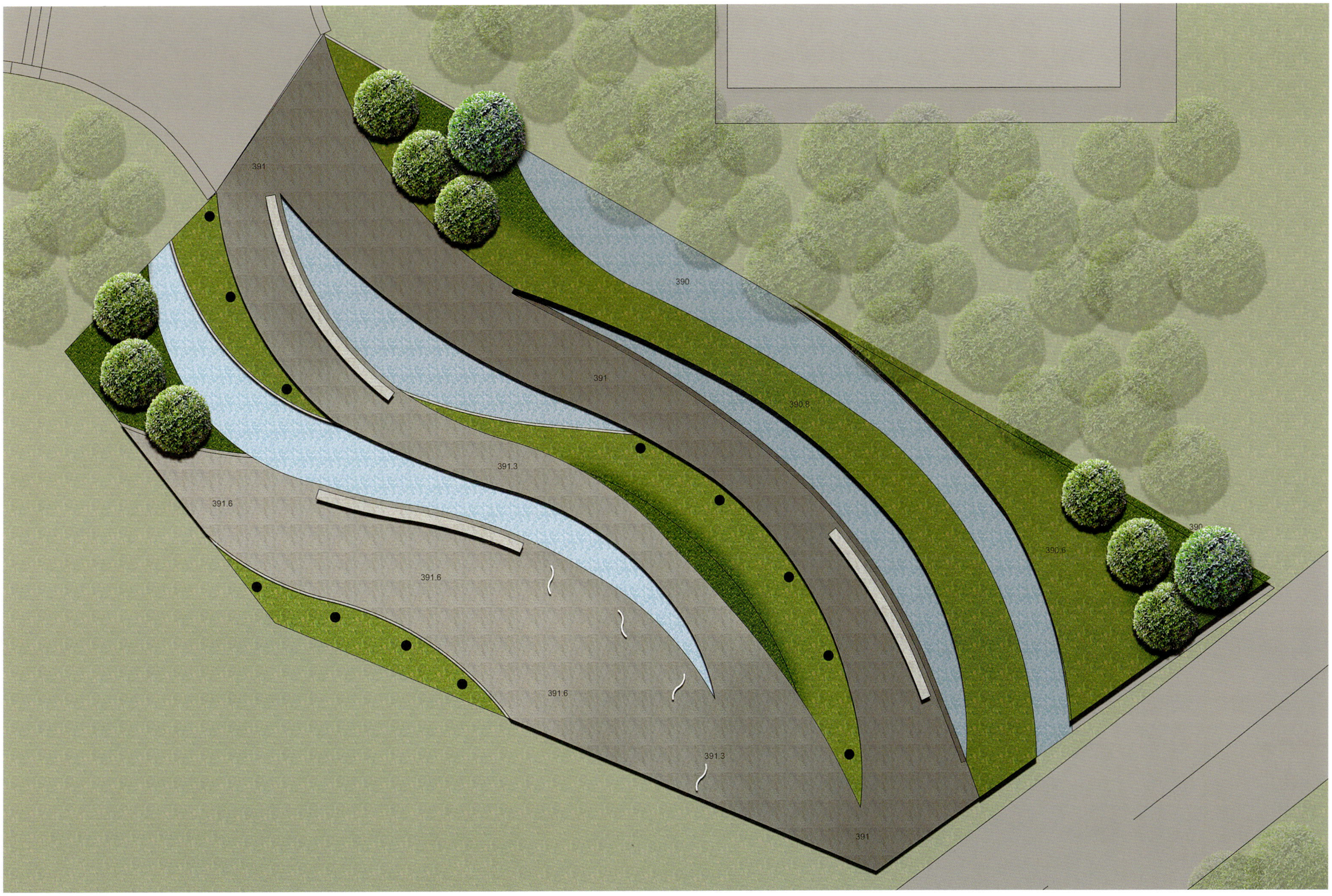

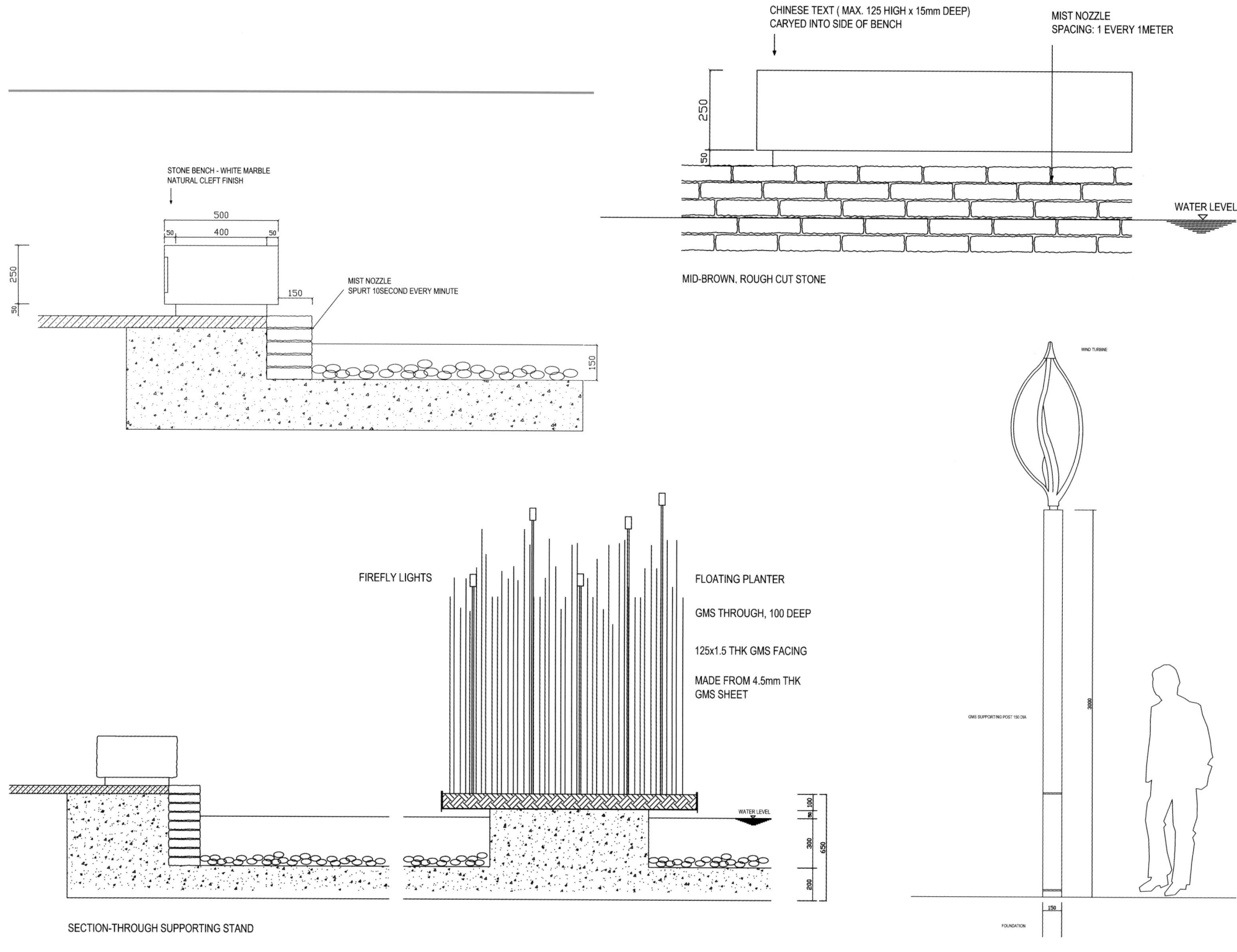

SECTION-MID SPAN

SPAN BETWEEN SUPPORT 1200mm

SECTION - GENERAL ARRANGEMENT

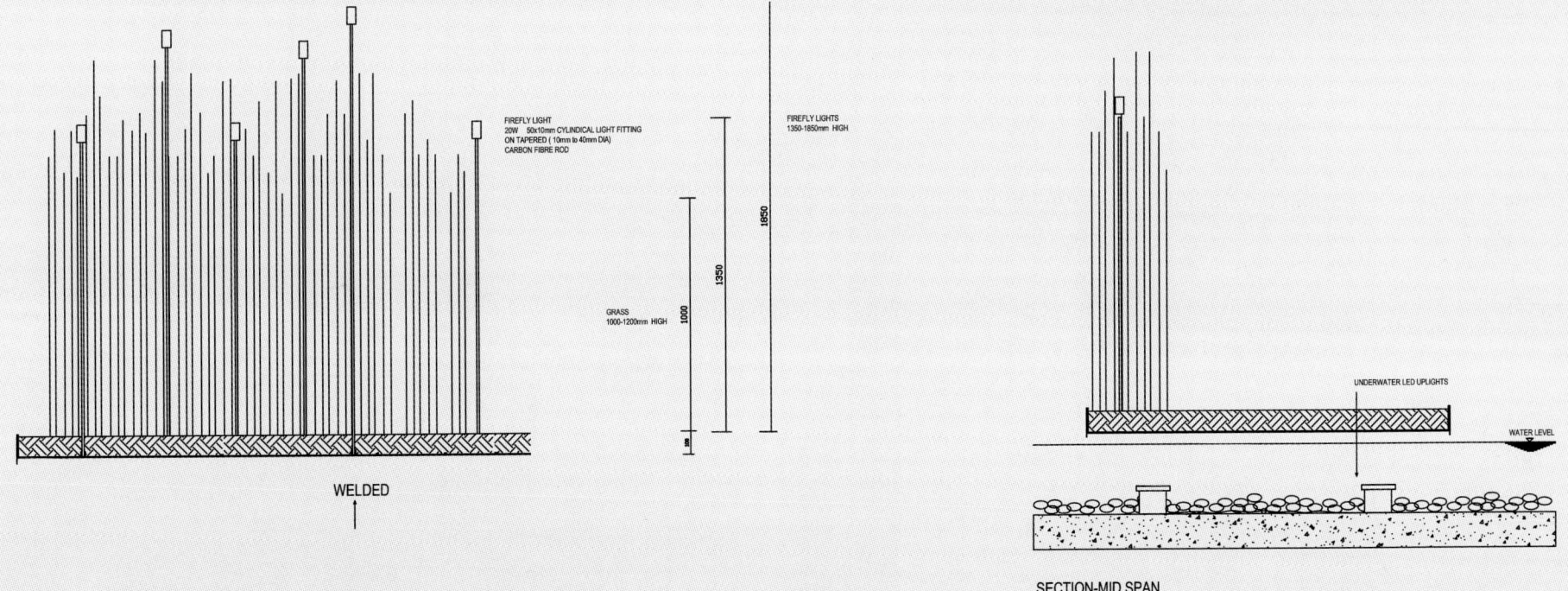

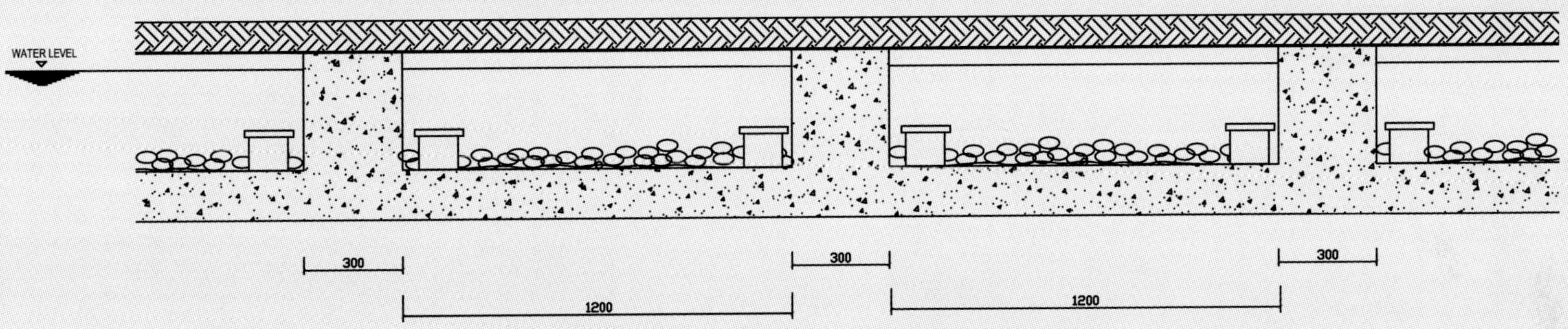

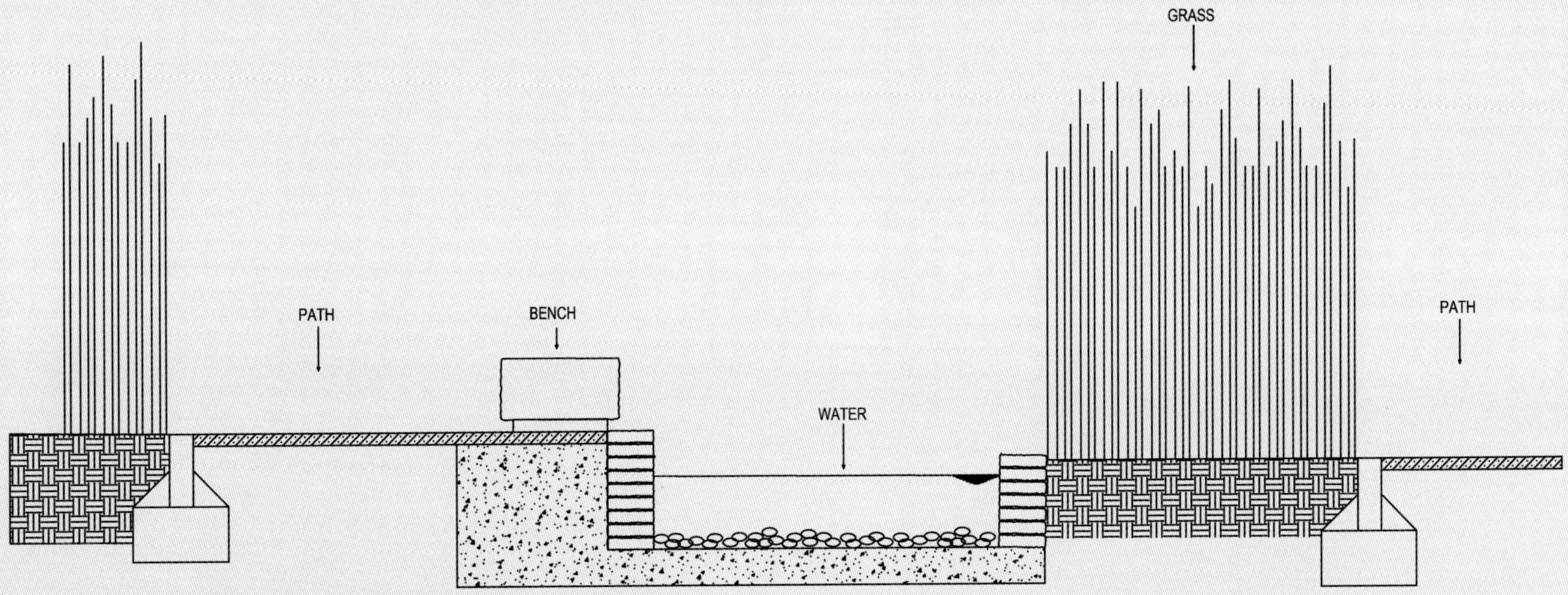

Aromatic Garden
创意园·芬香花园

展园位于世界园艺博览会创意园展区,由加拿大多伦多大学负责设计。设计以自然香气与人工香气相结合,在游客的嗅觉体验上填补了空白。除了栽种西安当地针叶林树种来营造属于西安的嗅觉外,更通过在园中放置多个"芳香柱"——内有来自中国不同地区的花粉香料,来丰富游客的嗅觉体验。"芬香花园"以科技手段,以中国含蓄内敛的审美情趣作为设计内涵,通过嗅觉这一相对间接的感觉,引发游客对景象无限的想像,扩充了景观的外延。在夜晚,夜景灯光让展园展现出令人难忘的温柔和绚丽。

The Garden is located at the Creativity Gardens Section, which is designed by University of Toronto. The integration of natural aroma and artificial aroma enriches visitors' sense of smell. It is planted with Xi'an local coniferous species in the Garden to create special Xi'an feeling, and also sets a series of aroma columns within pollen spices from different areas in China. The design connotation of this project is reflected by sci-tech means and Chinese special appreciation of beauty. Through the relatively indirect element - visitors' sense of smell, the designers would like to provide visitors with unlimited imagination of the scene in this limited landscape space. And the lighting decorations make the Garden much more glorious and charming at night.

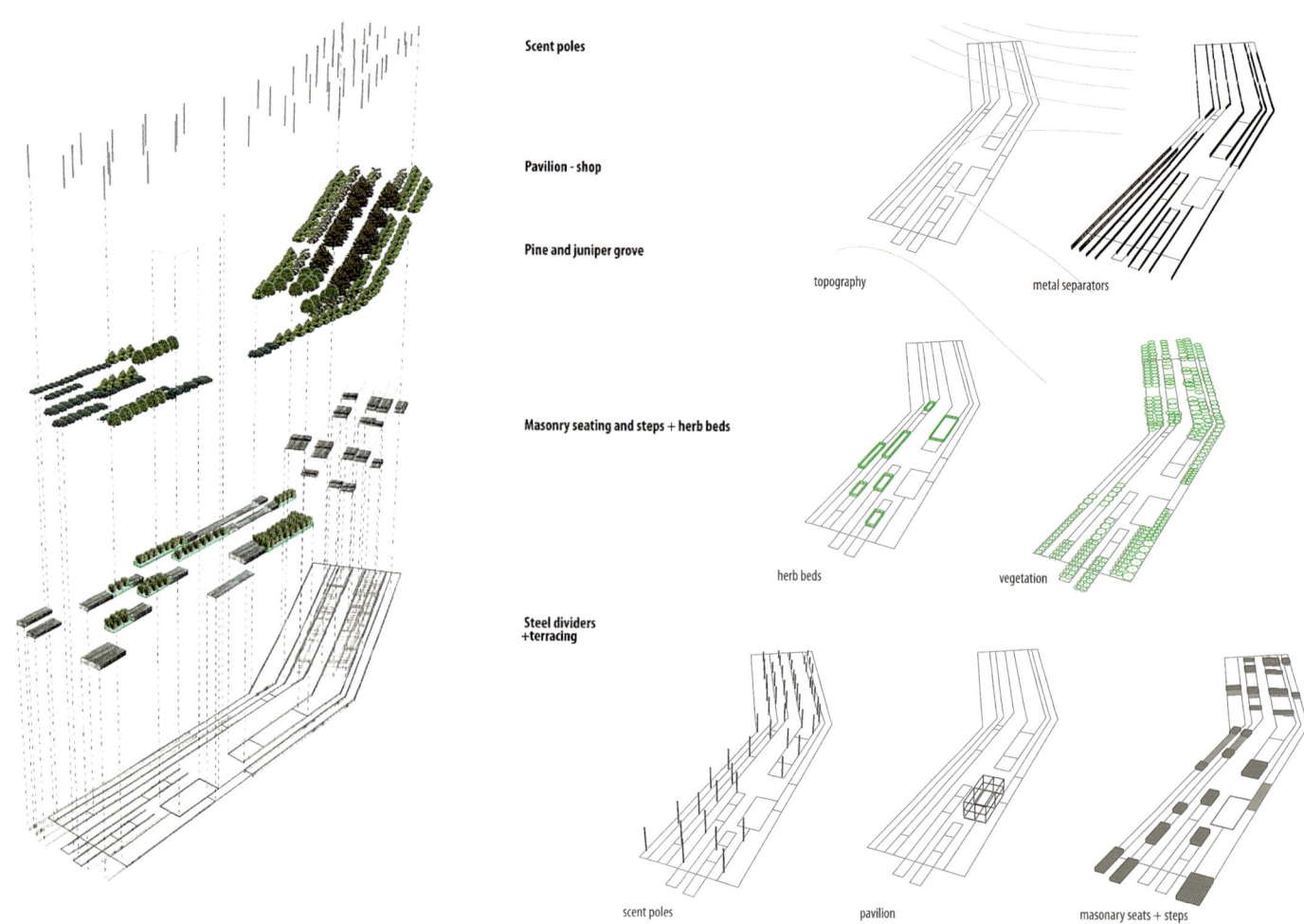

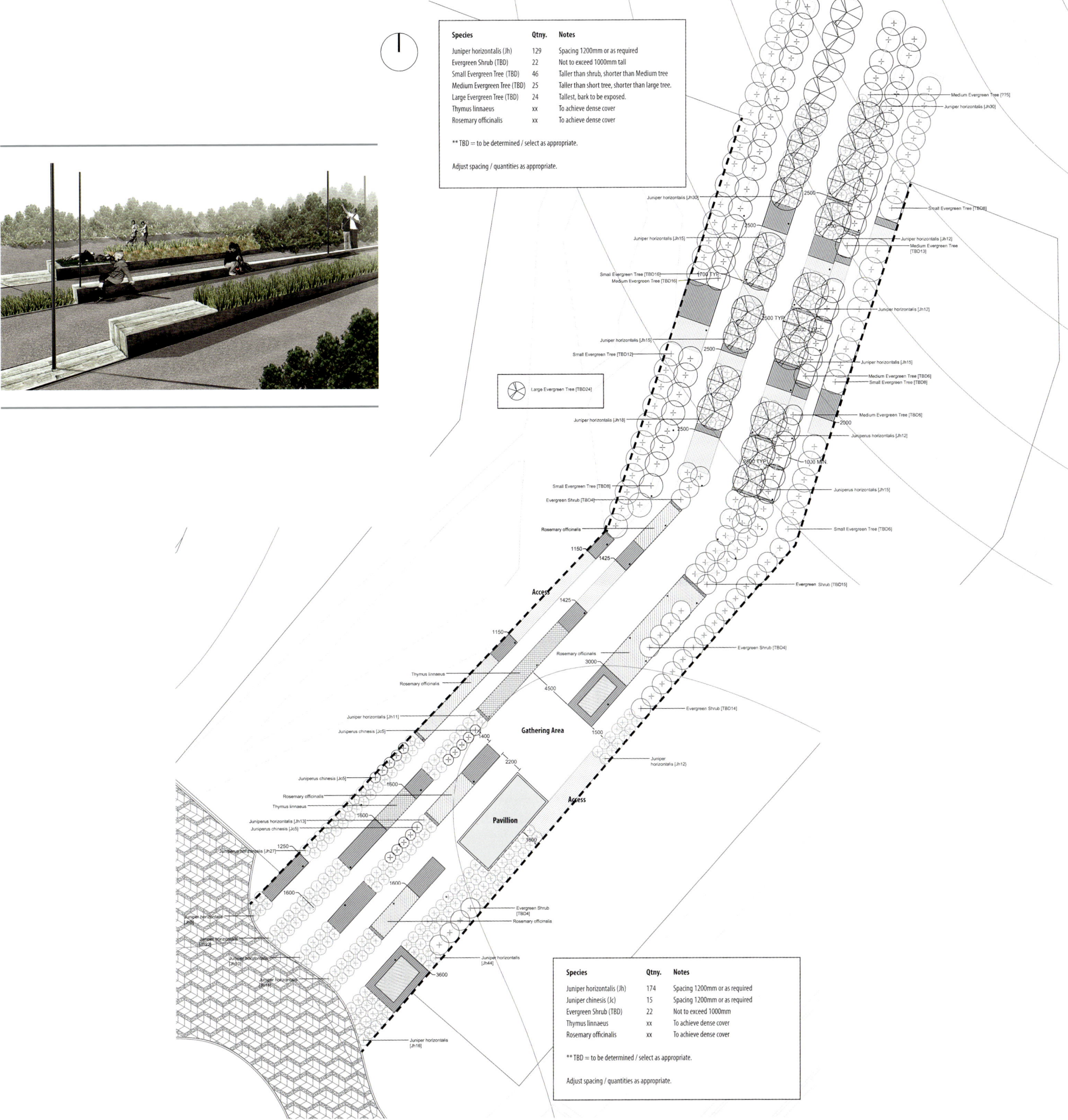

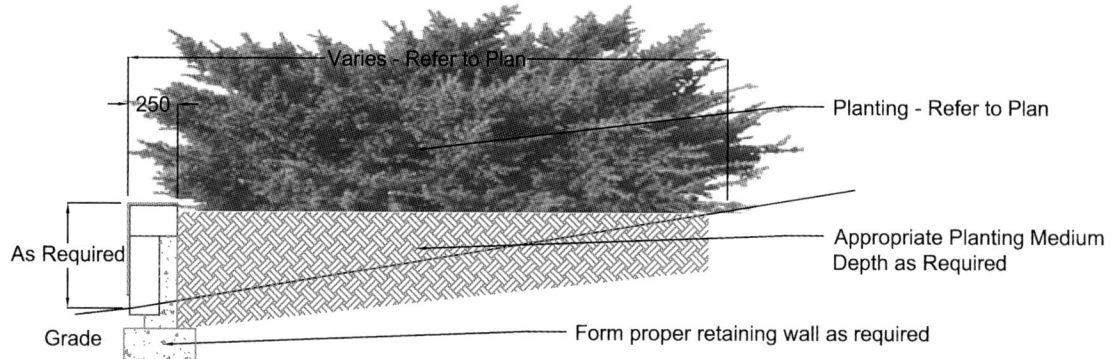

Section View

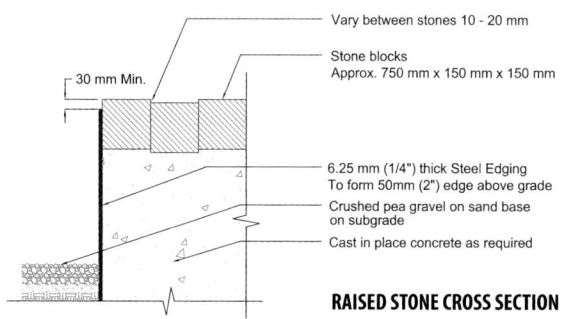

RAISED STONE CROSS SECTION

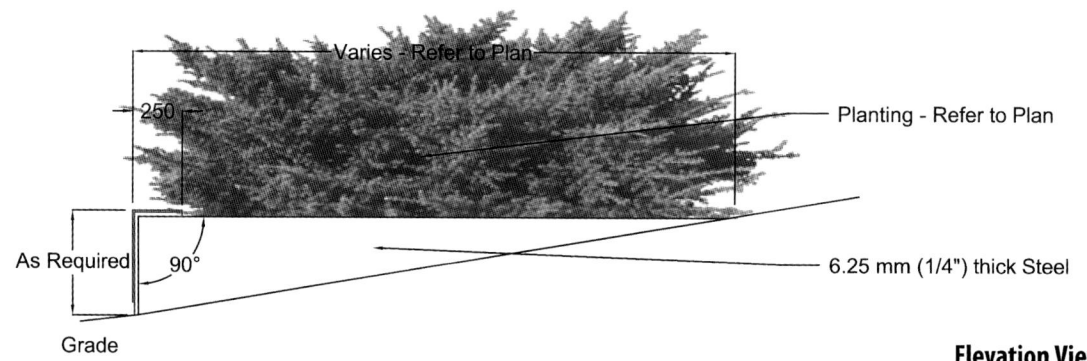

Elevation View

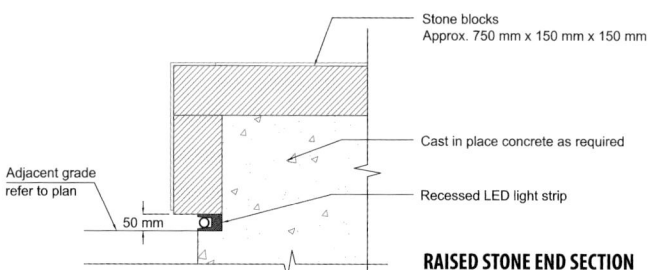

RAISED STONE END SECTION

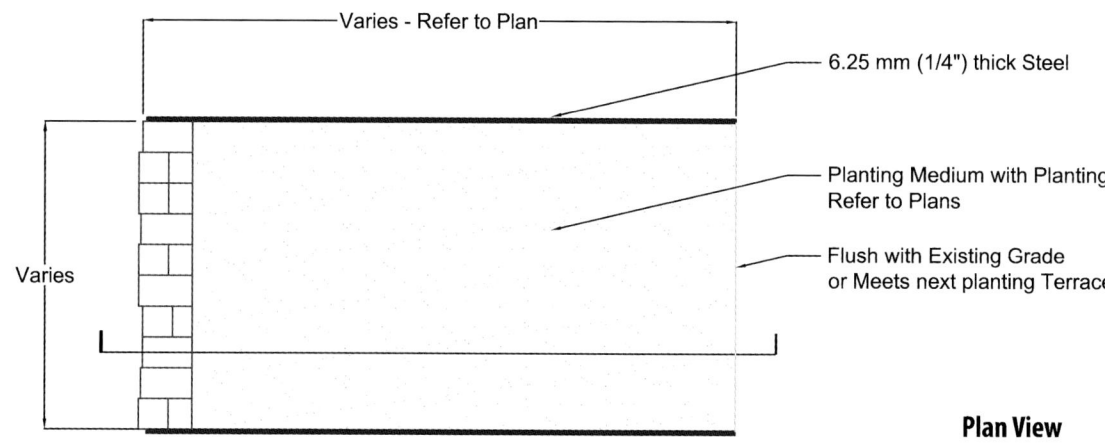

Plan View

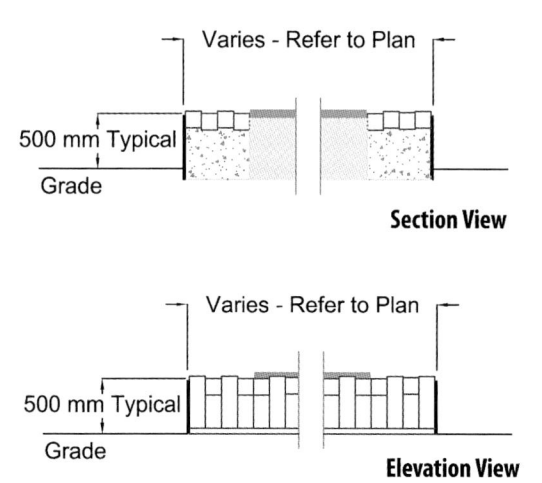

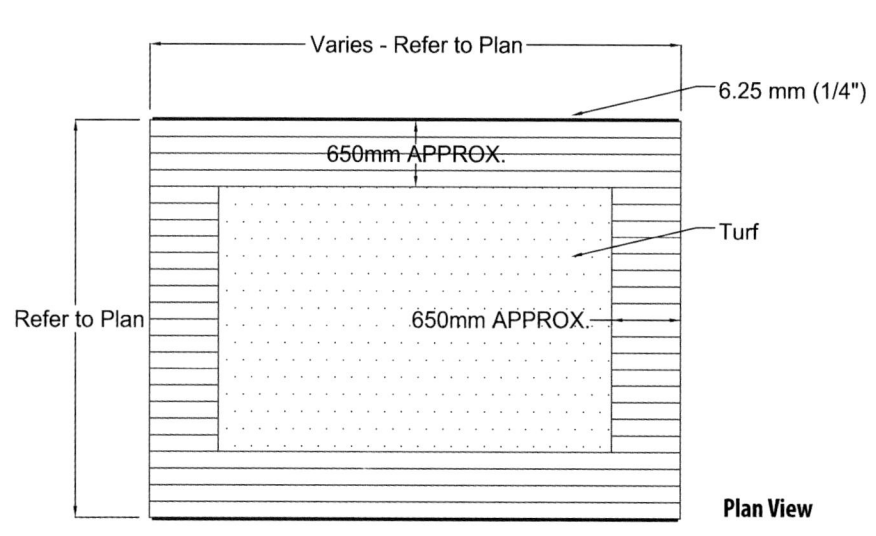

Plan View

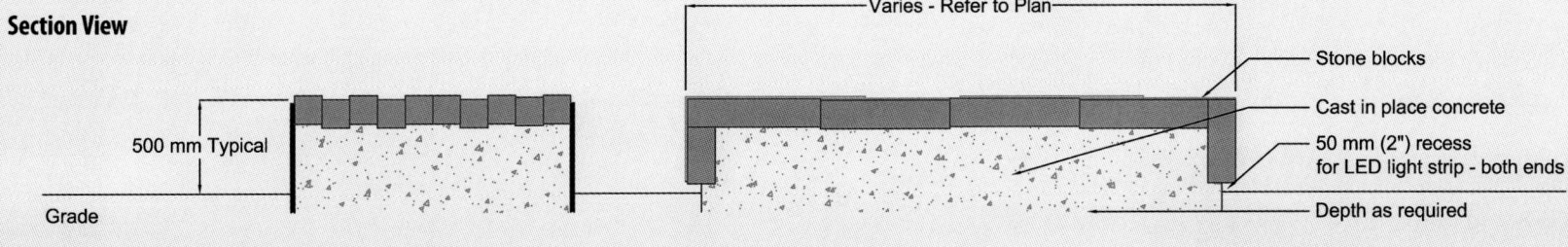
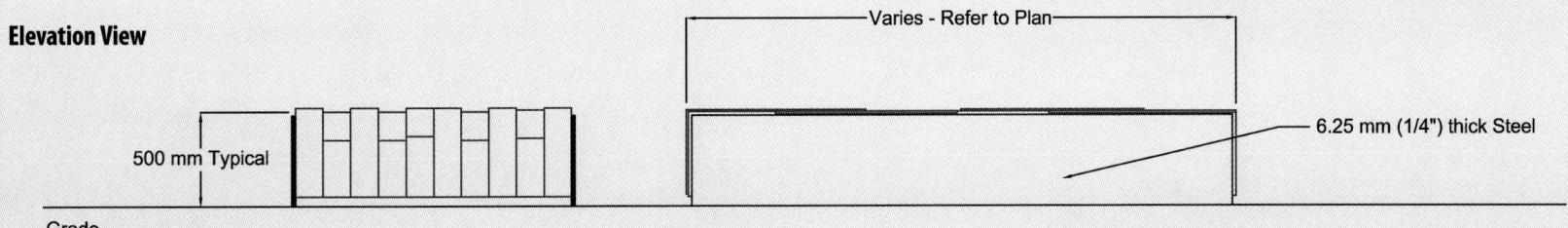
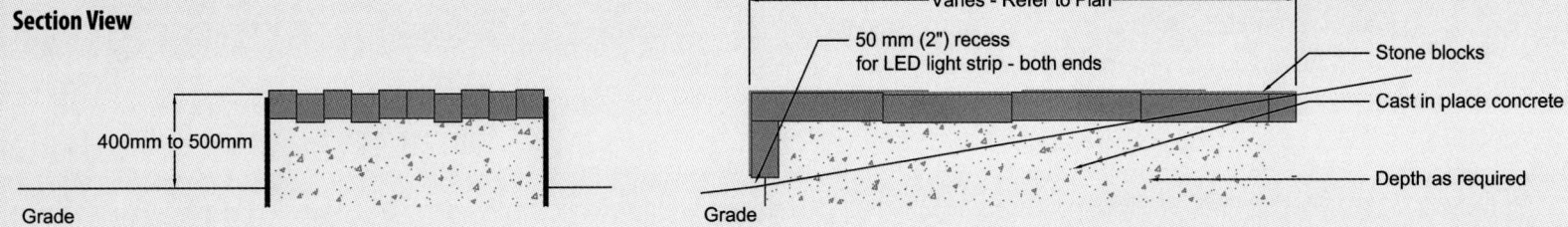
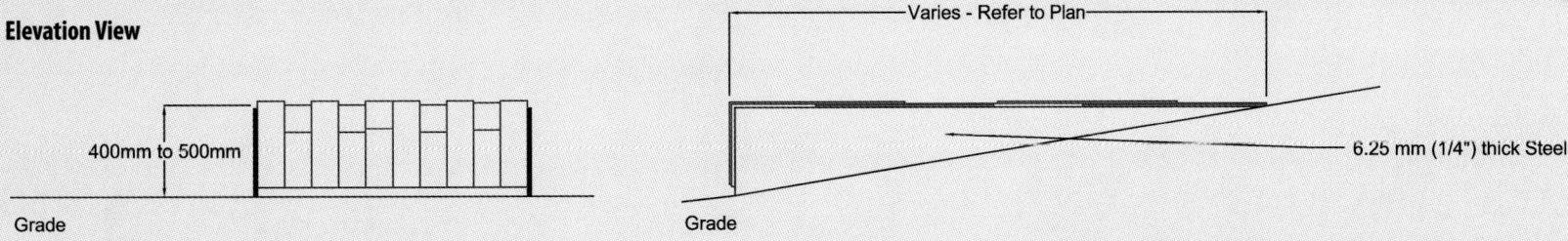
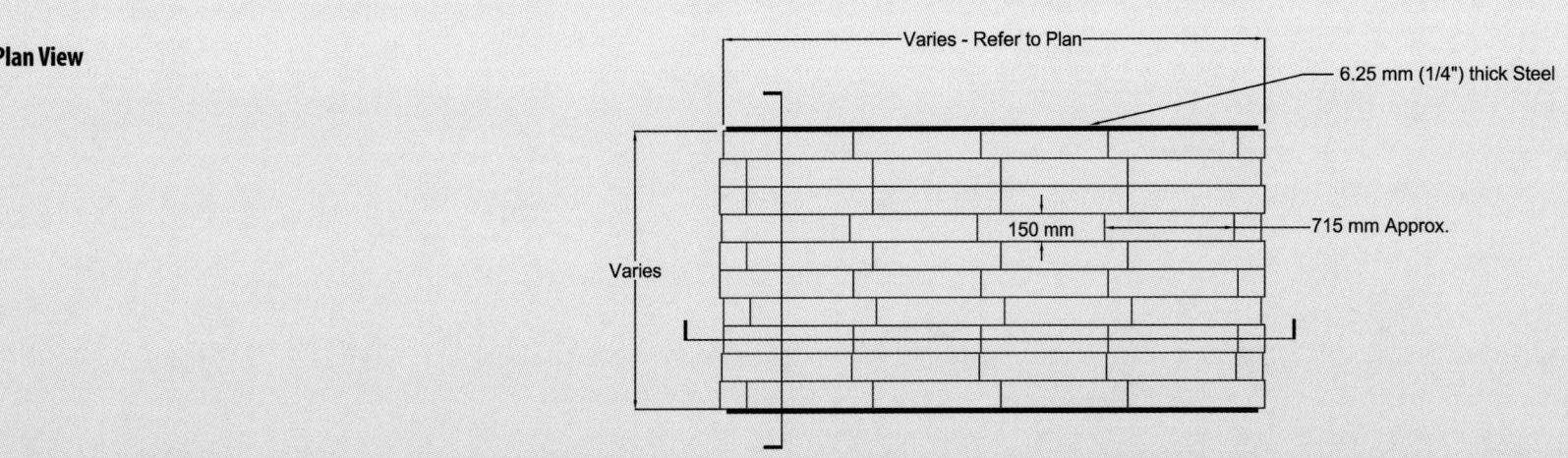

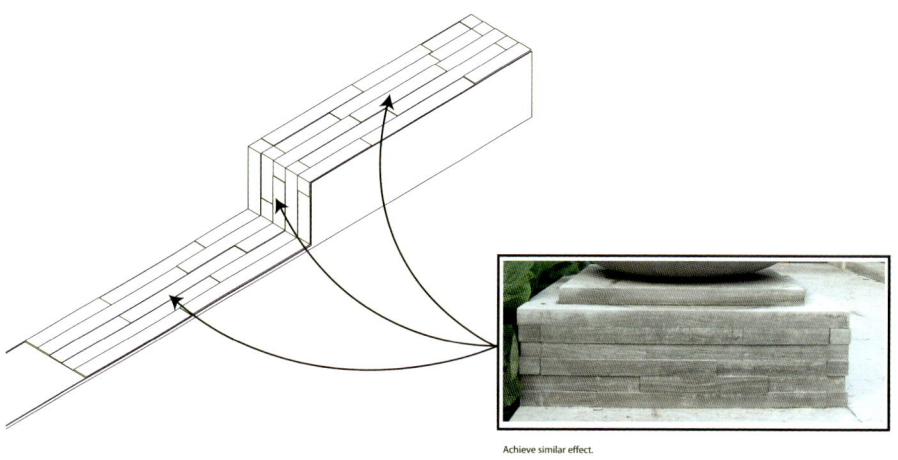

Achieve similar effect.
Side walls to be steel.

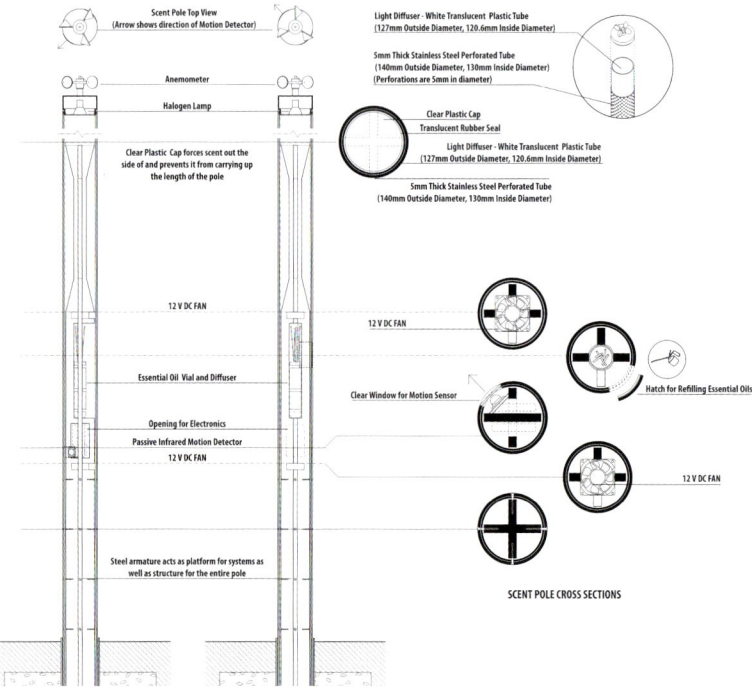

SCENT POLE CROSS SECTIONS

SCENT POLE VERTICAL SECTIONS

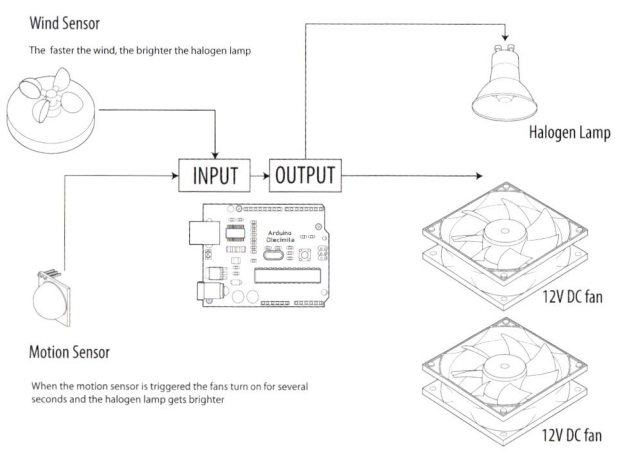

Wind Sensor
The faster the wind, the brighter the halogen lamp

Halogen Lamp

INPUT → OUTPUT

12V DC fan

12V DC fan

Motion Sensor
When the motion sensor is triggered the fans turn on for several seconds and the halogen lamp gets brighter

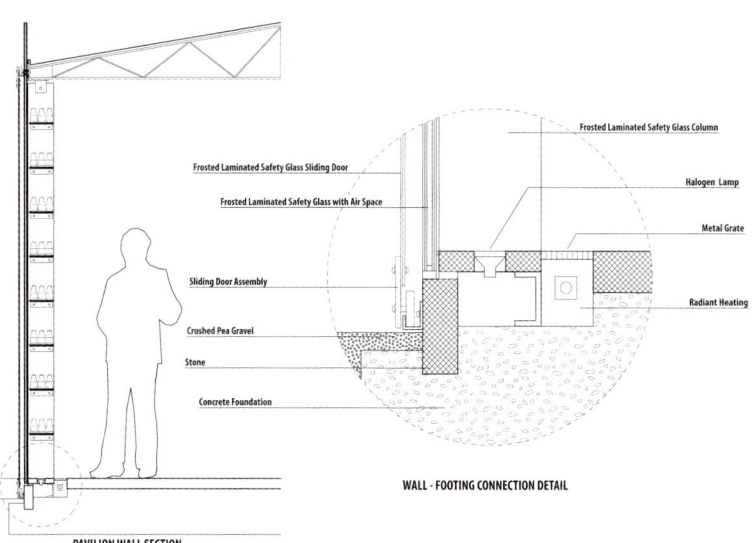

WALL - FOOTING CONNECTION DETAIL

PAVILION WALL SECTION

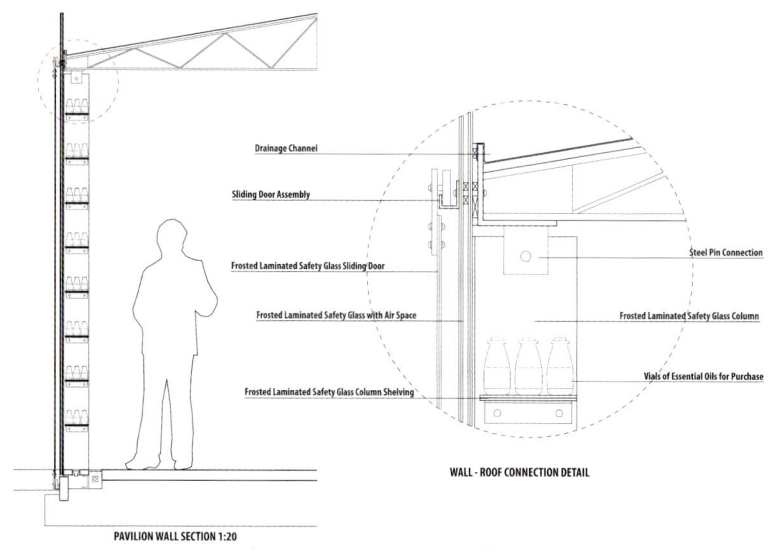

WALL - ROOF CONNECTION DETAIL

PAVILION WALL SECTION 1:20

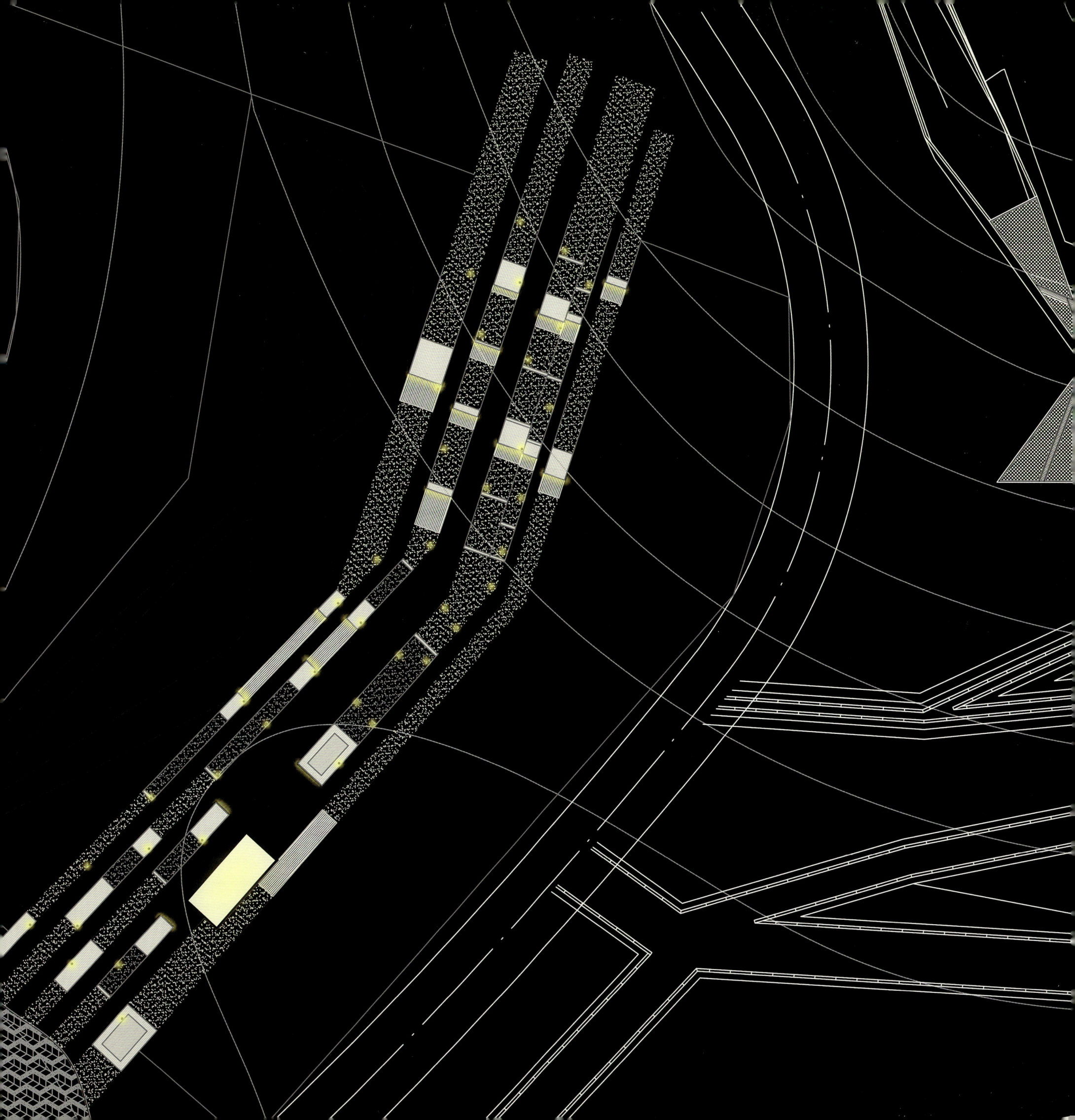

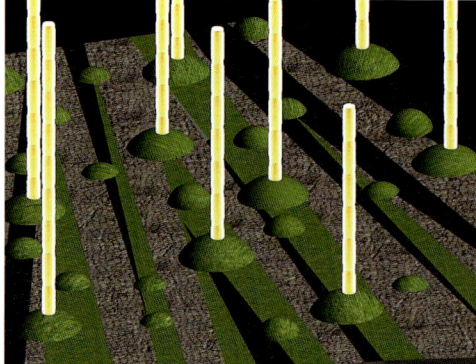
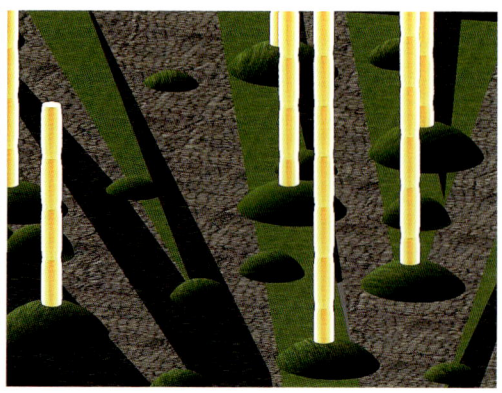

City Landscape Intervention
创意园·城市景观干预

展园位于世界园艺博览会创意园展区，由中国澳门圣若瑟大学负责设计。设计立足于澳门城市的转变，表现了在过去6个世纪中，澳门已经从一个依赖于自然资源的岛屿聚落转变成了国际化都市。原有可以代表澳门地域特色的渔村、棚屋、吊桥已被当今的高层建筑综合体所取代与主导。本案就是以放射状的绿轴，将参观流线与绿植交织，形成了互相干涉的格局，暗喻城市中景观与人类生活密不可分的关系，也呼应了世园会"创意自然"的主题。

Located at the Creativity Gardens Section, the Garden is designed by University of Saint Joseph. The theme of this garden is the changes of Macau in the past six centuries from an island living on natural resources into an international metropolis: the local features of Macau, such as fishing villages, shanties and hanging bridges, are replaced by high-rise complexes. This project applies actinomorphic green scrolls to interlace the visit routes with green plants, forming an interweaving layout to interpret the close relationship between urban landscape and human life, also echoing with the theme "Nature Creativity" of Expo 2011 Xi'an.

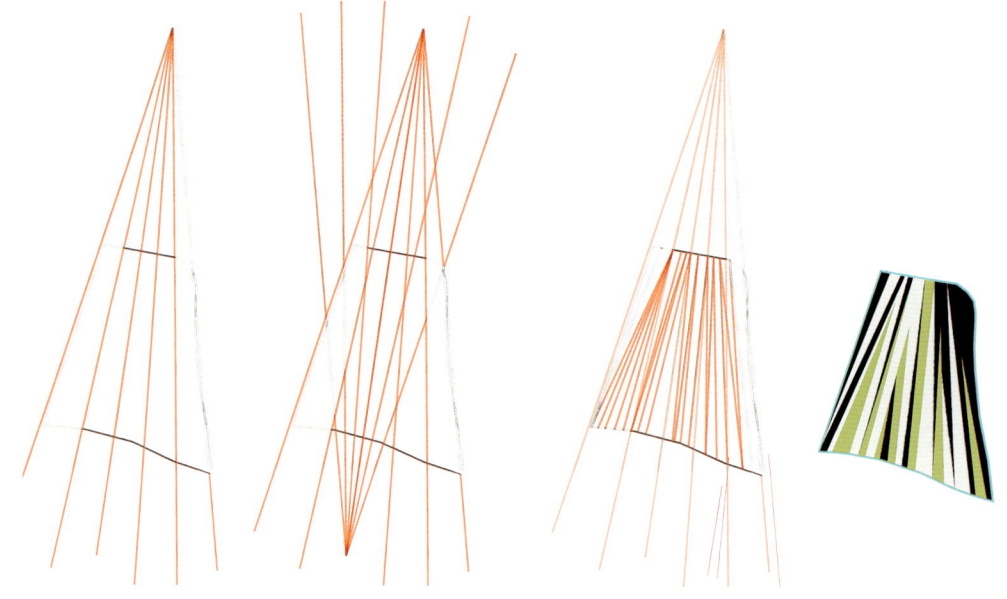
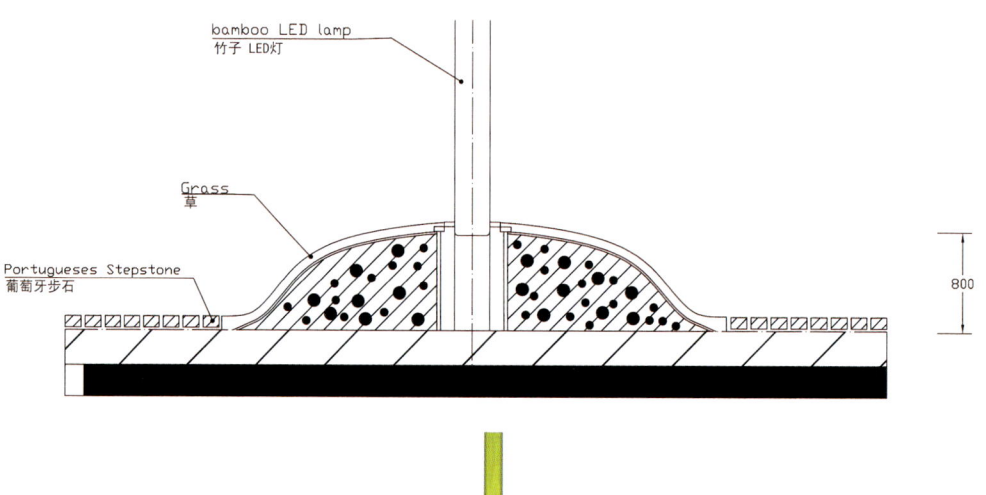
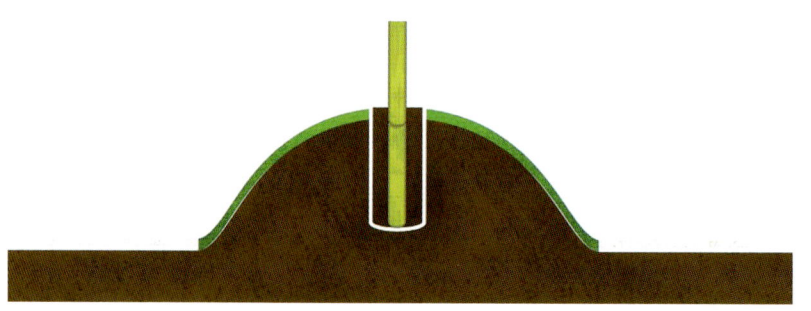

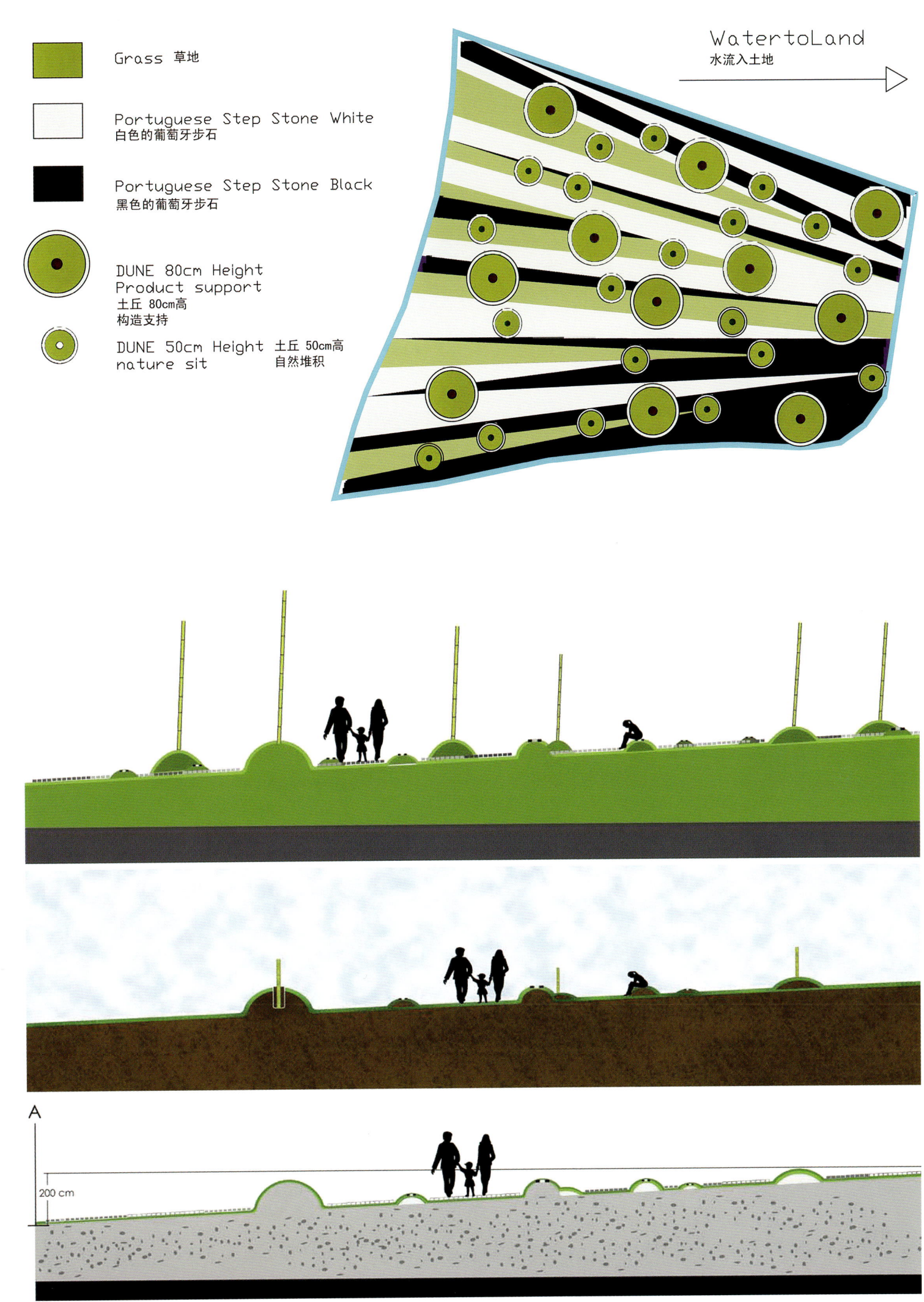

City of Sky
创意园·天空之城

展园位于世界园艺博览会创意园展区,由美国南加州大学负责设计。本案"天空之城"的设计理念立足于表达人们对于自然的美好憧憬,最能反映人们对于自然界中未知事物的无尽探知,是向着无法一眼看透的天空,永恒的追求。展园中的"日光中庭"、"反射花园"和"云雾园"分别反映了三种人类与天空的基本关系——直接感知、间接反射以及绿荫云雾萌蔽天空。"天空之城"想要带给游客的参观体验——通过喷雾系统加大了空气的湿度并且营造出云雾的观感,让游客在迷蒙中参与到整体景观之中,从而可以设身处地地反思人类与自然的关系。

The Garden is located at the Creativity Gardens Section, which is designed by University of Southern California. The design of this project expresses the endless curiosity and nice yearning of human for the nature and the mysterious sky. The sunshine atrium, reflection garden and cloud and mist garden in the "City of Sky" respectively reflect three basic relations between mankind and sky – direct perception, indirect reflection and roaming. "City of Sky" enables visitors to perceive the connotation in the bewildering sky and consider the relationship between mankind and nature.

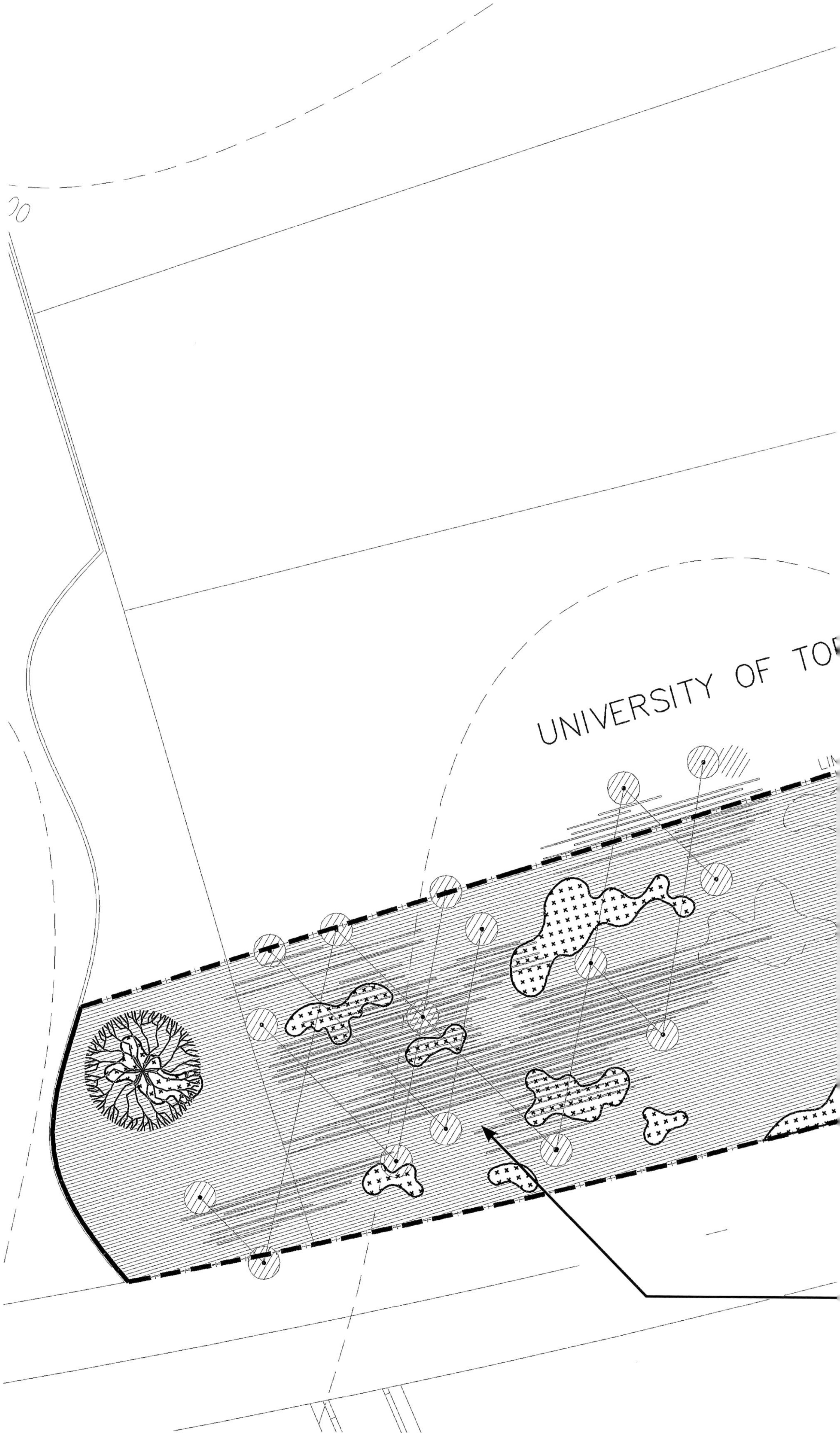

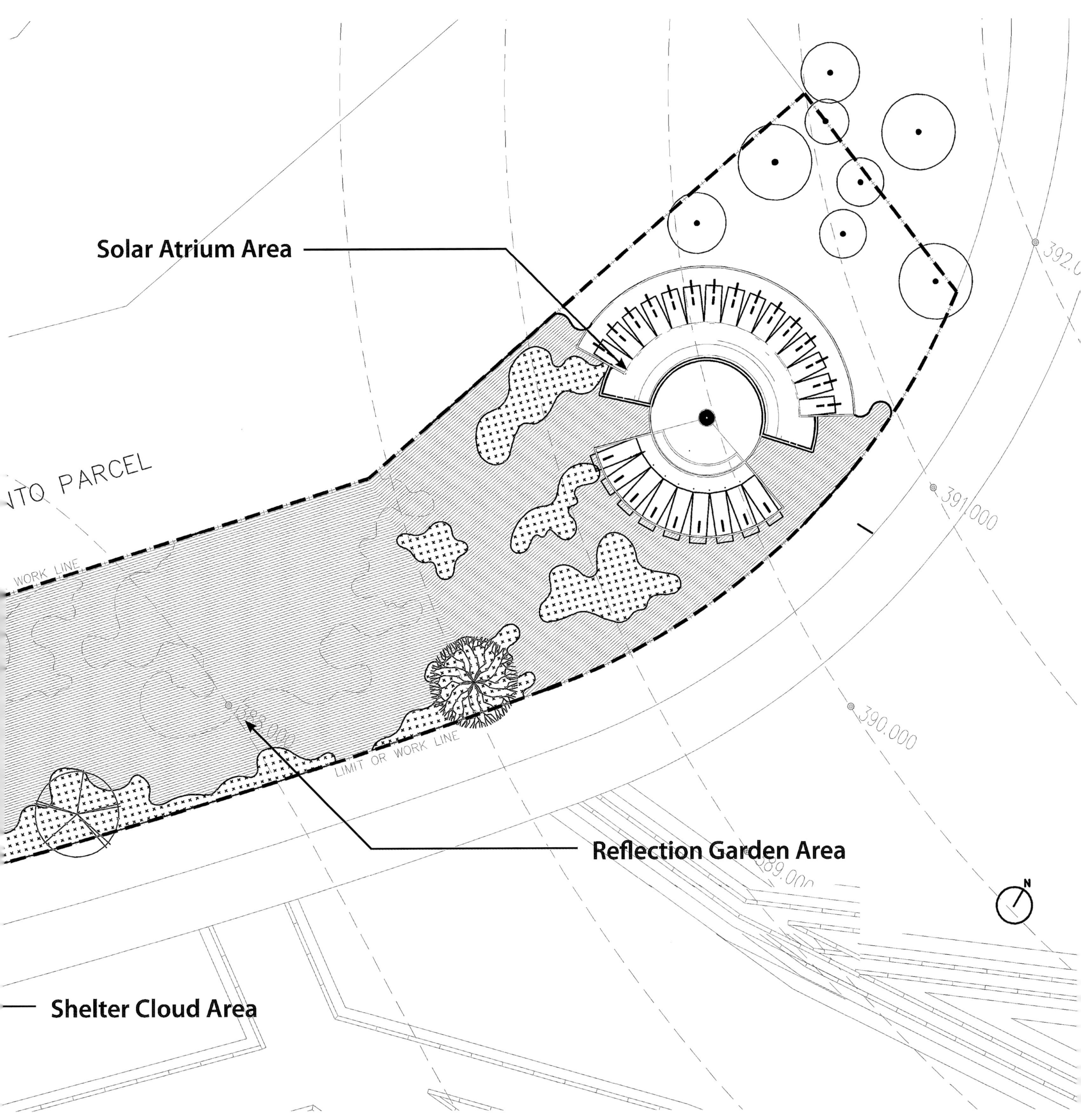

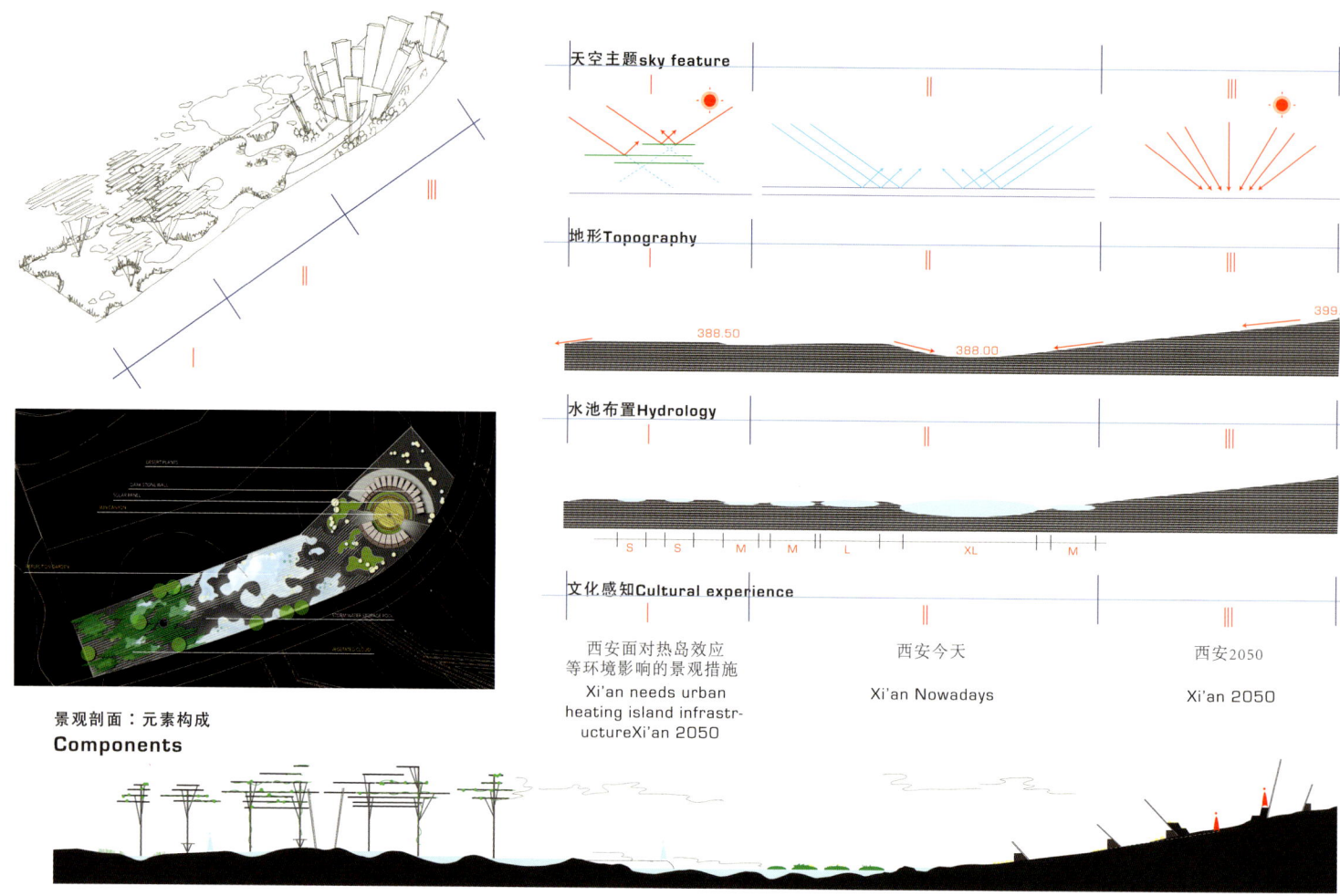

景观剖面：元素构成
Components

天空主题 sky feature

地形 Topography

水池布置 Hydrology

文化感知 Cultural experience

西安面对热岛效应等环境影响的景观措施
Xi'an needs urban heating island infrastructureXi'an 2050

西安今天
Xi'an Nowadays

西安2050
Xi'an 2050

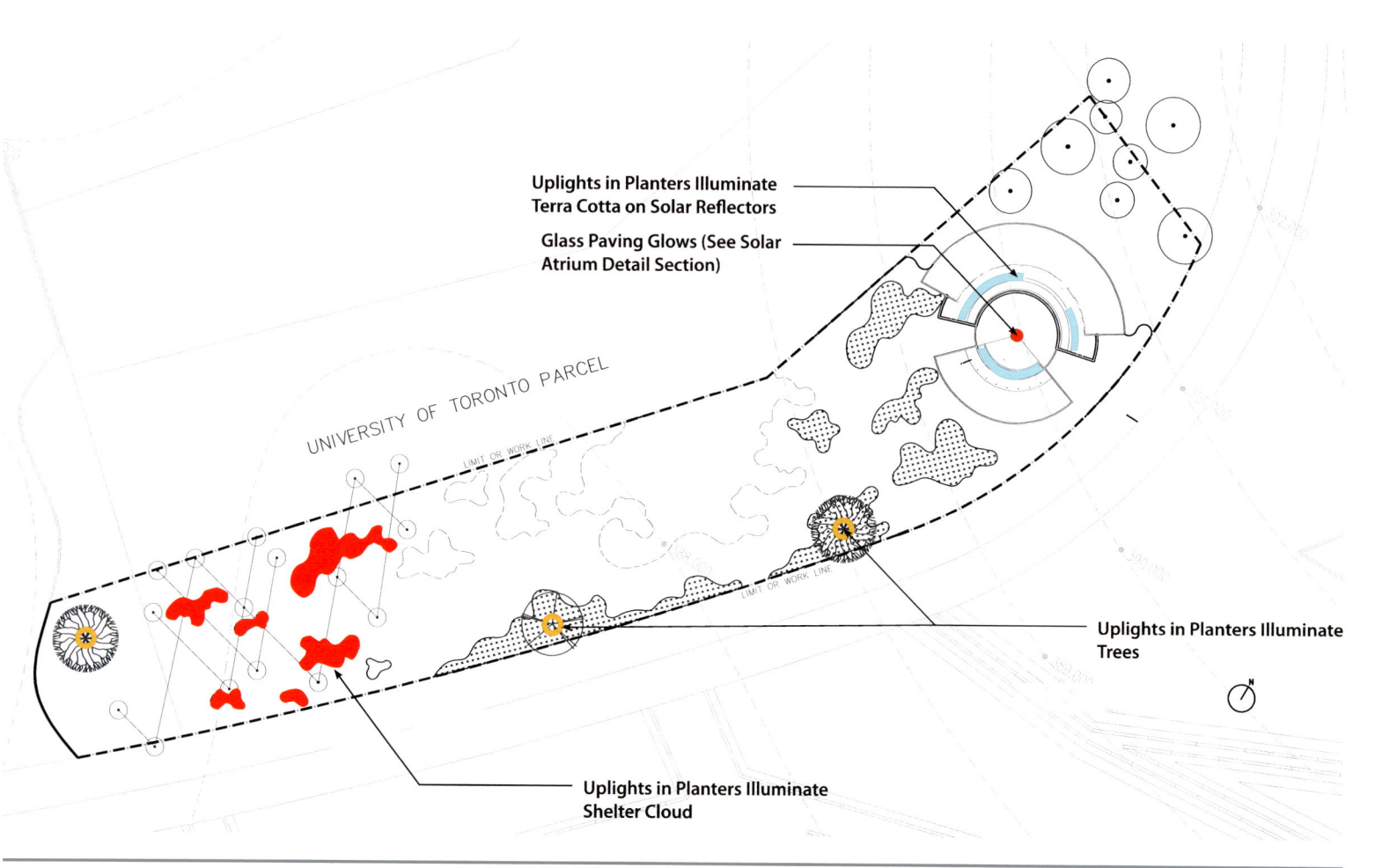

Uplights in Planters Illuminate Terra Cotta on Solar Reflectors

Glass Paving Glows (See Solar Atrium Detail Section)

UNIVERSITY OF TORONTO PARCEL

Uplights in Planters Illuminate Trees

Uplights in Planters Illuminate Shelter Cloud

Manual Drainage System
Visible Reflection Pools
Sub-Surface Inundated Area
Tree Planting Areas are Kept Dry
392.000
391.000
390.000
389.000

(PLANTING AREA TO MATCH
ADJACENT U.T. DESIGN)
SOIL CEMENT PAVING
SOLAR REFLECTOR, 5.6M
PINE TREE AS PER UT
PAVING EDGE
ADJUSTABLE REFLECTOR SUPPORT
PLANTING AREA WITH UPLIGHTS
SOIL CEMENT SEAT WALL
GLASS PAVING
WITH
LIGHTING
PLANTING
AREA WITH
UPLIGHTS
BASALT PAVING
SOLAR REFLECTOR, 3M
SOLAR
ATRIUM
SECTION
LIMIT OF WORK
1.00M
+391.83
+391.3
+390.79
3.34M
6.98M
8.7%
1.06M
8.7%
2.21M
390.0
0.98M
0.75M
10%
2.60M
+390.38
1.00M
390.0
KEYMAP
LEGEND
SOIL CEMENT PAVING
SOIL CEMENT SEATWALL
PLANTED AREA

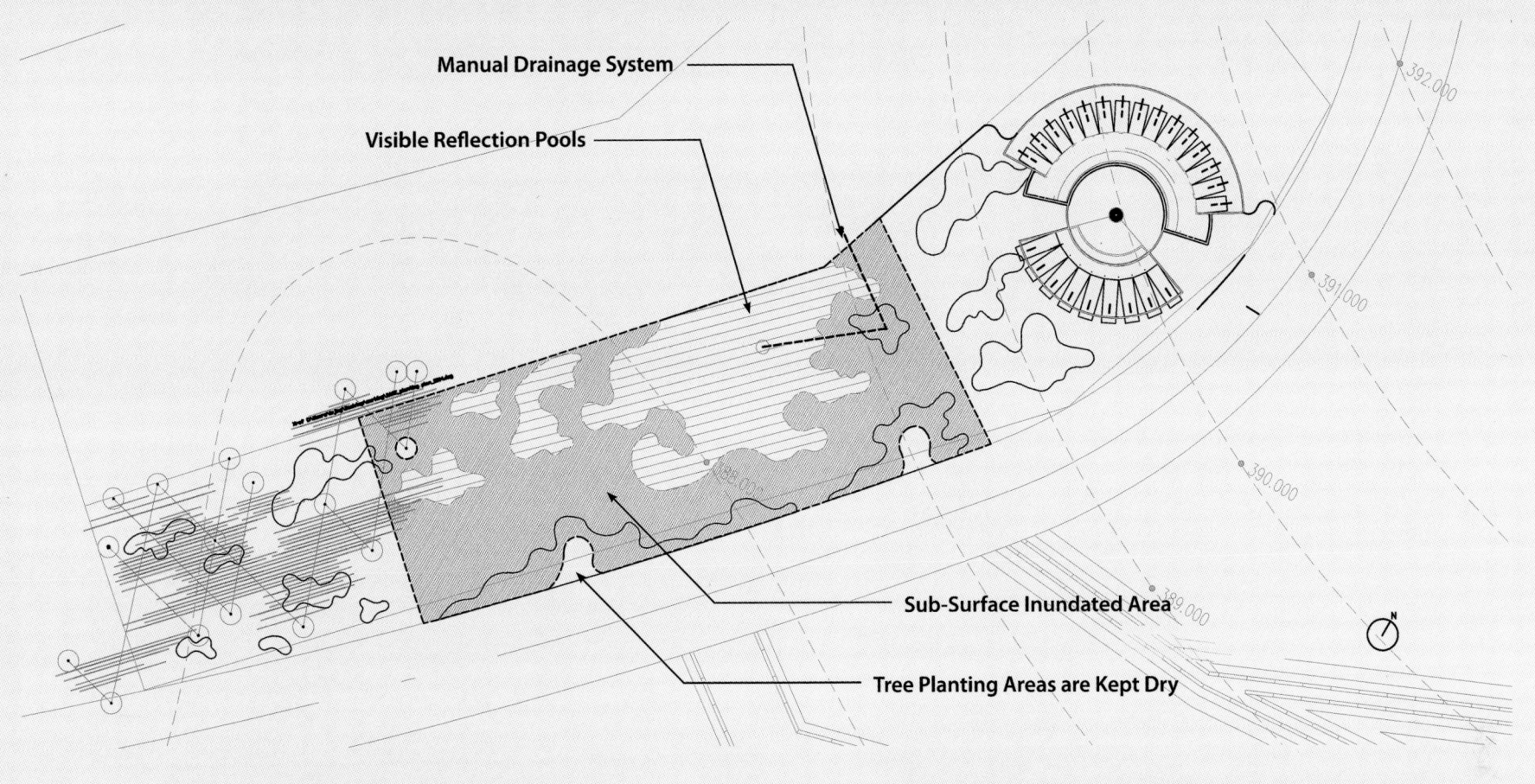

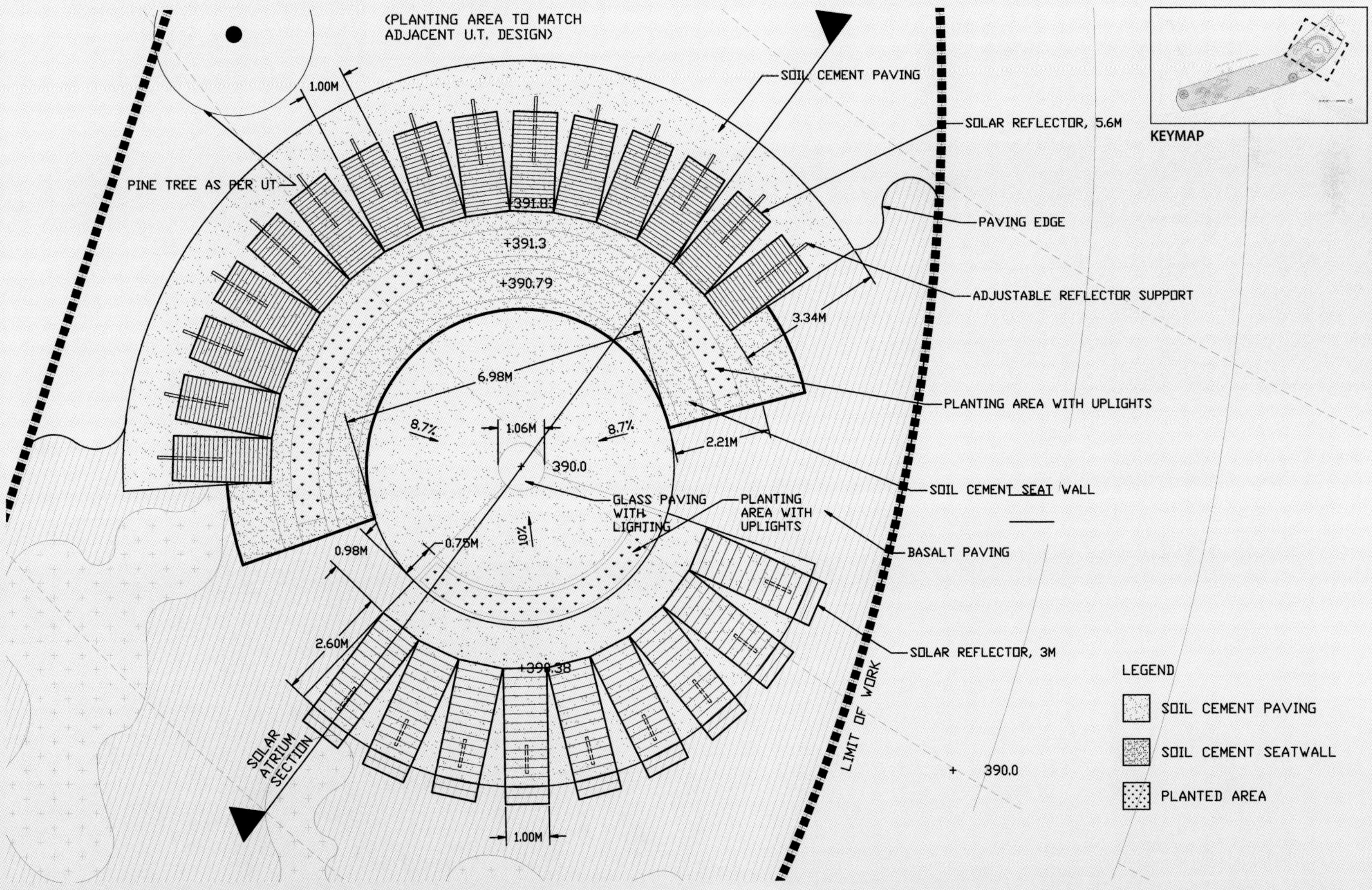

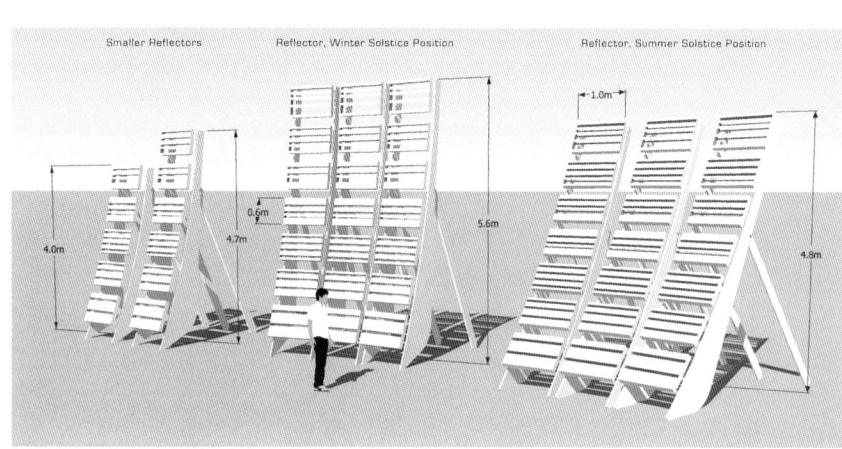

Conceptual Visualization of Reflector

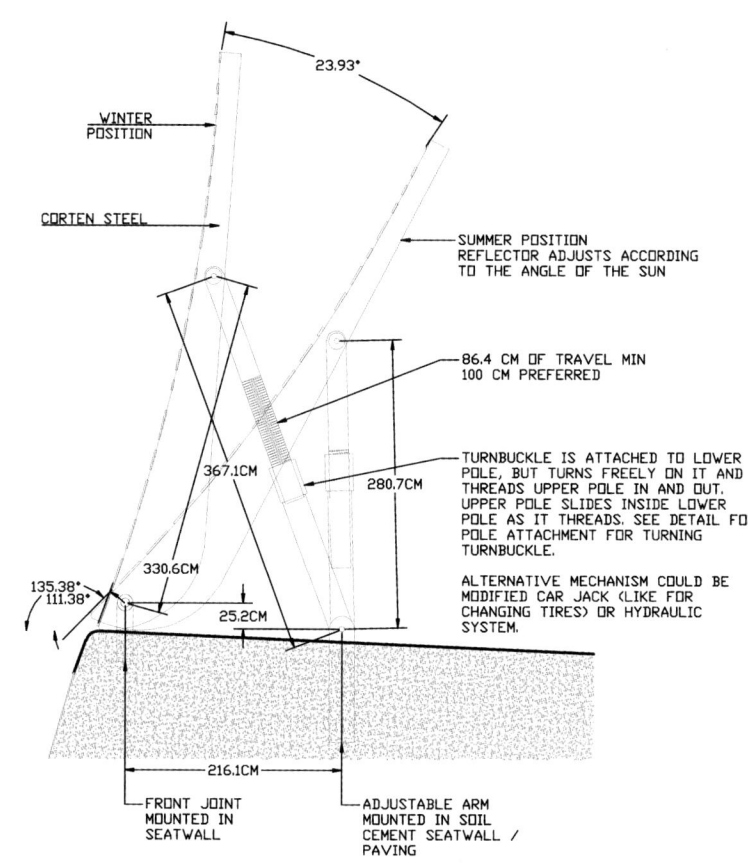

Large Solar Reflector Detail Side Section

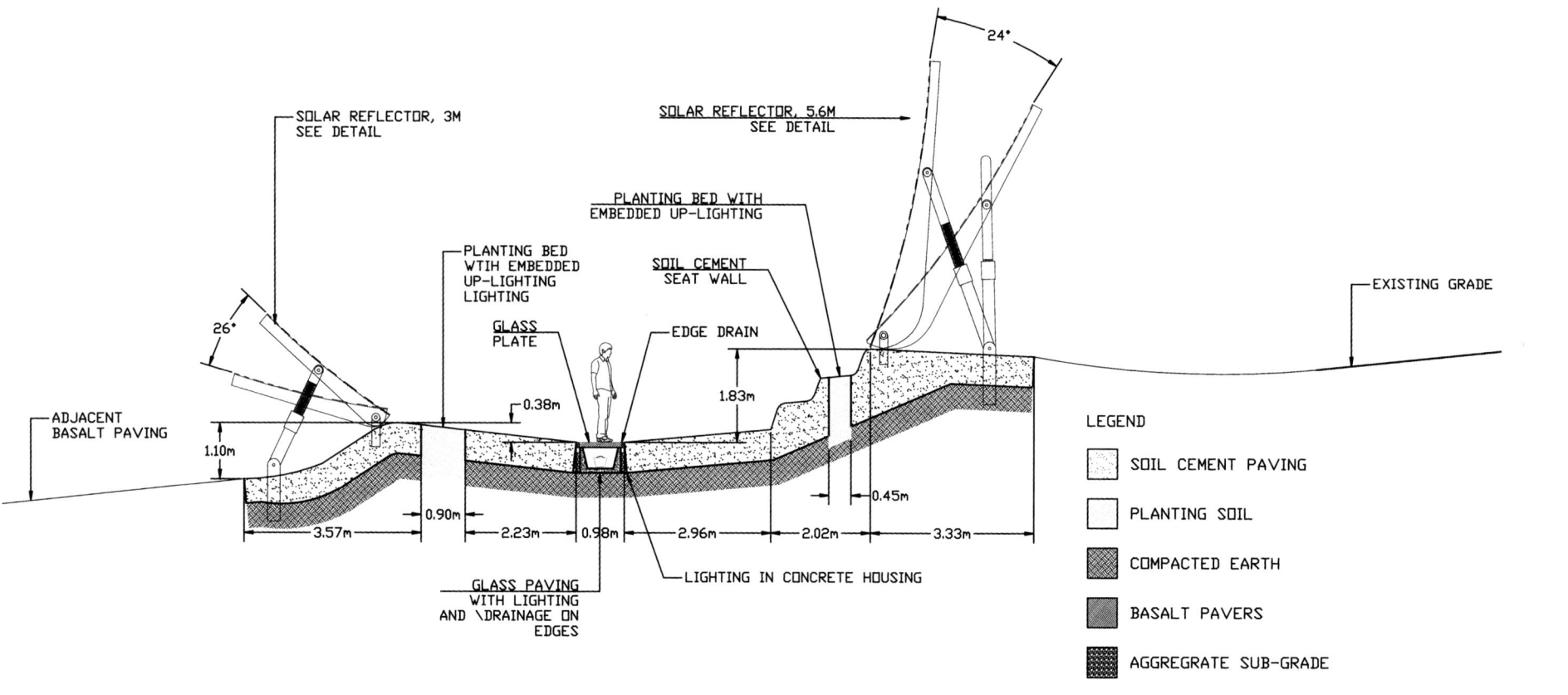

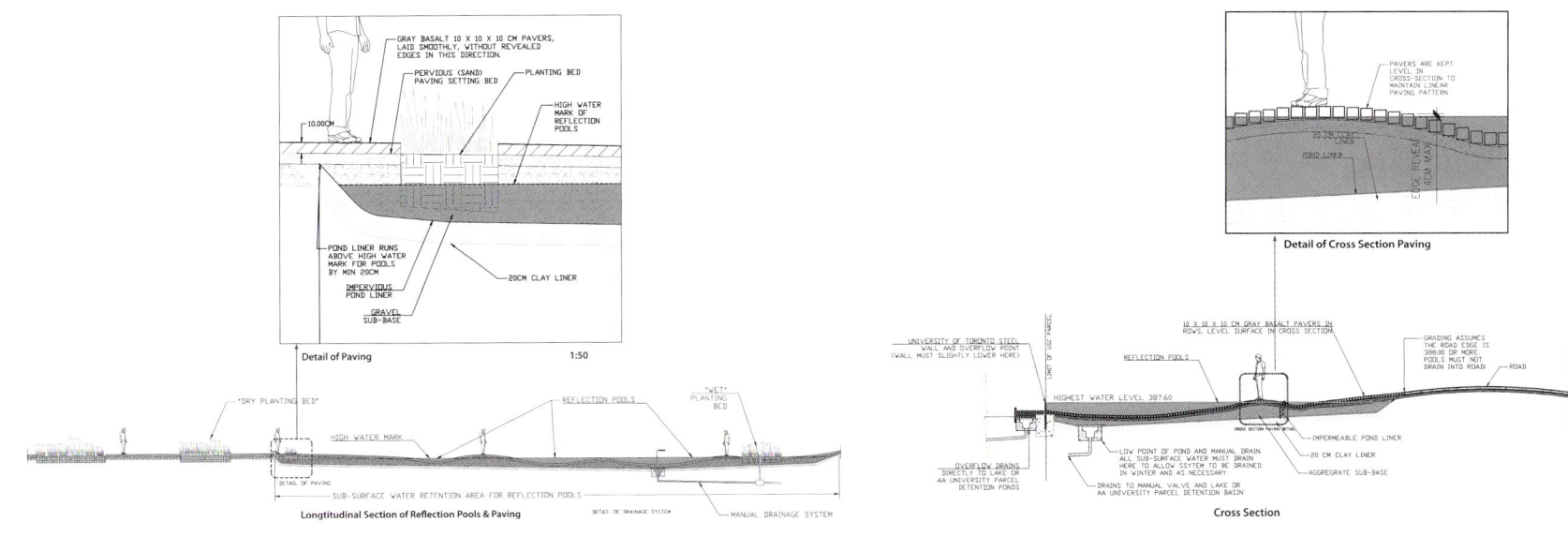

Layer Structure:
pole + rod + mister + planter + steel rope

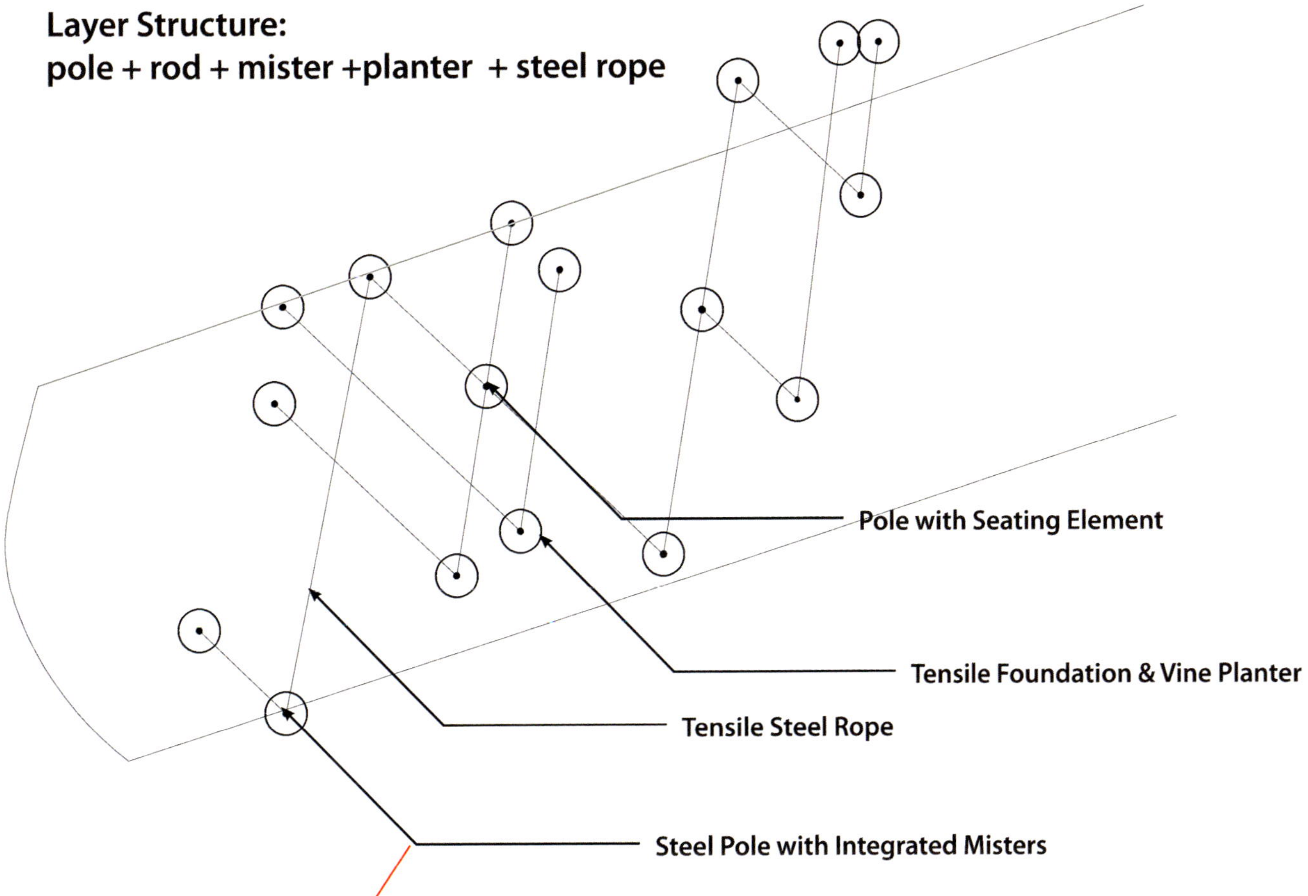

- Pole with Seating Element
- Tensile Foundation & Vine Planter
- Tensile Steel Rope
- Steel Pole with Integrated Misters

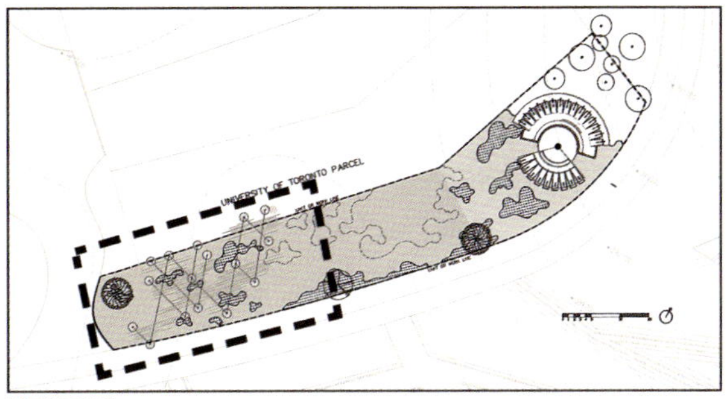

KEYMAP

Conceptural Visualization

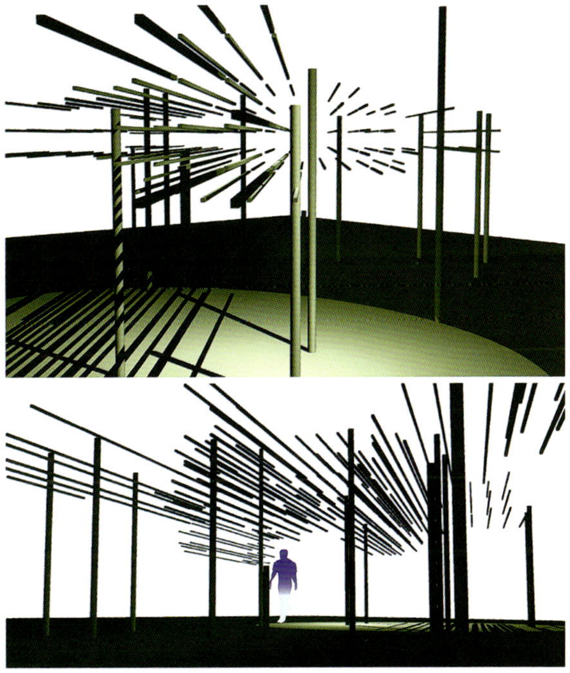

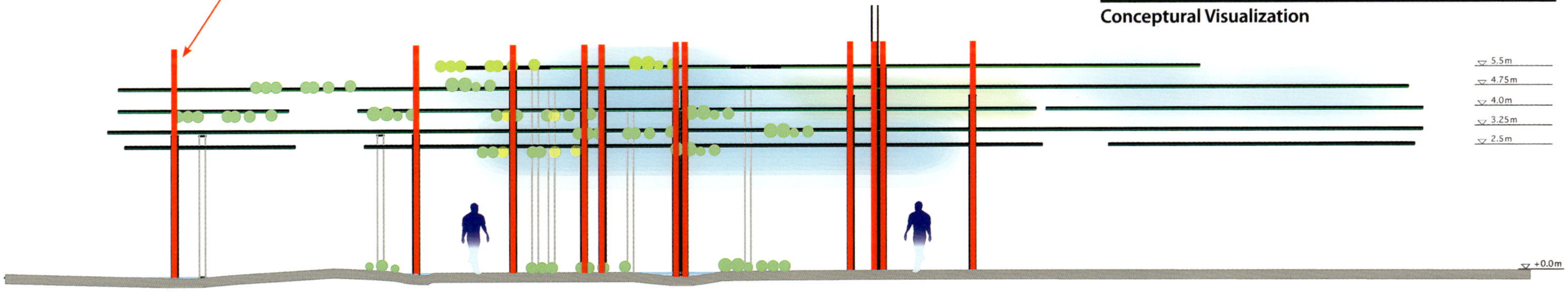

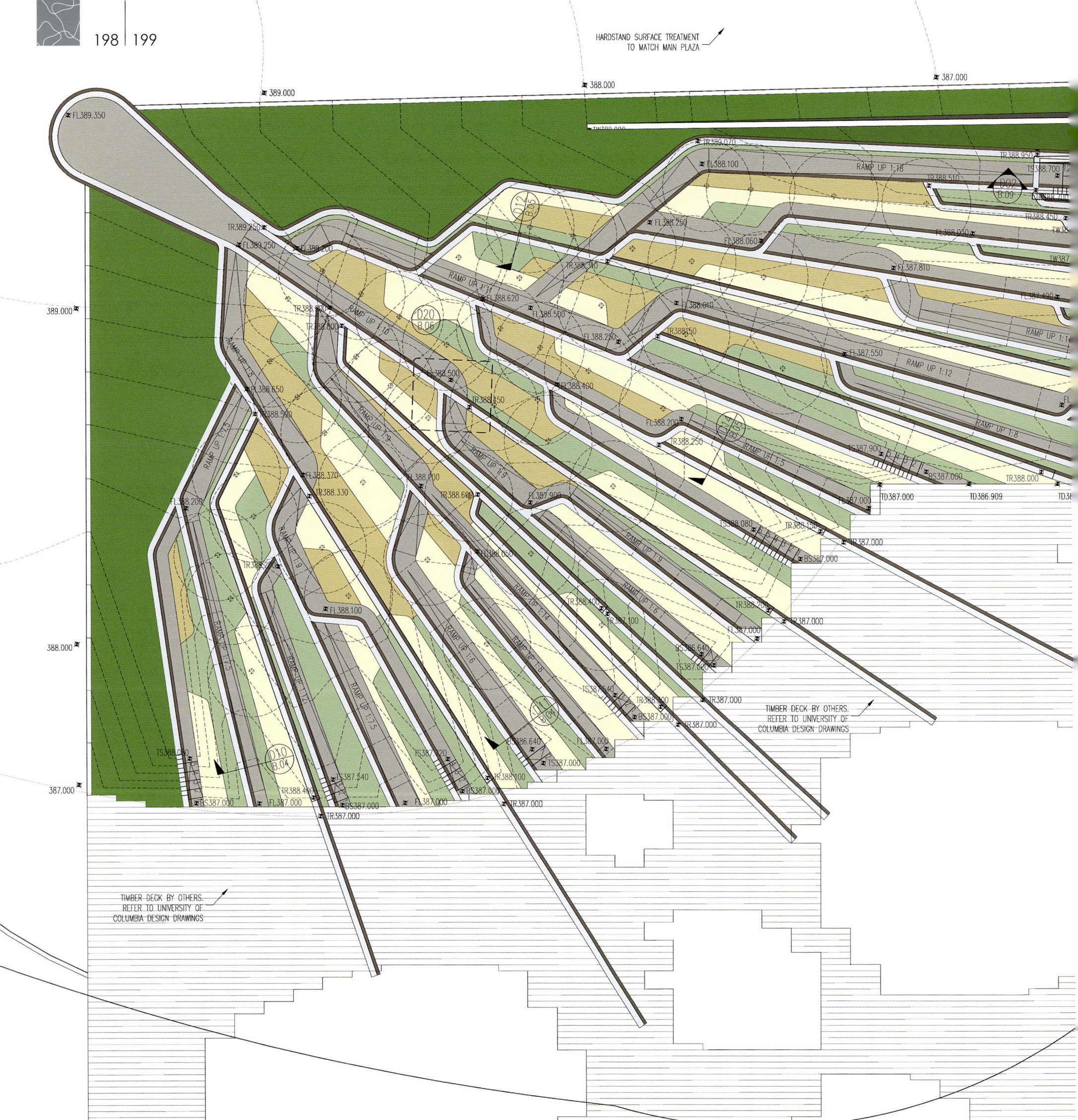

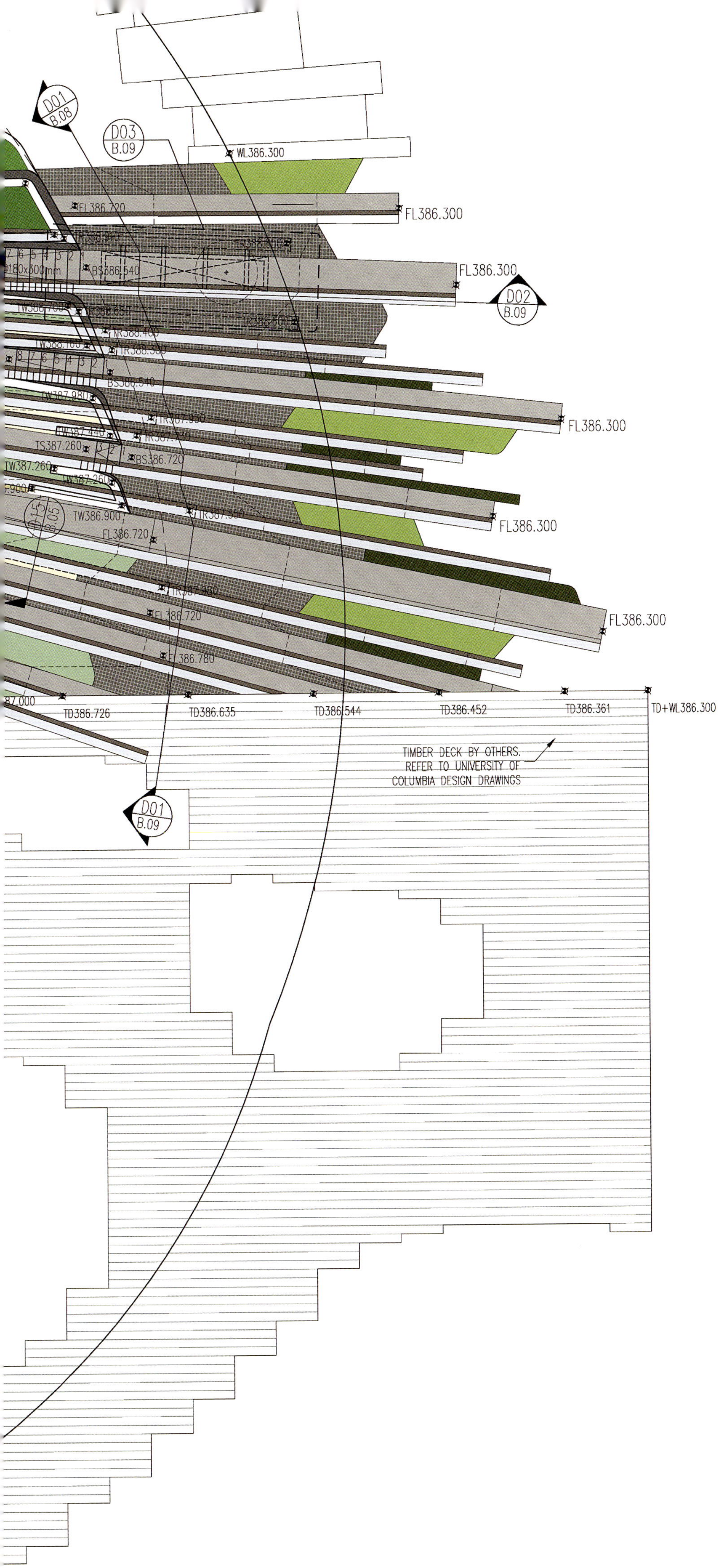

Floating Garden
创意园·流园

展园位于世界园艺博览会创意园展区，由美国加州大学伯克利分校负责设计。设计以"水流"为主题，将自然的水流分支产生的不确定性发挥到极致。游客进入展园可以自由选择参观的路线，不同的起点会给游客带来不同的经历和感受。路径分叉之间的逻辑关系通过水流来引导，参观者可以选择任何一种序列，任何一条游径作为自己的观赏线路。

Designed by University of California Berkeley, Floating Garden is built at the Creativity Gardens Section in the Expo Site. The Garden is themed with "water flow" to fully show the uncertainty of stream forking. Visitors could freely choose different routes in the Garden, among which the logical relations are led by water streams, to get different experiences.

1- Phyllostachys viridis
densly planted
reaching 10m in height

4- APSK
200 Litres

4- Acer palmatum 'Sango Kaku'
reaching 18-25' in decreasing height
23 count

2- Thysanolaena maxima
densly planted
canopy to be equal in elevation
with P. viridis

5- APB
200 Litres

5- Acer palmatum 'bloodgood'
reaching 13-20' in decreasing height
17 count

3- Bamboo Understory:
Liriope Spicata
plant in .3m grid

6- APBu
200 Litres

6- Acer palmatum 'Butterfly'
reaching 8-15' in decreasing height
10 count

7- Carex flacca
densly planted

8- Stipa arundinacea
densly planted

9- Carex buchanii
densly planted
trimmed to encourage upright
growth

10- Acorus gramineus 'Minimus Aureus'
densly planted

11- Scirpus tabernaemontani 'Gmelin'
densly planted

12- CC
30 Litres

12- Bonzai: Carpinus coreana
Specimen Tree
approximately 1m tall, .5m wide
needs careful maintenance

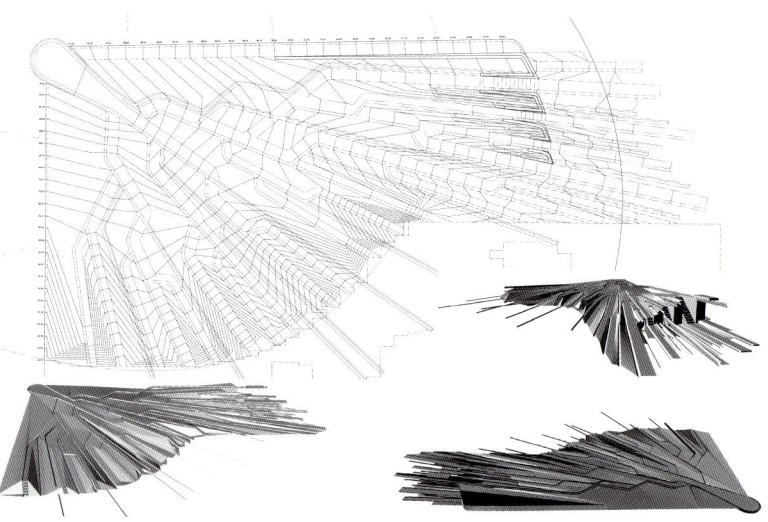

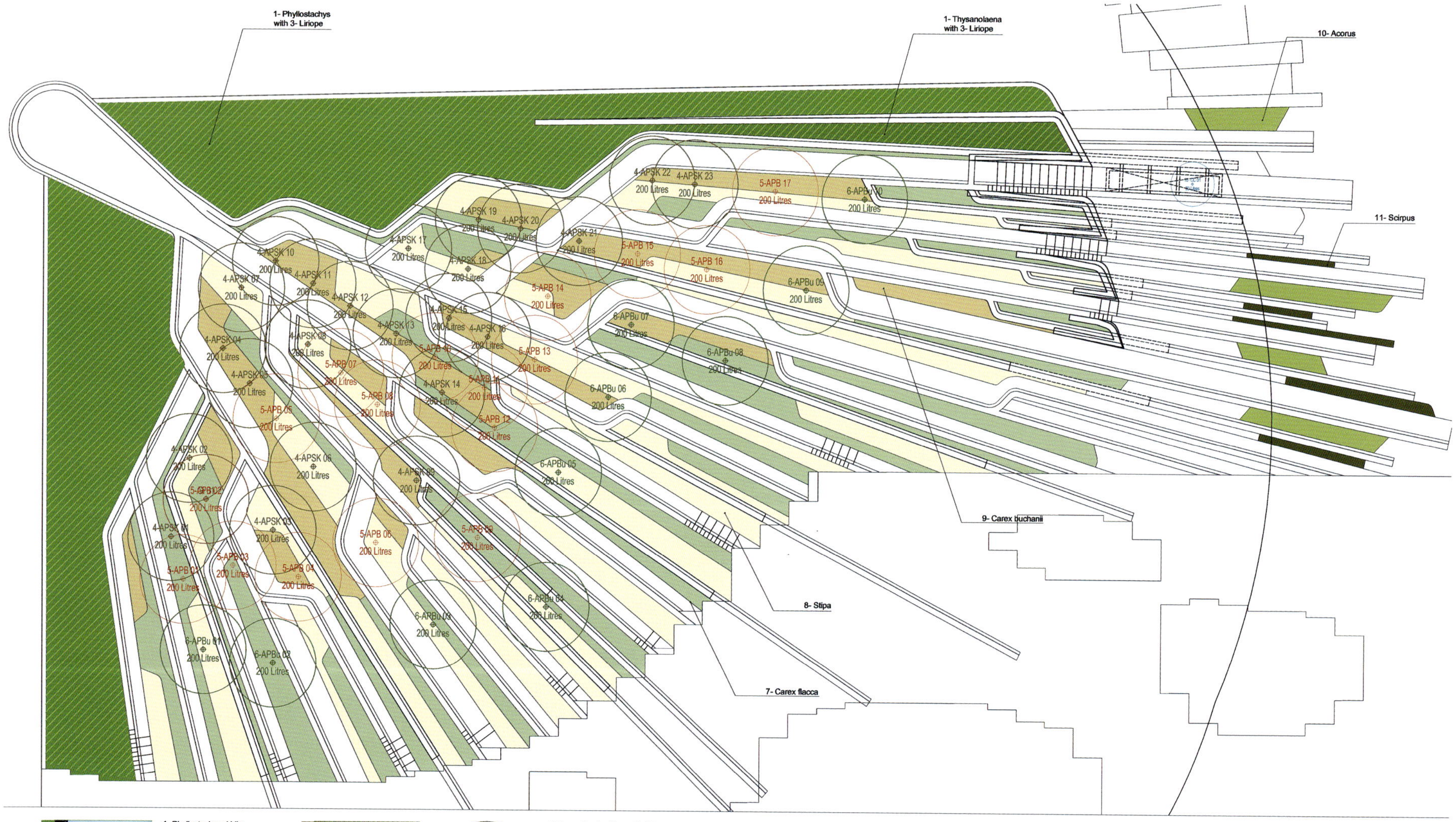

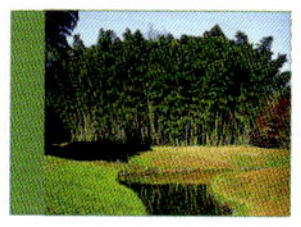

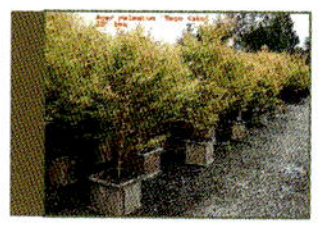

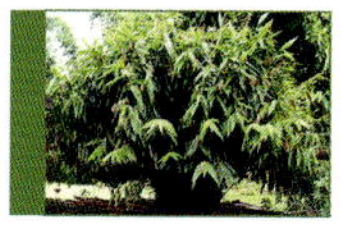

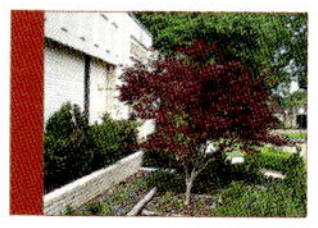

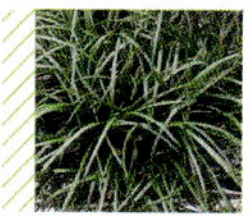

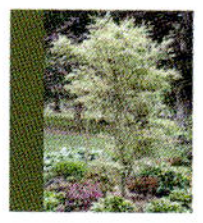

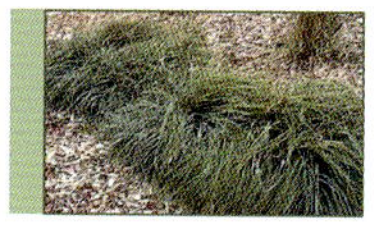

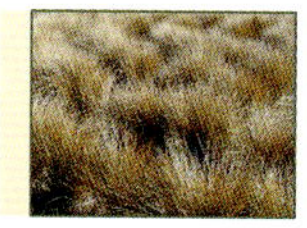

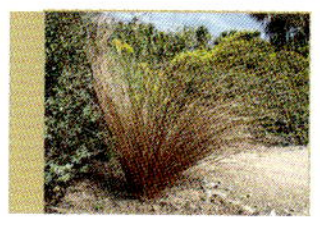

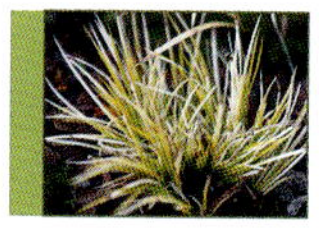

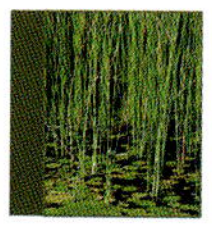

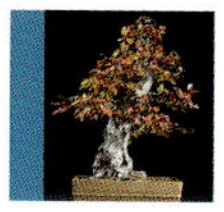

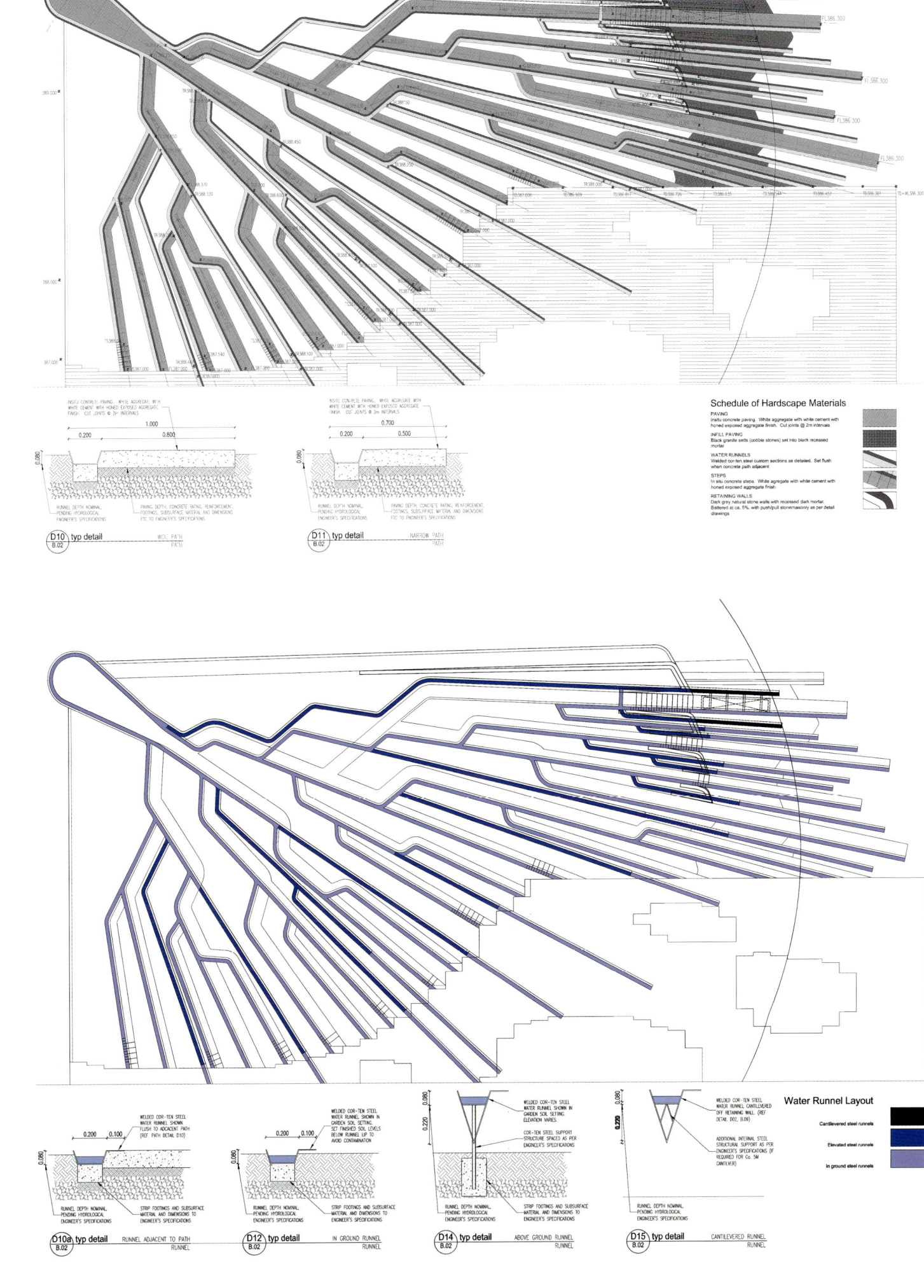

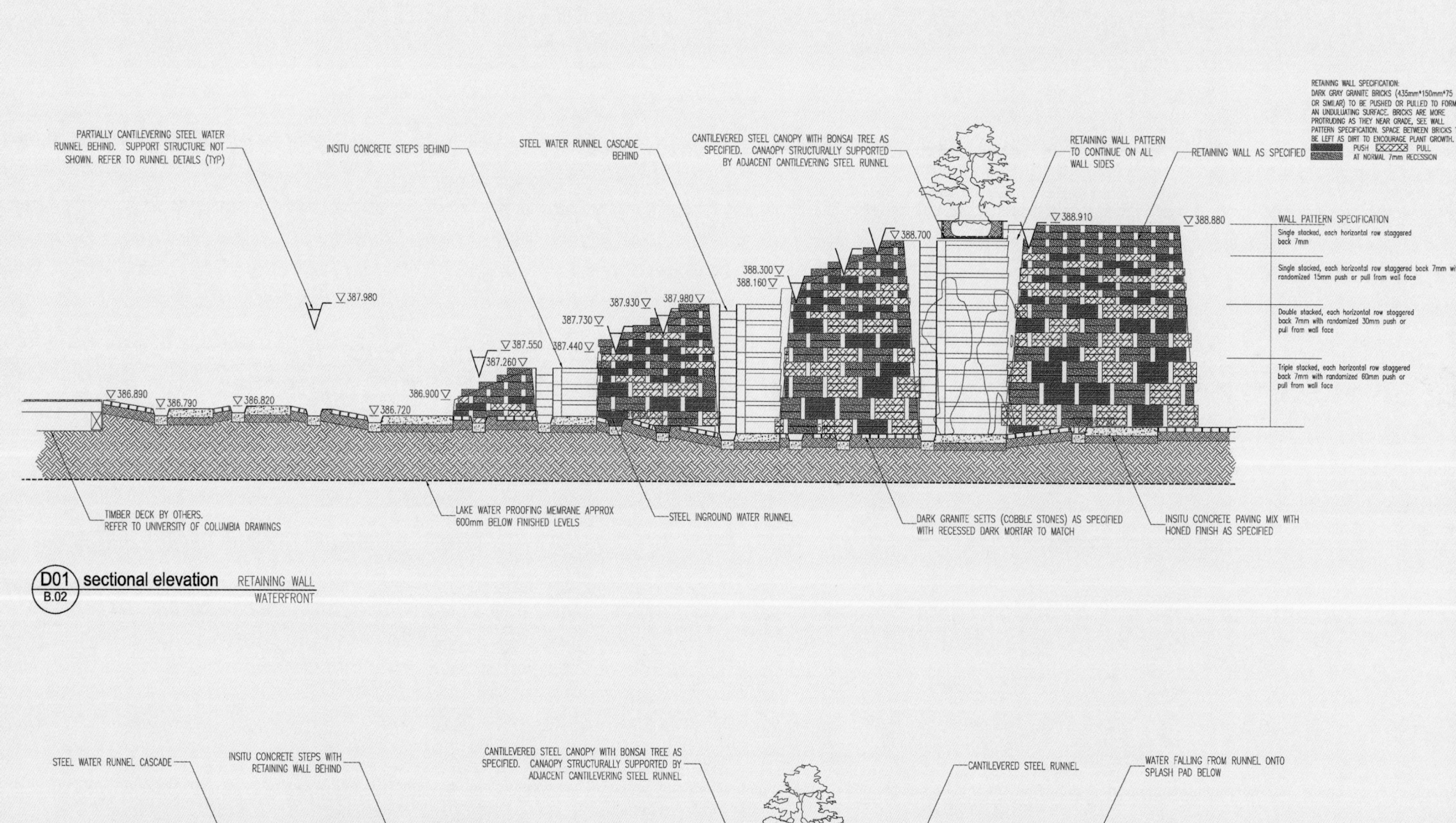

D03 B.02 roof plan CANTILEVERED CANOPY WATERFRONT

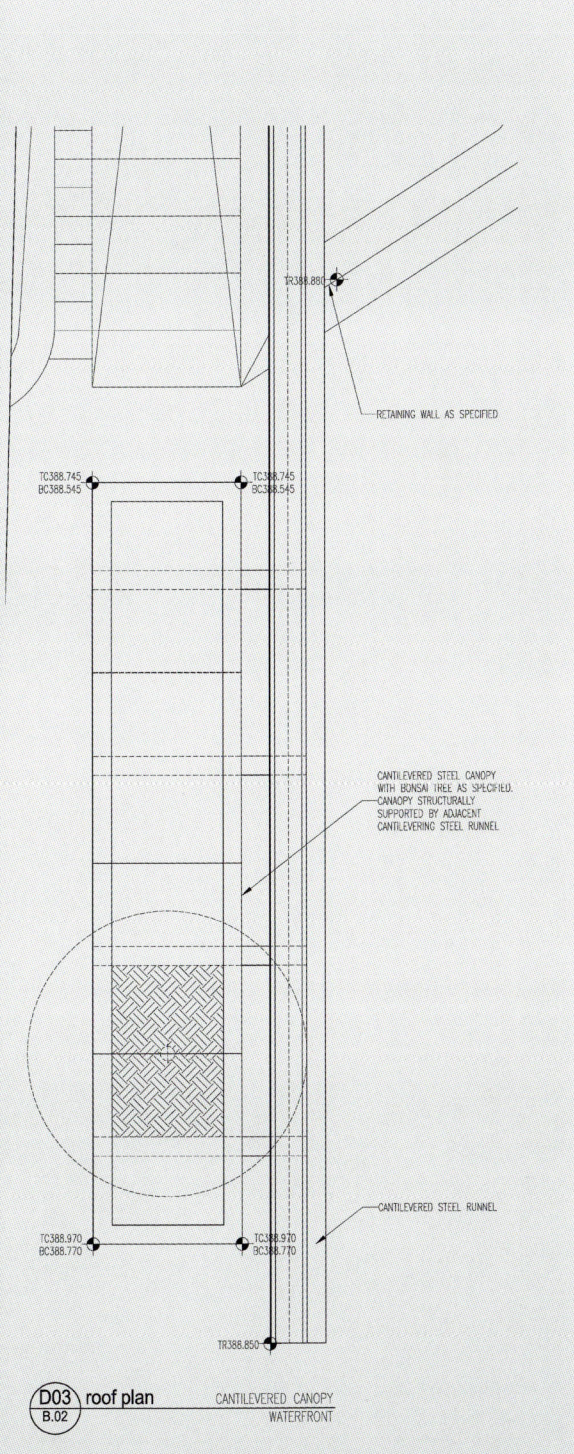

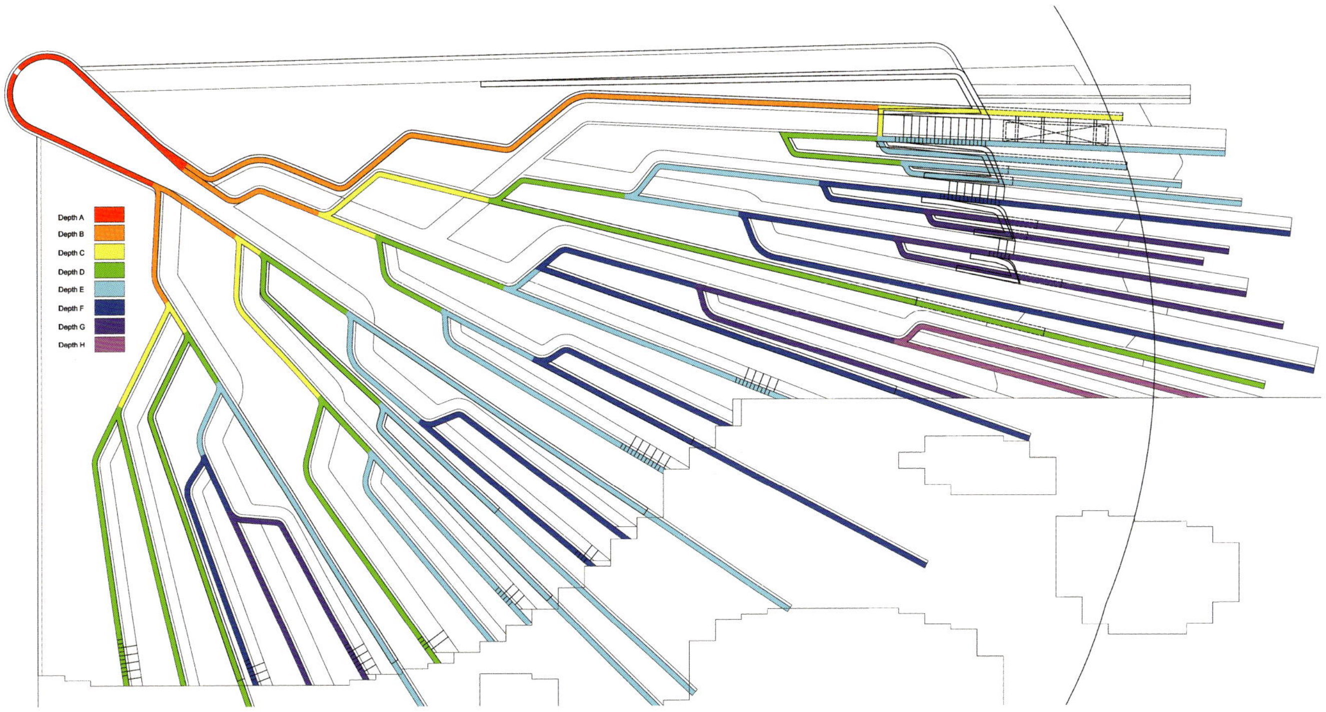

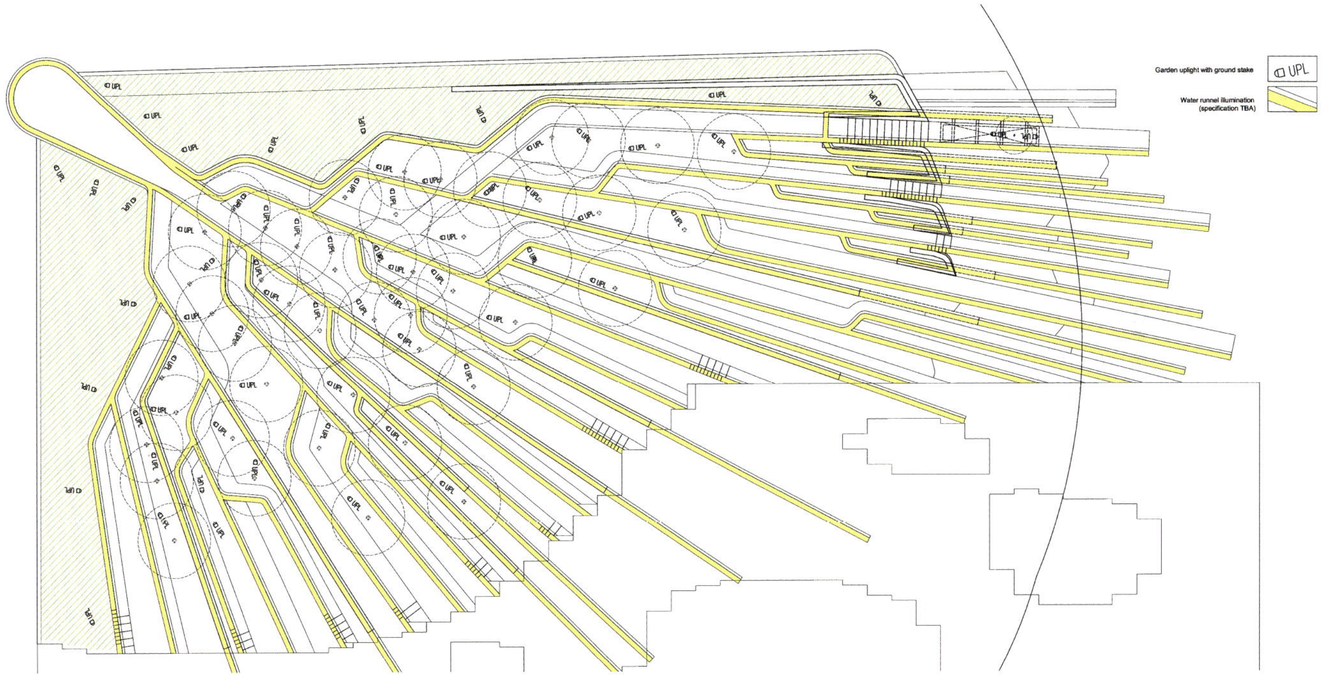

Pampas Impression
创意园·潘帕斯印象

展园位于世界园艺博览会创意园展区，由阿根廷托尔夸托迪特加大学负责设计。设计以潘帕斯草原景象为入手点，试图以现代景观设计手法，在物理、地理、文化风韵与阿根廷不尽相同的西安创造出局部的潘帕斯草原的感觉，再现阿根廷优美的自然景观。潘帕斯草原是阿根廷一片非常广阔肥沃的低地，看起来似乎与中国的庭院完全没有任何共同之处。但是了解它们的人却能从中获得一种共有的影像——一种基于记忆本源而非仅仅对现实存在的景象的理解。展园通过设置简单的游径，在路径两旁栽种一人左右高的灌木植被，同时在植物上加入了能够发光的纤维体，两者互为交织，微风起，枝摇曳，就像潘帕斯草原上金光闪闪的麦田。

Located at the Creativity Gardens Section, the Garden is designed by Universidad Torcuato Di Tella. The designers attempt to apply the physical, geographic and cultural epitome of Pampas, which is a vast steppe of Argentina, to create a sense of Pampas in the Expo Site. It seems that there is no relation between the Pampas Steppe and traditional Chinese garden. However, for those who deeply know that, both of traditional Chinese Garden and Pampas Steppe share a mutual image which is based on the memory origin of human beings, but not only on the understanding of the real landscape. In the Garden there set simple visit routes, on both sides of which are planted with shrubs of average height and vegetations with luminous fibrous to represent the clinquant cornfield on the Pampas Steppe.

PLAN

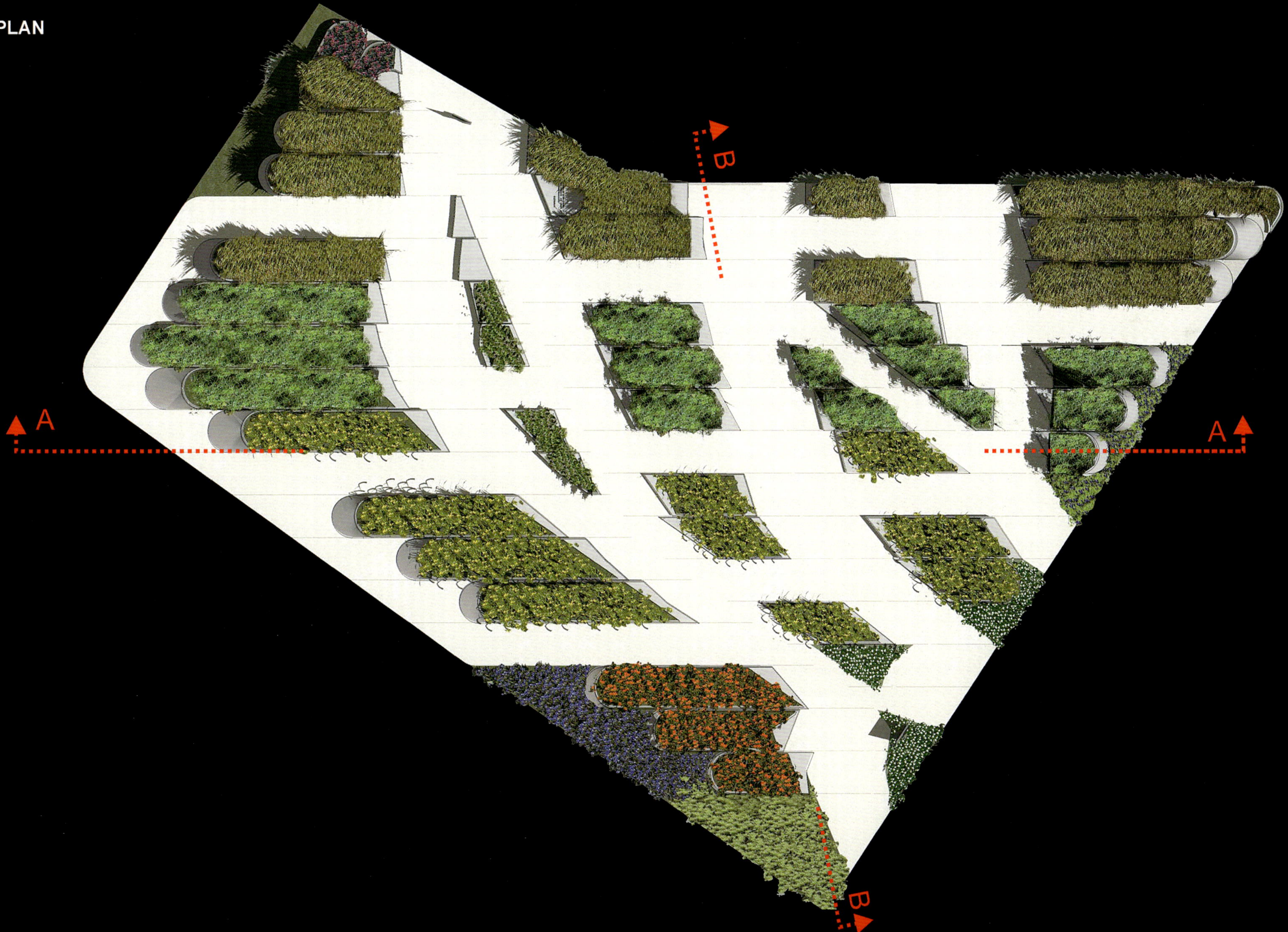

Illustrative Master Plan

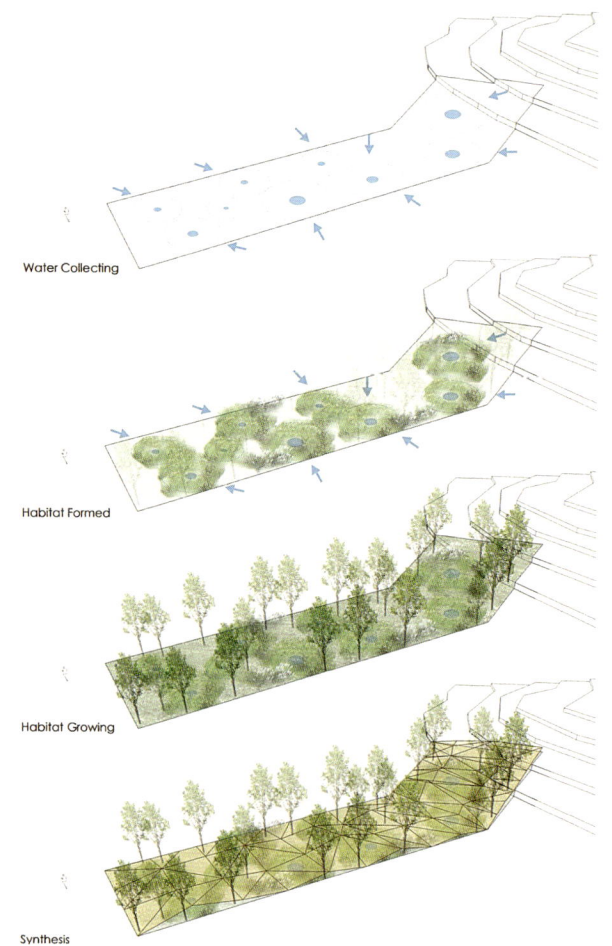

Water Collecting
Habitat Formed
Habitat Growing
Synthesis

Weaving Nature
创意园·编织自然

展园位于世界园艺博览会创意园展区，由中国北京大学负责设计。设计者在花园上铺了一层多功能的薄网，形成了一个存在于植物与参观者之间的层面。这个由薄网所组成的平面把人们在花园中的活动转移到上层，照明系统被安置其下，绿色景观系统穿梭其中。设计者通过使用不同尺寸大小的网格，为游客提供了多种活动的可能性，包括休息区、步行路径和休闲区。本案的设计主旨，主要是为了让参观者与自然能够彼此无障碍地互动交流，同时减少人类活动对景观的负面影响，使景观自然地成长发展。

The Garden is located at the Creativity Gardens Section, which is designed by Peking University. The designers set a multi-functional knitted flimsy slight web in the garden to interpret a special way of interaction and communication between visitors and plants. There are two layers in the web. The upper layer is where visitors are and the lower layer is where illumination system is installed with plants to form the interaction. There are also several functional spaces with different sizes in the web for rest, walk, and leisure. The design of the flimsy web greatly reduces the destruction to nature by human beings and creates a harmony of co-existence between mankind and nature.

Exploring in the Wood

Jumping in the crops

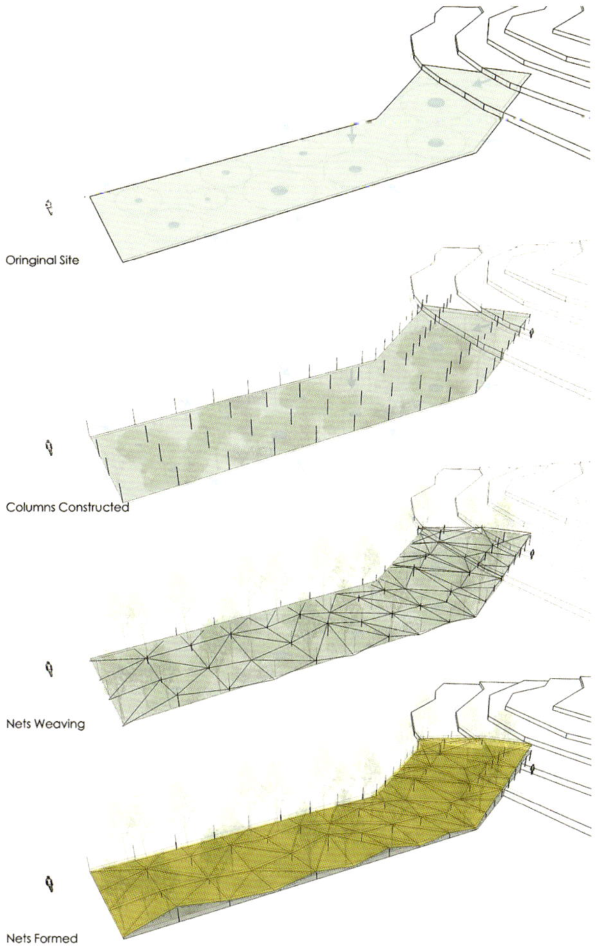

Oringinal Site

Columns Constructed

Nets Weaving

Nets Formed

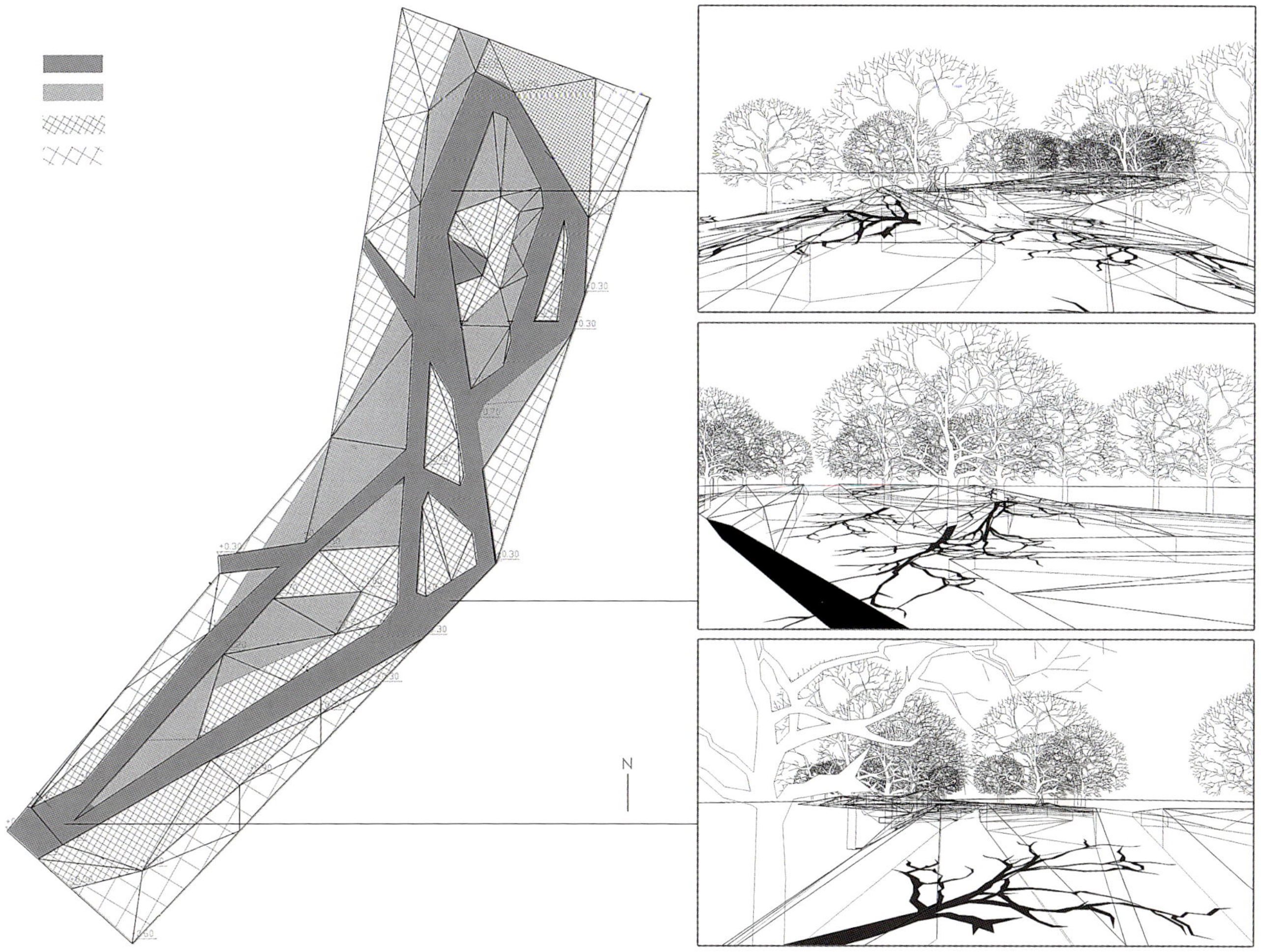

Lying in the flowers

Mesh Size
- 3*3 cm
- 6*6 cm
- 24*24 cm
- 96*96 cm

Different Types of Weaving Density and programs

Density A

Density B

Density C

Density D

Density E

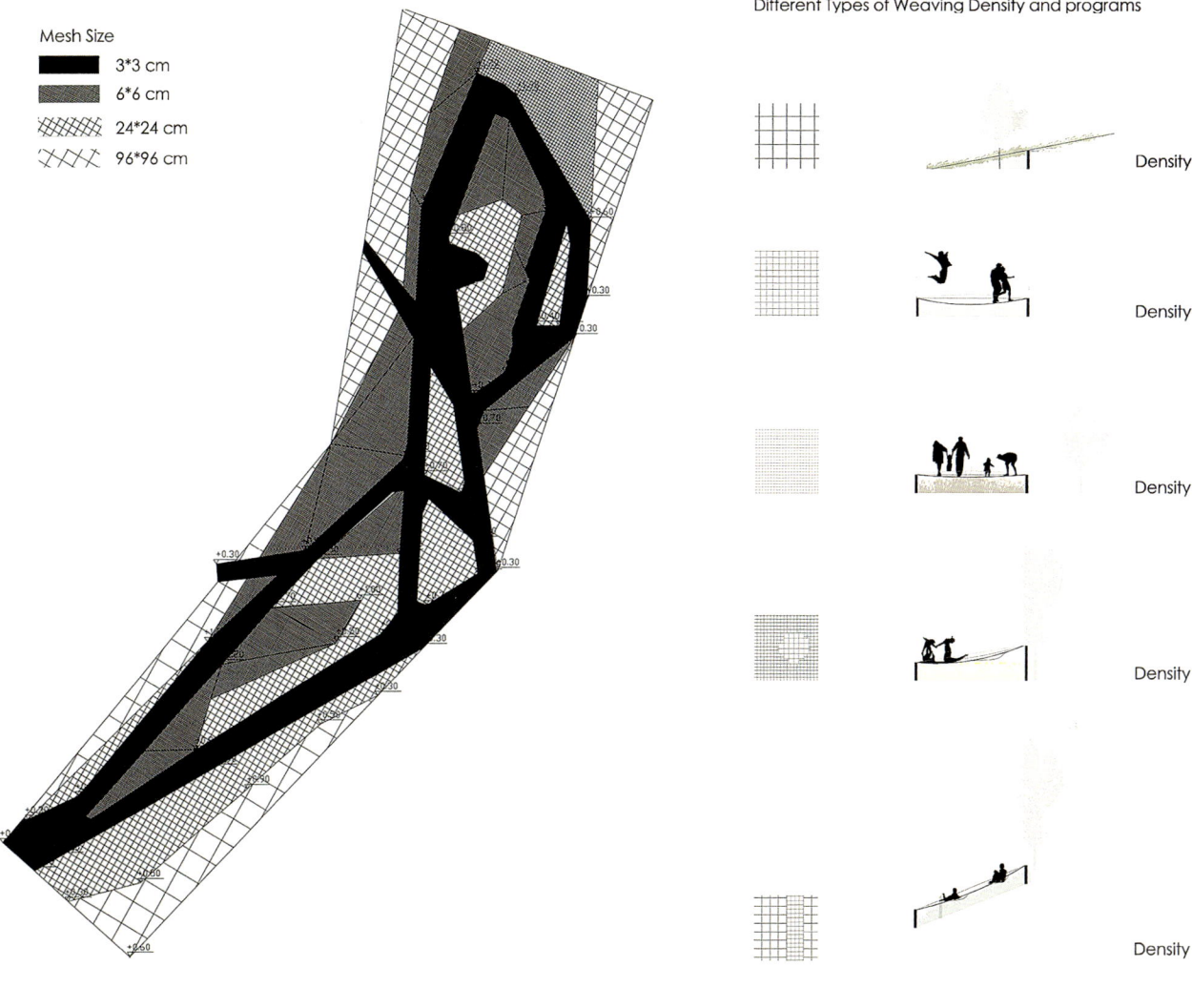

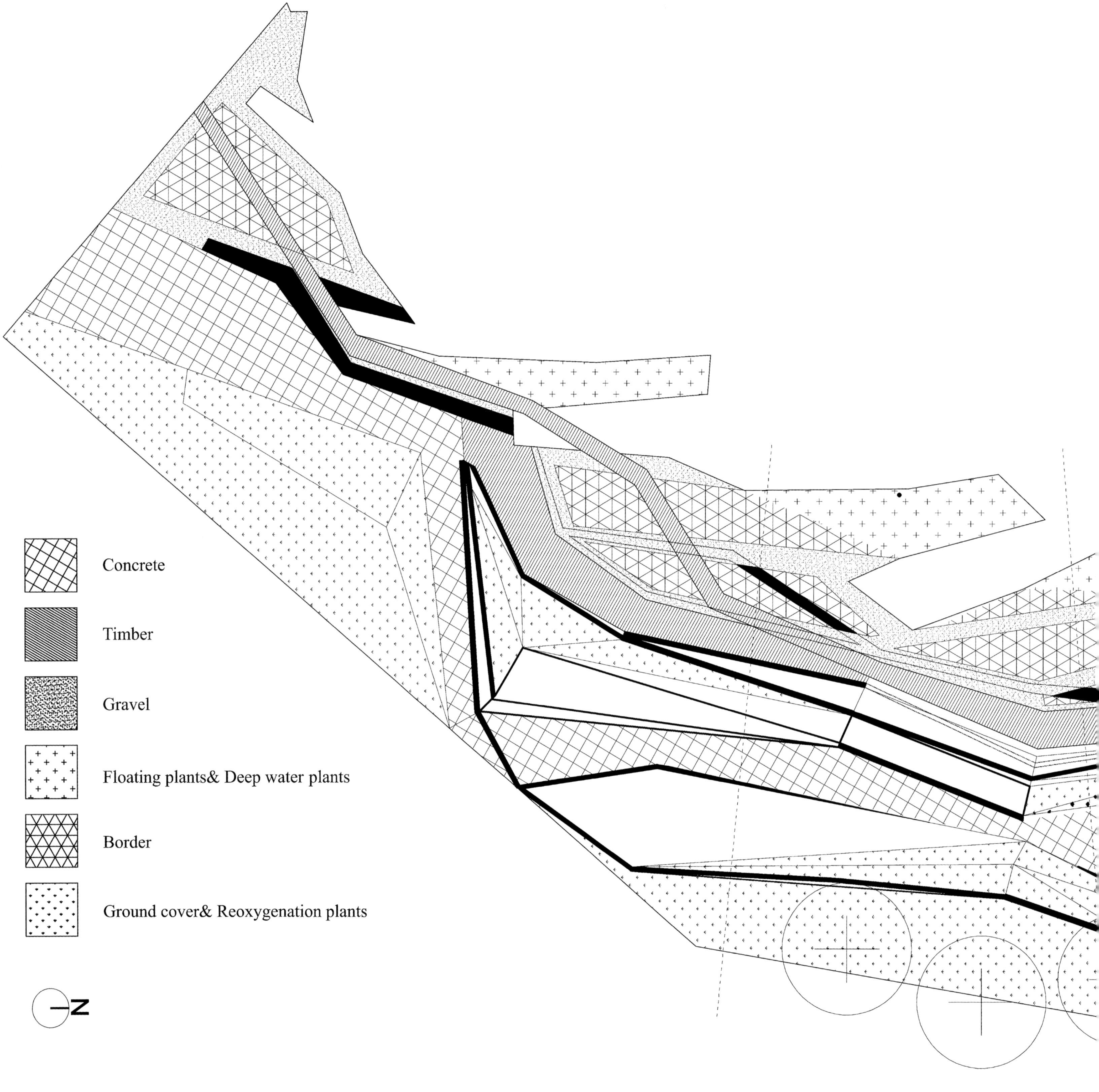

Concrete

Timber

Gravel

Floating plants & Deep water plants

Border

Ground cover & Reoxygenation plants

N

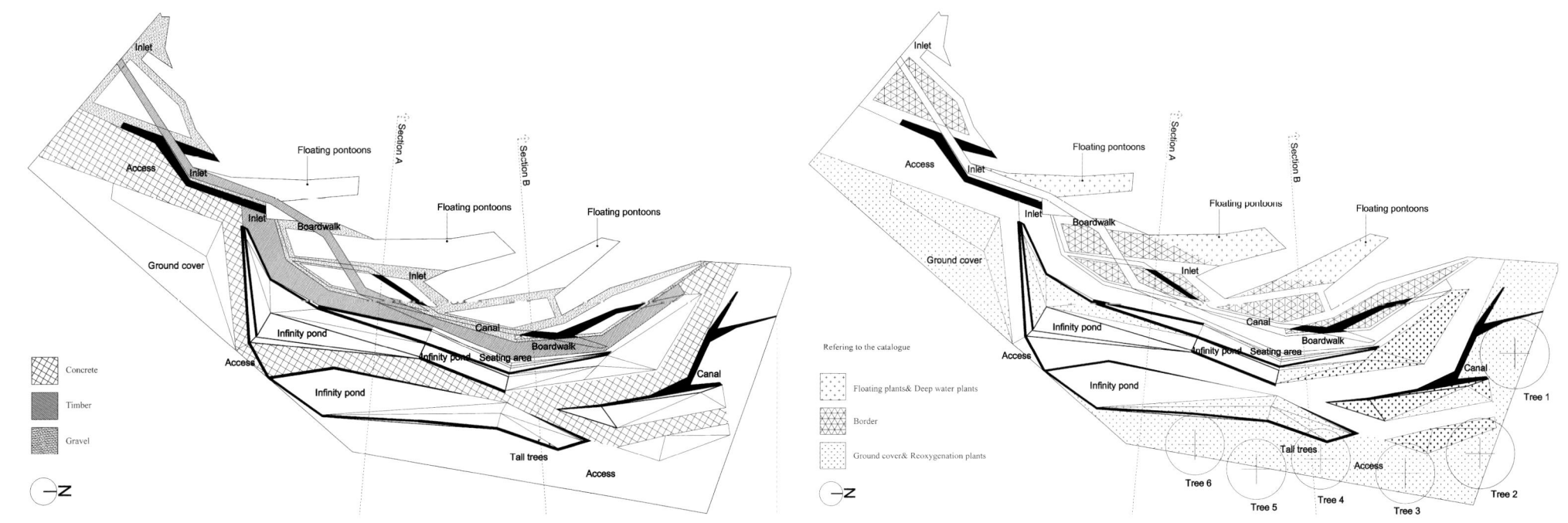

Reinforced Riverside
创意园·强化的河畔

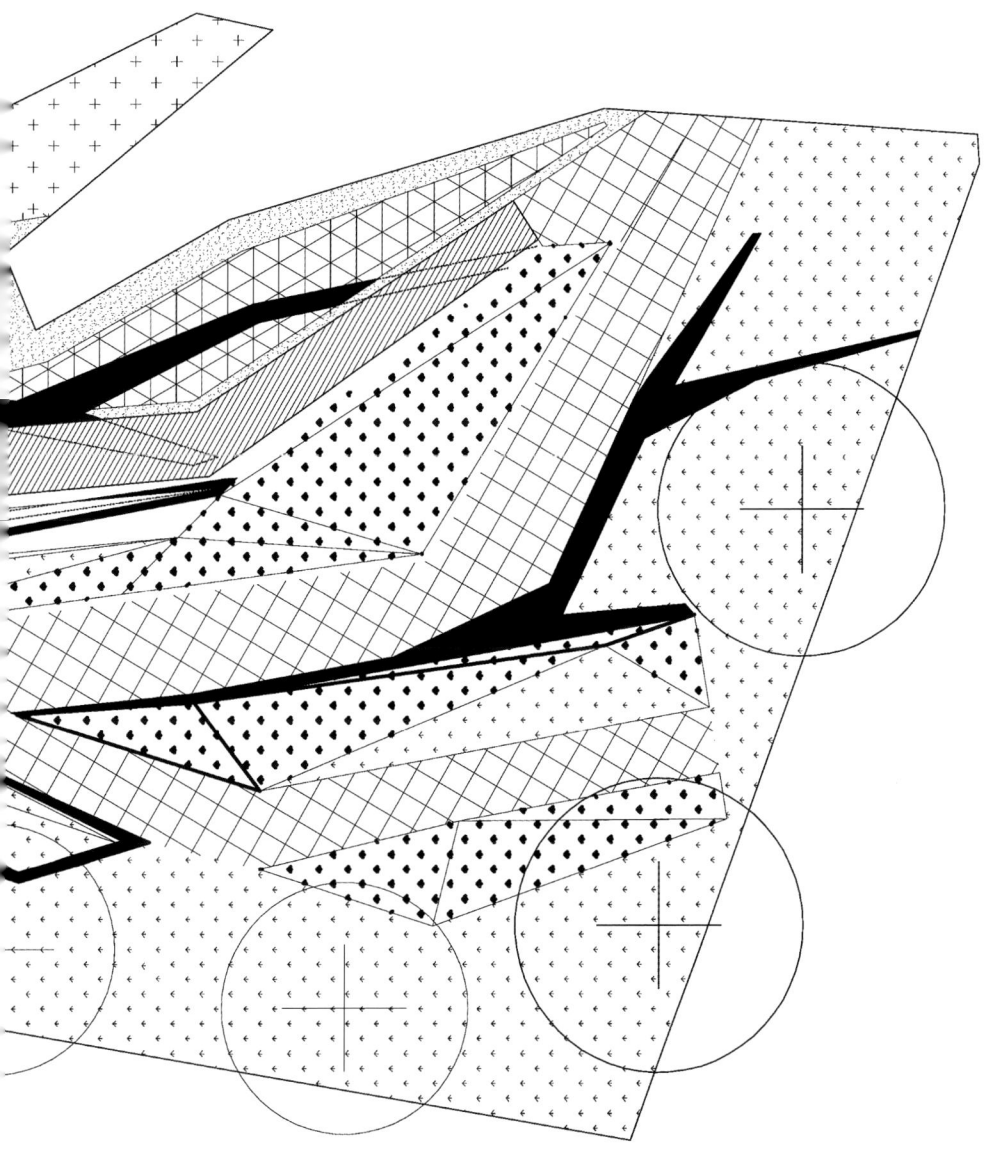

展园位于世界园艺博览会创意园展区，由英国伦敦建筑联盟学院负责设计。展园以"强化的河畔"作为设计主题，利用了地处水边的优势和特点，探讨边缘的处理手法。"边缘"在本案中不是一个水陆相交处的严格线形空间，而是一段液体、固体轮流交替转化的过渡地带。混凝土和木材的基础设施，浮桥、陆地和水生植物，别致的水景，包括池塘和水渠模糊了湖水与庭院的边界——倾斜的空间形式打破了人们所熟悉的水平和纵向空间维度，并突出了整体的不明确性、模糊性和透明性。

Located at the Creativity Gardens Section, the Garden is designed by the Architectural Association, which is themed with "Reinforced Riverside". The advantages and characteristics of riverside are applied to probe into the way of handling the common boundary between water and land, which is reflected by the transition belt built by the alternation of solid and liquid. The infrastructure built of concrete and timber as well as the unique landscape created by pontoon bridge, land and aquatic plants blurs the boundary between lake area and the garden. The acclivitous space form breaks the horizontal and vertical space dimensions that people are familiar with to highlight the uncertainty, fuzziness and transparency of the Garden.

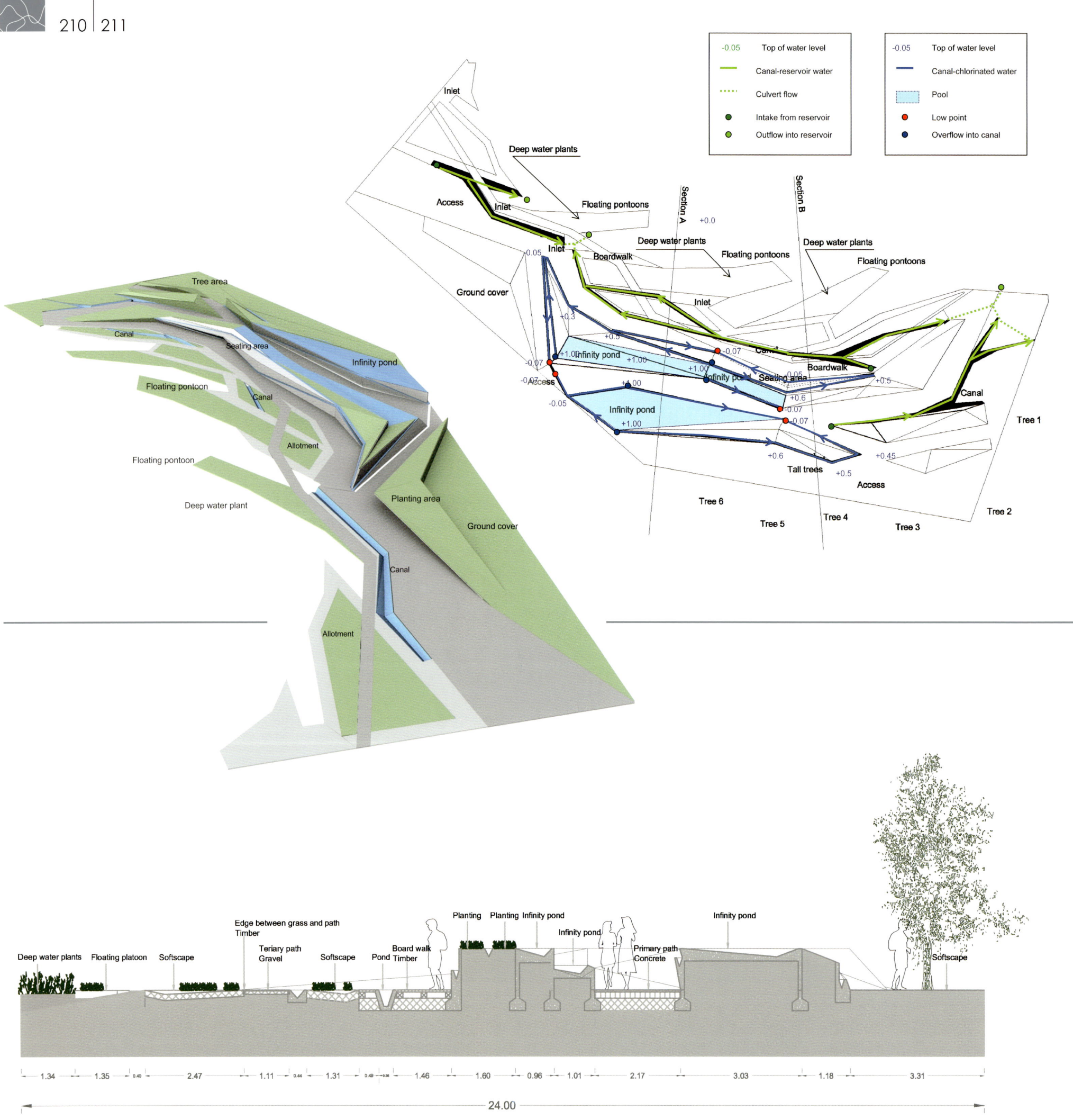

CHANEL	P1	P2	h1	h2	Dh	Length	Slope	Cooment
			m	m	m	m	%	
1	2	1	-0.05	-0.07	0.02	11	0.2%	Stepped finishing
2	3	2	1	-0.05	1.05	10.5	10.0%	Stepped finishing
3	4	2	0.3	-0.05	0.35	5.7	6.1%	Stepped finishing
4	5	4	0.5	0.3	0.2	5.4	3.7%	Stepped finishing
5	8	5	1	0.5	0.5	11.2	4.5%	Stepped finishing
6	5	9	0.5	-0.07	0.57	11.7	4.9%	Stepped finishing
7	6	7	1	-0.07	1.07	8.6	12.4%	Stepped finishing
8	8	10	1	0.6	0.4	8.9	4.5%	Stepped finishing
9	11	9	0.3	-0.07	0.37	8.1	4.6%	Stepped finishing
10	10	12	0.6	0.5	0.1	9	1.1%	Rough bottom canal
11	12	11	0.5	0.3	0.2	9.1	2.2%	Rough bottom canal
12	14	13	-0.05	-0.07	0.02	4.1	0.5%	Rough bottom canal
13	15	14	1	-0.05	1.05	4.9	21.4%	Stepped finishing
14	16	14	1	-0.05	1.05	6.4	16.4%	Stepped finishing
15	16	17	1	-0.07	1.07	17.4	6.1%	Stepped finishing
16	15	18	1	0.6	0.4	17.4	2.3%	Rough bottom canal
17	20	17	0.45	-0.07	0.52	8.7	6.0%	Stepped finishing
18	18	19	0.6	0.5	0.1	6	1.7%	Rough bottom canal
19	19	20	0.5	0.45	0.05	2.8	1.8%	Rough bottom canal

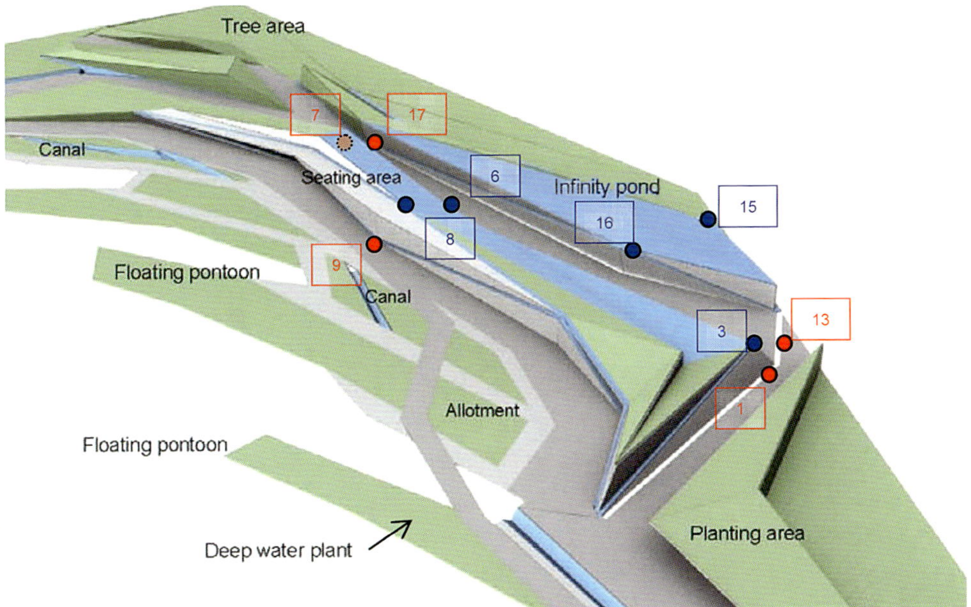

OUTFLOWS FROM INFINITY POOLS AND LOW POINTS

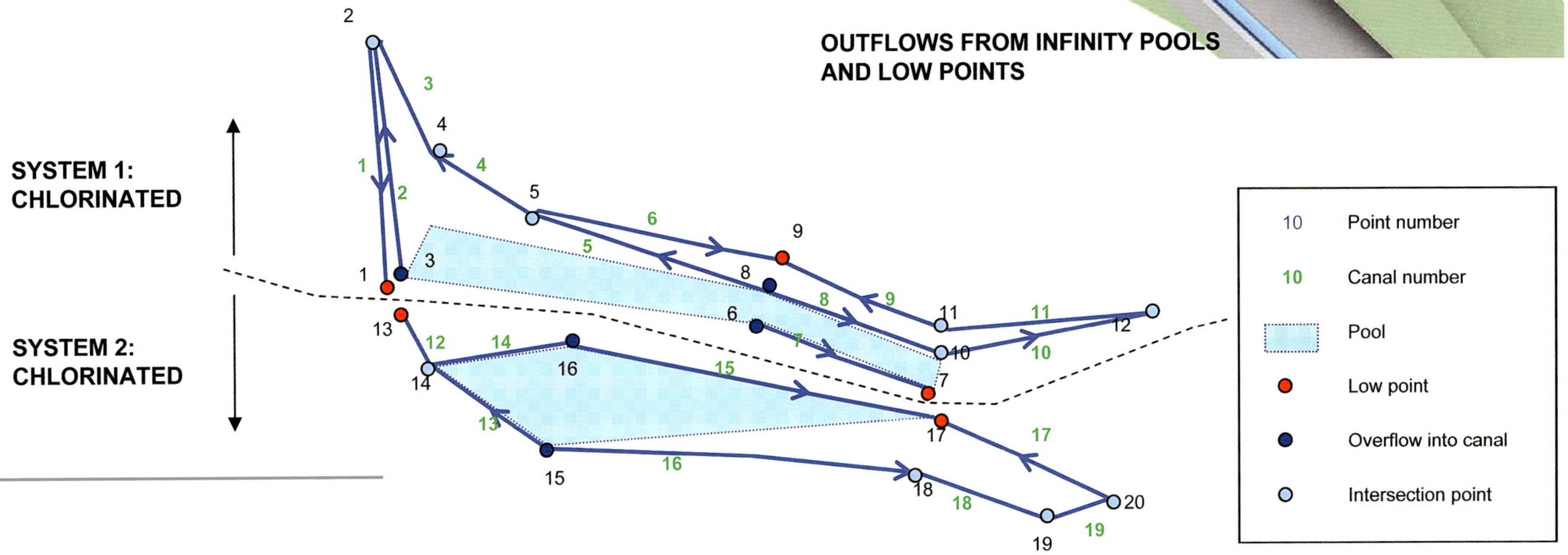

SYSTEM 1: CHLORINATED

SYSTEM 2: CHLORINATED

POINT AND CHANNEL NUMBERING

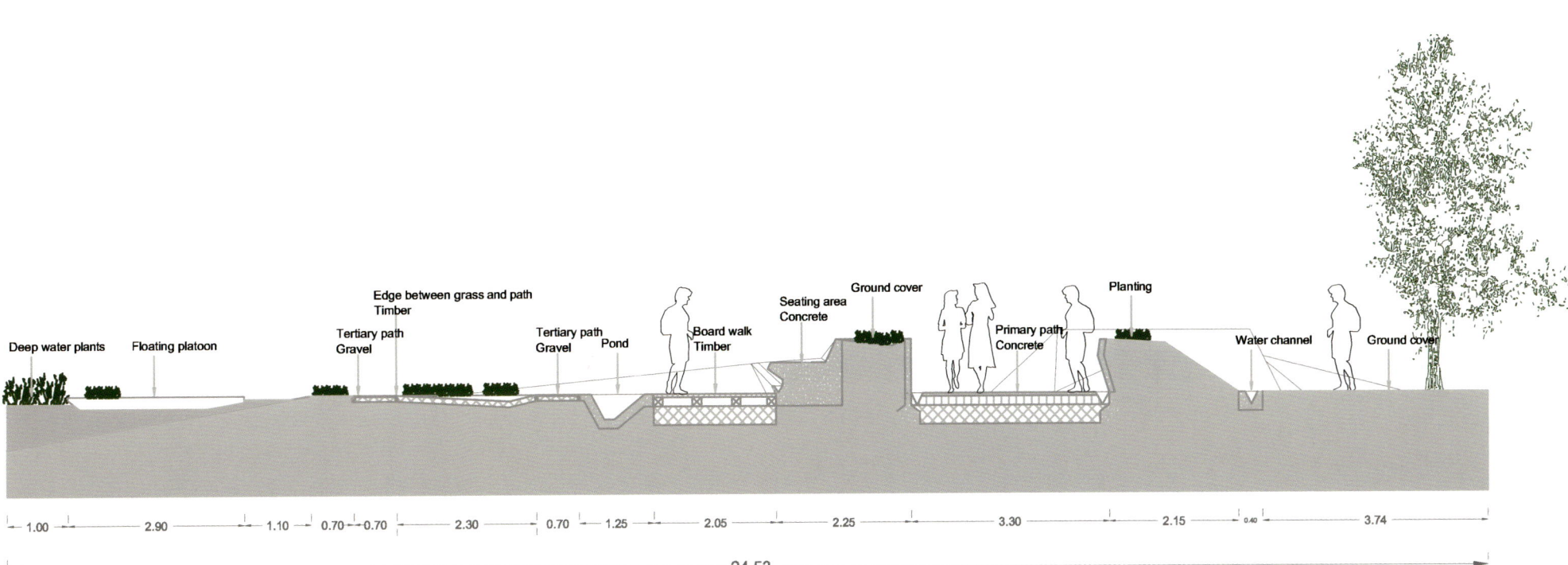

Section B

Ecological Platform
创意园·生态平台

展园位于世界园艺博览会创意园展区，由哥伦比亚大学负责设计。设计者从基地的不确定因素产生的变化出发，将之作为设计的重要概念。设计者通过构建一个宜居的水滨平台，创造生物系统多样性最为丰富的湿地。随着季节变化，水位会上升和下降，营造水与平台间不同的边界形式。这种"顺其自然"的设计态度，与过去设计师们都试图"驯服"自然不同。设计者希望营造一种环境，让生物群落都能在不断改造的过程中适应，最终以一种极为平和的手段拥有该地块的"控制权"。

The Garden, designed by Columbia University, is located at the Creativity Gardens Section in the Expo Site. Designers start their design with the changes of uncertain factors in the design scope, and make "change" an important design concept in this project. They create a multidimensional ecological swamp by building a riverside platform and display different landscape effects by seasonal plants change as well as the rise and fall of water level. The concept of "let nature take its course" in this project is very different from the idea of "overcome the nature", which was popular in the past. Designers would like to create a natural local Xi'an environment to make biocenosis adapt to it bit by bit and finally to be the masters of this environment naturally.

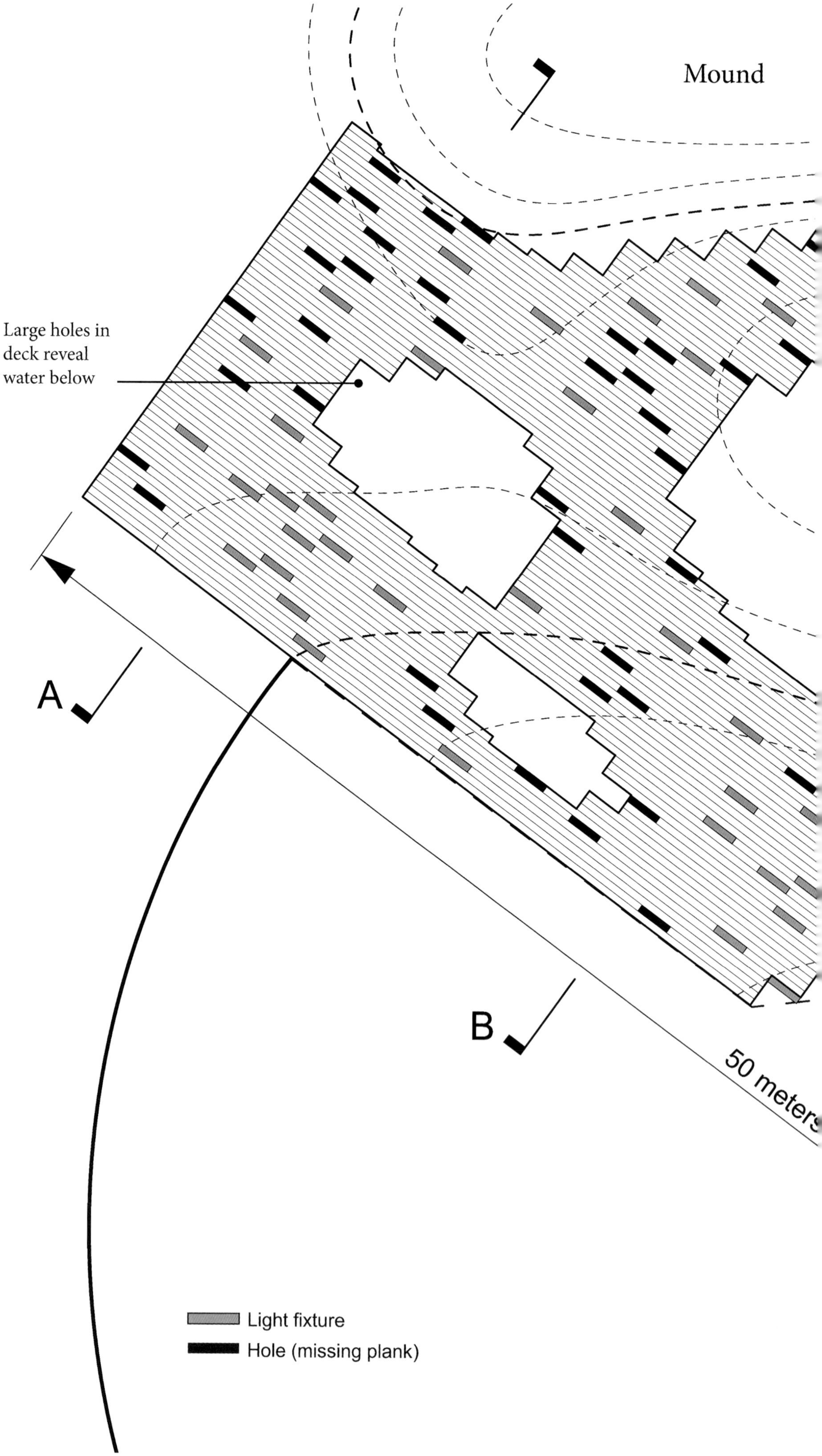

Mound

Large holes in deck reveal water below

A

B

50 meters

Light fixture
Hole (missing plank)

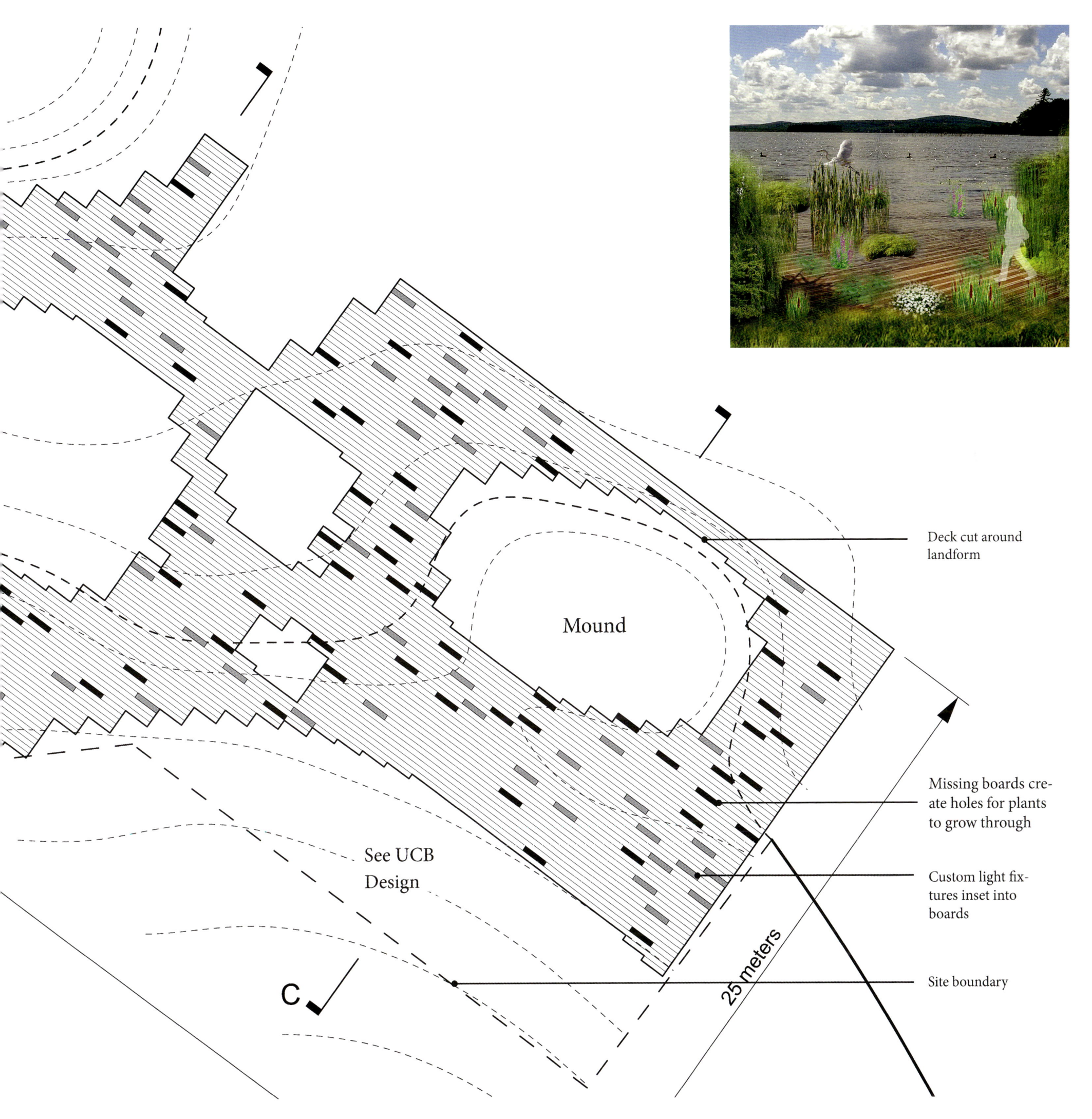

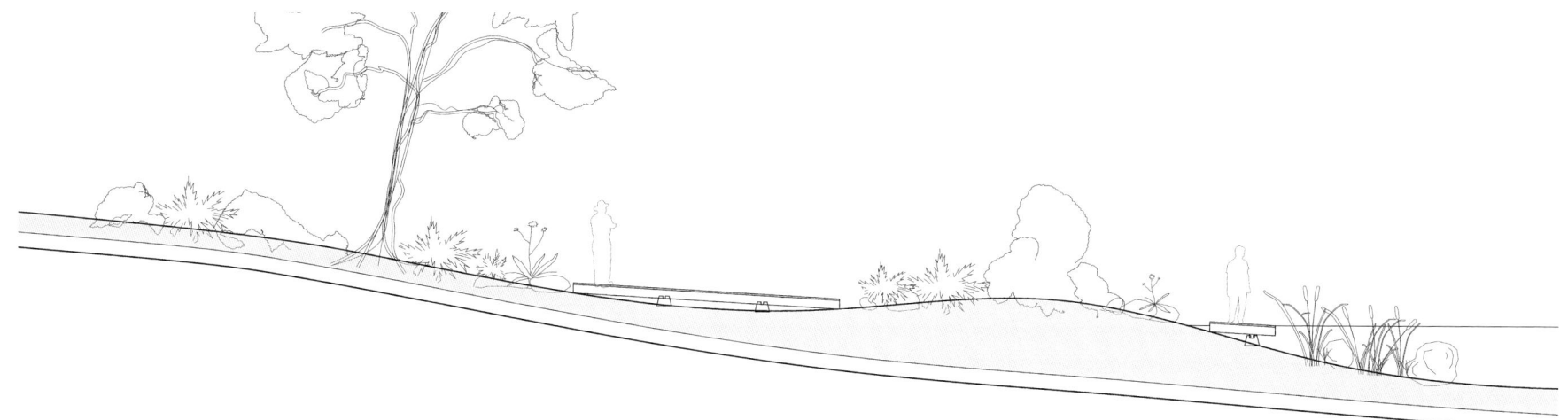

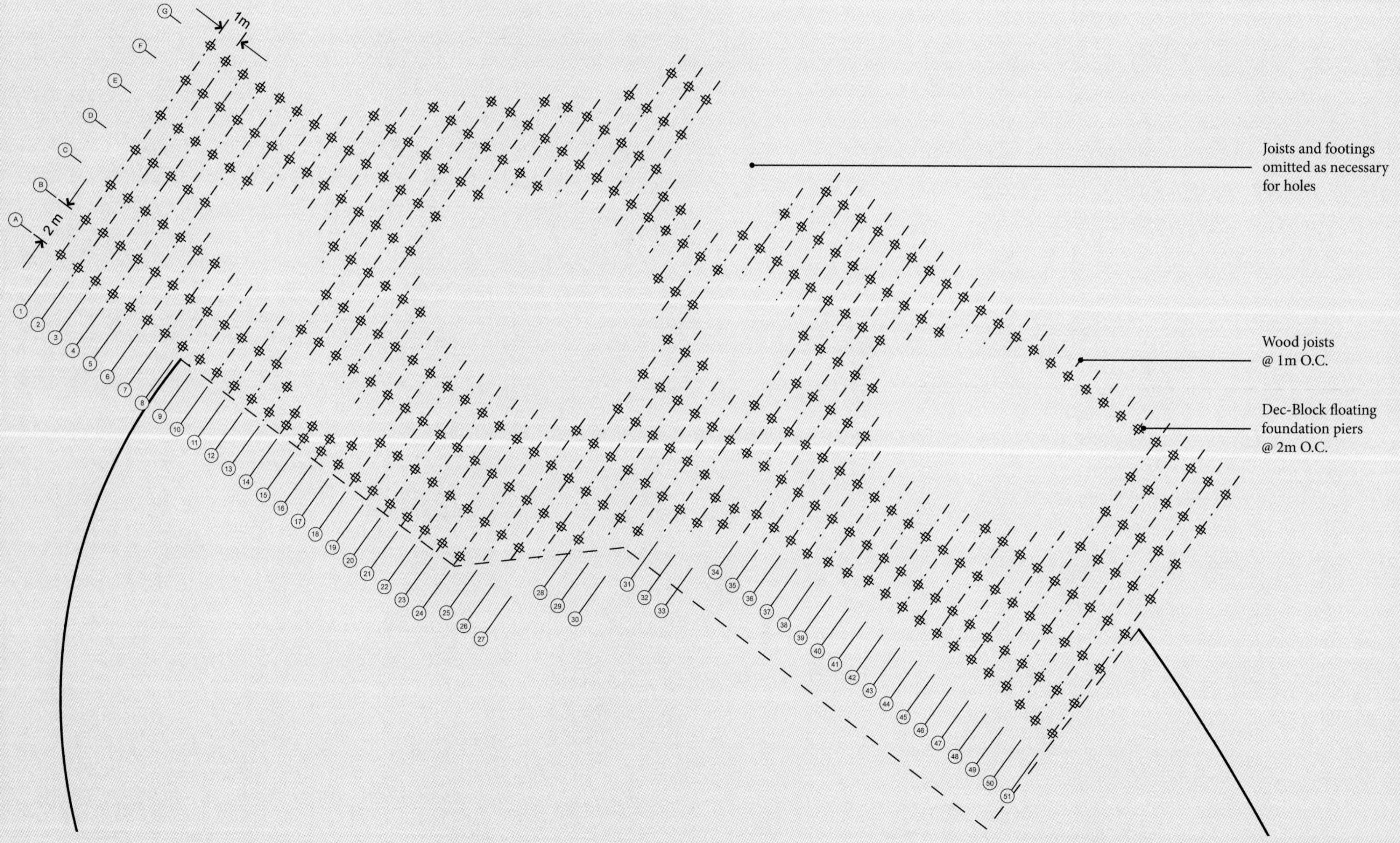
1m
2 m
A
B
C
D
E
F
G
1 2 3 4 5 6 7 8 9 10 11 12 13 14 15 16 17 18 19 20 21 22 23 24 25 26 27 28 29 30 31 32 33 34 35 36 37 38 39 40 41 42 43 44 45 46 47 48 49 50 51
Joists and footings omitted as necessary for holes
Wood joists @ 1m O.C.
Dec-Block floating foundation piers @ 2m O.C.

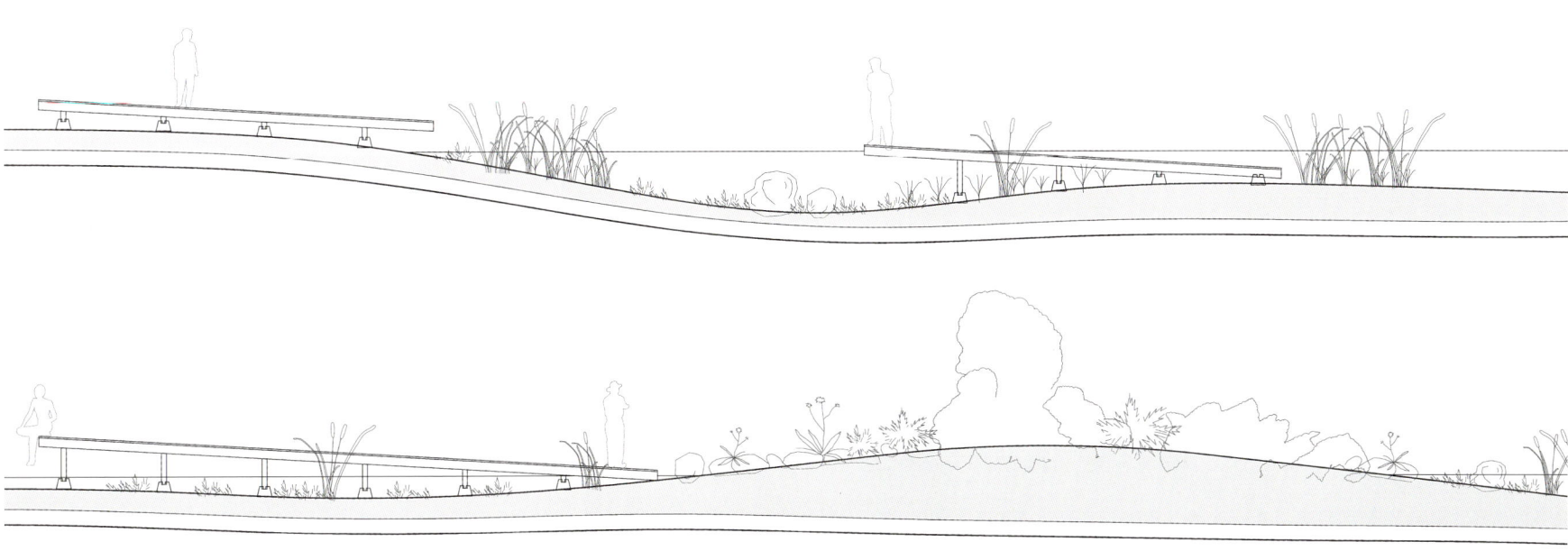

Notes:
Please refer to the *Plant Catalogue* for plant types.

Marsh Plants are to be planted in assigned regions with a mixture of plants M1-M5.

Mound Plants are to be planted in assigned regions as indicated. Plants MD1 and MD2 should be planted with a spacing of 45 cm. As specified in the *Plant Catalogue*, MD3-MD5 are dwarf types.

Water Plants are to be planted in assigned regions with a mixture of plants W1-W5.

REF: BERKLEY

Ecology-Solar Term 创意园·生态——节气

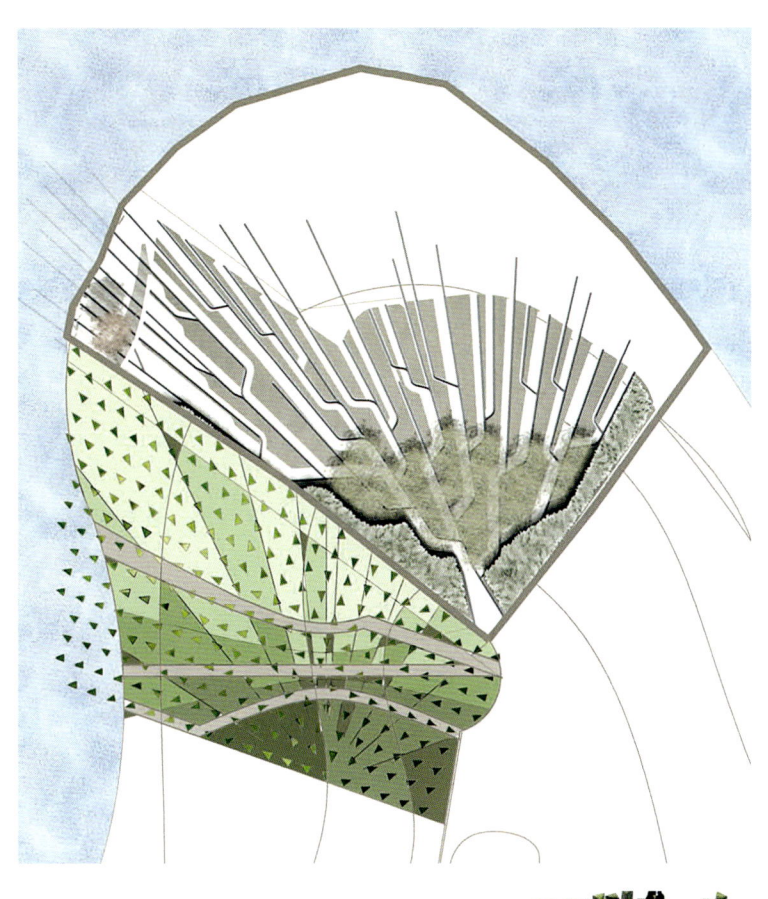

展园位于世界园艺博览会创意园展区，由台湾逢甲大学负责设计。设计从"二十四节气"出发，将"日晷"的概念引入了具体设计手法，来定义西安的位置，从而在基地上得到一组有关时序周期的形态。本案以高密度等间距的形态作为布局的基础，根据节气原理、基地环境限制、地形、路径、植栽多样性、气候条件和参观者行为的不可预测性，创造出一个持续变化的微气候环境，来丰富参观者的体验。

Designed by Feng Chia University, the Garden is located at the Creativity Gardens Section in the Expo Site. According to the "twenty-four solar terms", the designers apply the concept of "sundial" to define the sequential period pattern of Xi'an. Making the layout based on the situation of high density and same space, designers create a continuous changing micro climate to feast visitors' experience in the light of principles of solar terms, environmental constraint of the base, terrain, routes, the variety of plants, weather conditions and the unpredictability of visitors' behavior.

Hong Kong, Macau & Taipei Gardens

港澳台北园

香港园 Hong Kong Garden

展园设计以"天人合德"作为主题，以风(风能树)、山(地质公园)、水(循环不息之水)、喜(触动喜乐之心灵)配上蝴蝶起舞，雀鸟悦耳歌唱声及三环绿化带为主要手段。总面积716平方米。在空间排布上，设计利用高科技的手段将动植物、风、山、水等自然元素引入园内，与地块形成的气场，一同牵引着人们内在心灵世界的喜悦。

Hong Kong Garden, occupying an area of 716 m², is themed with "harmony and virtue between man and nature", applying the elements of Wind (wind-energy tree), Mountain (geological park), Water (circulating water), Joy (touching the joyful mind), together with dancing butterflies, beautiful songs of birds, and the green belt along the third-ring road. The animals, plants, wind, mountains, water and other natural elements introduced in the Garden arouse the joyful feeling in our inner world.

澳门园 Macau Garden

展园设计以澳门历史积淀作为主题，主要展现了澳门历史城区所拥有的四百多年中西文化交流的历史。总面积697平方米。澳门历史城区内留存了大量历史古迹，既有欧陆式古典建筑物，也有颇具东方美的中国建筑。这种东西方文化的碰撞，就是澳门的魅力所在。园区在南北向设置中西式回廊，回廊之间以黑白波浪形葡国石铺面串联。简洁的设计语言加上丰富的材质使用，让澳门园呈现出华南地区景观设计的特色。

Macau Garden occupies an area of 697 m². The design of the Garden is inspired by the historical connotation of Macau to represent the unique local Sino-West cultural integration in the Historic Centre of Macau in the past 400 years. There are many historical places in the Historic Centre of Macau, most of which are classical Euro-style architectures and traditional Chinese buildings. In the north-south Garden sets a winding corridor with both Chinese and Western style. The black-and-white waved stone paved in Portuguese style decorates the floor of the corridor. Simple design with rich materials in the Garden shows distinct landscape in the region of South China.

台北园 Taipei Garden

总面积572平方米的台北园以全面展现台北风情为主题，在基地两侧竖向栽植了热带、温带和寒带植物，表现了台湾丰富的自然环境。本案使用现代建筑材料所创造的代表台北地形的地景雕塑和装置，与自然资源一同，表达了台湾的丰饶。园中等比例缩小的台北101大楼雕塑旁，栽植西安当地处处可见的银杏树，代表台湾对本届西安世园会的支持和友好。在展园的空间排布上，本案大量使用的圆形元素，与中国传统的"圆融通和"相呼应，既表达了两岸在文化上的交流，同时也说明了两岸人民血脉同源。

Taipei Garden, with an area of 572 m², is designed with the theme of Taipei local manners and feelings. On either side of the Garden plot vertically planted a lot of tropical plants and plants from Variable Zone and Frigid Zone to represent the fascinating natural scenery in Taiwan. All landscape decorations in the Garden are created by the modern construction materials, showing visitors the rich and fertile Taiwan. The common ginkgoes in Xi'an planted around the model sculpture of towering 101 building in the Garden give expression to the friendship between the two cities and Taipei's support to Expo 2011 Xi'an. The space arrangement of the Garden is circle-oriented to respond to the blood lineage between Taiwan and Mainland China.

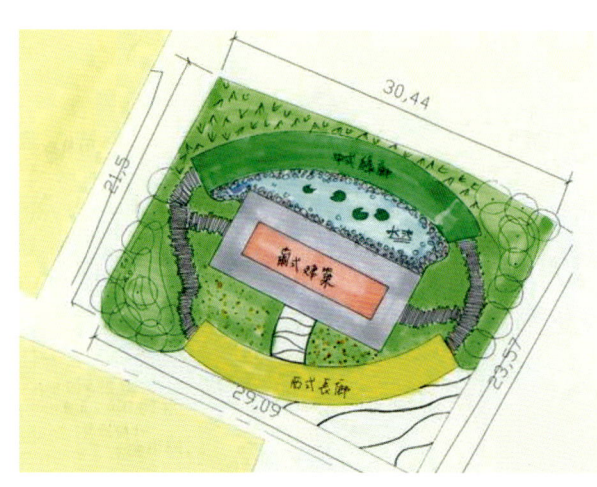

上海园 Shanghai Garden

以"海上华庭"为创作主题的上海园，总面积1,126平方米，将绿色、科技、低碳、融入项目之中，以此来象征今日上海正步入一个新的环保世纪。漫步于弧形步道，在漏斗形装置旁，乐队演绎出悠扬旋律，让参观者仿佛进入了聆听、冥想、信步的心灵之旅，体验着和谐的城市生活。

Shanghai Garden occupies an area of 1,126 m². The main part of the Garden is a sunken water garden integrated with the concepts of ecology, science and technology. The whole garden represents that Shanghai is stepping into a brand new eco epoch. Along the arc path, an orchestra plays melodious music beside an infundibulate landscape device, creating a tranquil musing environment for visitors to experience the harmonious life in metropolis.

江苏扬州园 Yangzhou Garden, Jiangsu

总面积1,049平方米，继承和发扬中国传统园林精髓——"山水"主题，采用自然式丘壑布局手法"挖池堆山"、"延山理水"，将江苏悠久的历史文化融入景观之中，发挥了造园特长，从而展现了江苏园林的独特魅力。

Yangzhou Garden, Jiangsu occupies an area of 1,049 m². Inheriting and developing the essence of Chinese painting, which is featured with mountains and waters, designers take the natural style of the hills and valleys to make the layout of the garden. What's more, they bring the history and culture of Jiangsu Province to the landscape design to present the charm of Jiangsu gardens.

Domestic Gardens
内地园

新疆园 Xinjiang Garden

总面积977平方米的展园将"和谐新疆"定为主题，以西域风情为表现手法，突出体现新疆当地的民俗、文化、景观建筑。

The 977 m² Xinjiang Garden is themed with "Harmonious Xinjiang", showing the scenery of the Western Region to highly embody the local custom, culture and landscape architecture in Xinjiang.

河北园 Hebei Garden

总面积1,003平方米的河北展园，将"书院"概念作为创作入手点，以建筑、山石、水体、道路和平台构成河北园的骨架，以绘画、雕刻和书法为表现手法，从而把河北园打造成具有丰富历史内涵的古典庭院。

Hebei Garden, covering an area of 1,003 m², is designed as an academy of classical learning, creating a framework with buildings, hillstones, water, paths and pavilions. Taking various artistic gardening techniques, such as painting, carving and calligraphy, it is built as a classical garden with rich historical connotation.

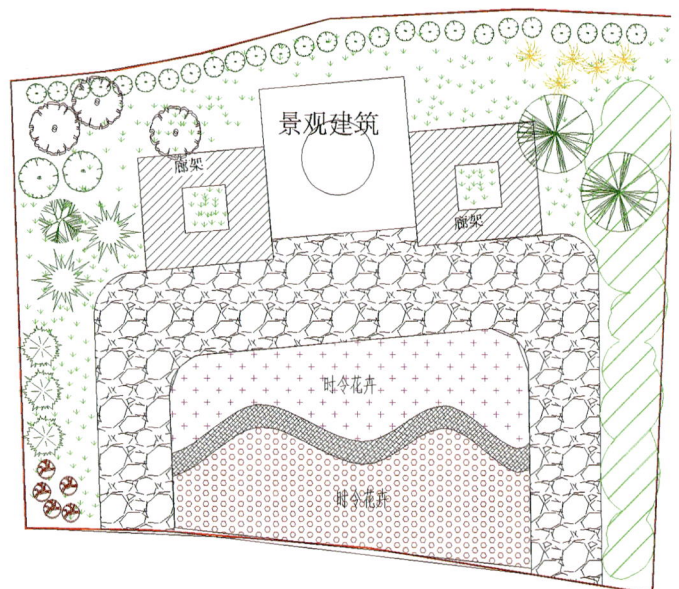

吉林长春园 Changchun Garden, Jilin

总面积843平方米,以圆形和圆弧线构图来展现"圆融和谐"的设计内涵,以市花君子兰和城市风光作为表现特色,形成"通透大气、疏朗开放"的整体空间外貌,展现了北方园林的特色。

Covering an area of 843 m², Changchun Garden, Jilin is dominated by compositions of circles and circular curves to represent the design concept of "harmony". The Garden also applies kaffir lily, the city flower of Changchun, and its city view to present local characteristics. The overall appearance of the garden space is "transparent, grand and open", which showcases distinguishing features of northern gardens in China.

辽宁沈阳园 Shenyang Garden, Liaoning

总面积932平方米的辽宁沈阳园设计以"沈阳流年"作为主题,透过"沈阳历史"、"沈阳味道"、"沈阳印象"三大景观区,来表现沈阳辉煌的历史、发达的现代都市以及和谐的城市文明。

Shenyang Garden, Liaoning occupies an area of 932 m². The theme of this Garden is "Shenyang in the flux of time" and it is divided into three landscape exhibition parts: "Shenyang History", "Shenyang Flavor" and "Shenyang Impression", to present glorious history, modern city view and harmonious city life in Shenyang.

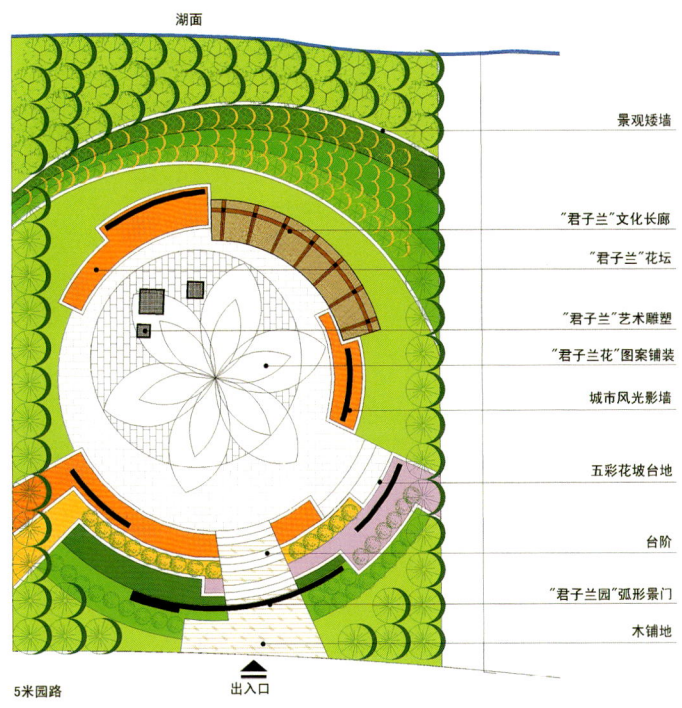

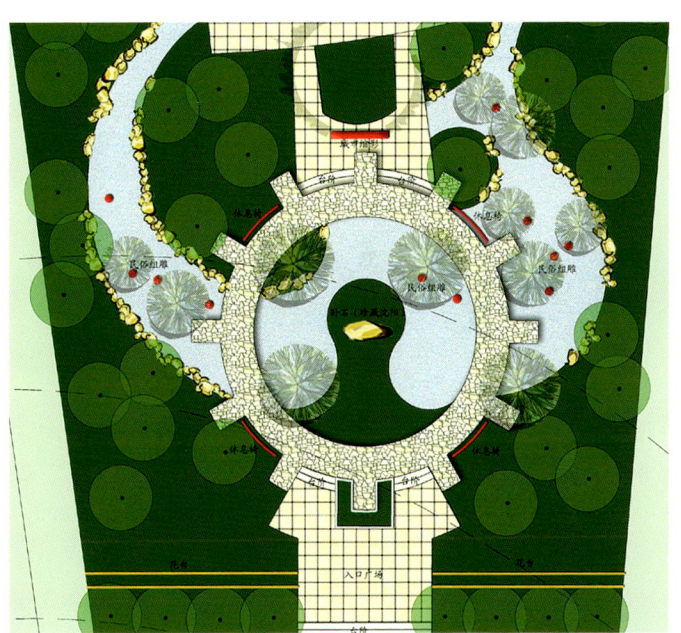

黑龙江园 Heilongjiang Garden

以"冰雪胜地、生态龙江"作为设计主题,通过植入具有东北地域特色的建筑、原始森林缩影和祥和的人文理念,共同营造一个具有北疆冰雪、自然生态的黑龙江园。总面积2,594平方米。

Heilongjiang Garden occupies an area of 2,594 m². The theme of the Garden is "Ice and Snow Land, Ecological Heilongjiang". It is built into a landscape space characterized by ice and snow and natural ecology of northern frontier by the constructions of unique northeast feature, the miniature of the virgin forest and the humanism of auspiciousness.

山东园 Shandong Garden

展园切合世界园艺博览会的主题,挖掘、提取山东历史文化及地域文化,采用写意的手法,在有限的空间内通过景观营造来体现"中庸、和谐"的儒家文化思想核心与精髓,将人与自然和谐共生的理念融入园艺设计之中,形成磅礴大气、文化丰富、景观优美的整体效果。总面积920平方米。

Shandong Garden occupies an area of 920 m². The design concept of this garden corresponds to the theme of the Expo. With the excavation and extraction of Shandong historical and local cultures, the garden creates enjoyable space to embody the essence of harmonious Confucian Culture by landscaping in the limited area. It is a successful practice in the integration of nature, harmony, and garden design, which forms a majestic, exquisite landscape space of Confucianism.

山西园 Shanxi Garden

总面积917平方米的山西园，设计构思源于山西的皇城相府，紧扣世园会主题"天人长安，创意自然"，通过建筑与园林的完美结合，体现出天人合一的最高境界：人与自然的和谐，并且勾画出山西未来"和谐共荣、共同发展"的美好蓝图。

Shanxi Garden covers an area of 917 m². The design of this garden is inspired by the Royal Prime Minister's Palace in Shanxi. It is closely in accordance with "Nature and People in One in Chang'an, Nature Creativity-A City for Nature, Co-existing in Peace", the theme of Expo 2011 Xi'an. The Garden achieves the perfect combination of landscape and architecture, representing the eternal peace and harmony between nature and mankind, and showing a glorious future for Shanxi's development.

福建厦门园 Xiamen Garden, Fujian

总面积1,100平方米的福建厦门园，以厦门传统石文化和海洋文化作为设计的出发点，将自然景观与闽南传统红砖建筑结合，以现代的设计方法对展园进行合理的流线排布，将厦门园打造成具有古典审美情趣的现代园林。

Xiamen Garden, Fujian occupies an area of 1,100 m². This garden features traditional stone culture and special ocean culture of Xiamen. It applies the unique style of traditional redbrick building in the south of Fujian with the natural landscape design, to provide a logical flow line layout for the Garden by modern techniques. Xiamen Garden creates a contemporary landscape space with classical beauty and flavor.

重庆园 Chongqing Garden

总面积1,150平方米的重庆园，其设计从"人人重庆"出发，将代表重庆的"火凤凰"图腾意象以及重庆在新时代的水上门户"朝天门码头"融入平面布局之中，从而展现了重庆独具特色的地域文化和新时期的城市形象，进一步宣扬了和谐共生的城市品格。

Chongqing Garden covers an area of 1,150 m². The theme of the Garden is "People's Chongqing". The whole planning features the "Phoenix" totem image and Chaotianmen Port, the landmark of Chongqing, showing the unique local culture and city identity in the new era. The harmonious atmosphere of the city is also highlighted in the Garden.

天津园 Tianjin Garden

总面积2,231平方米的天津园，设计理念立足于表现天津作为渤海湾的一颗明珠，同时要凸显天津城市蓬勃发展的现代风采。设计结合了天津的历史与传统文化，以"金樽"突出天津的深厚工业城市内涵，而柔美的彩带将展园各分区之间轻轻串联起来，并且意指天津拥有的海运、漕运的文化历史遗产。

Tianjin Garden occupies an area of 2,231 m². The design concept of the Garden is based on showing Tianjin as a bright pearl on Bohai Bay while highlighting flourishing city image of Tianjin. Integrating with local history and culture of Tianjin, the garden gives prominence to the profound connotation of an industrial city by a "Jin Zun" (golden wine jar). Mellow colored ribbons gently join various functional sections together in the Garden to represent rich cultural and historical heritage of ocean and river shipping in Tianjin.

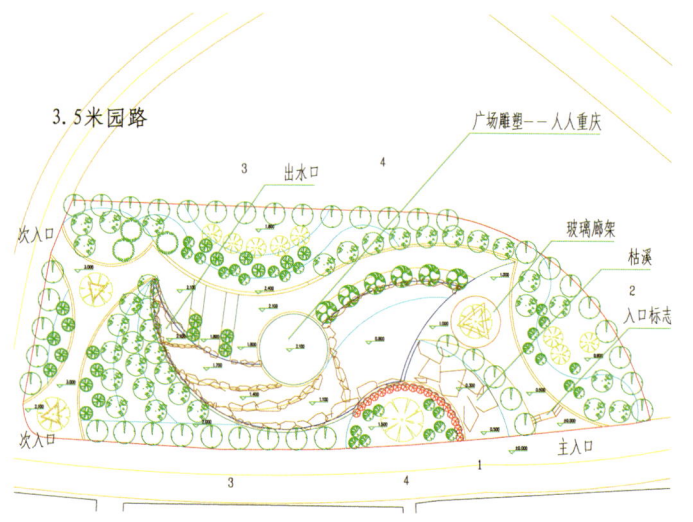
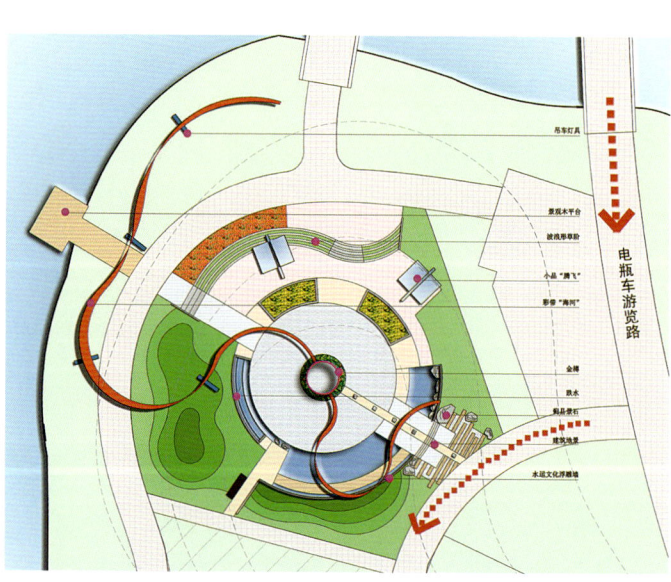

汶川园 Wenchuan Garden

展园以"汶川一路走来"作为项目主题,以羌族建筑和水景特色为主要内容,具体包括:羌碉和羌寨、山岩瀑流、锅庄广场、索桥、白石神图腾、羊图腾的雕塑造型等,同时展园还移栽了汶川特色植物,例如珙桐、甜樱桃树等。总面积751平方米。

Wenchuan Garden occupies an area of 751 m². The Garden is themed with "Wenchuan on its way", mainly exhibiting unique Qiang buildings and waterscapes, including Qiang stockade villages, waterfalls from hill rocks, Guo Zhuang Plaza, a rope bridge, Baishi God Totem and Sheep Totem Sculptures. Also the Garden is transplanted with various local plants in Wenchuan, such as Chinese dove tree and sweet cherry trees.

青海西宁园 Xining Garden, Qinghai

展园的设计以"唐蕃古道"为主线,总面积1,515平方米。沿线安排"大美青海"、"河湟文化"、"锦绣西宁"三个景点由东向西贯穿整个园区,强调河湟文化及通过唐蕃古道传播的多民族文化,进一步表达西宁所具有的多元文化内涵,更表达了我国各族之间友谊牢不可破的传统。

Xining Garden, Qinghai covers an area of 1,515 m². The Garden features a Tang court-Tibet route, along which there are three scenic spots including Beautiful Qinghai, Hehuang Culture and Colorful Xining from east to west. These designs aim at highlighting Hehuang Culture and multi-ethnic cultures to give further expression to the specific multicultural connotation and to show the unbreakable friendship among all peoples in China.

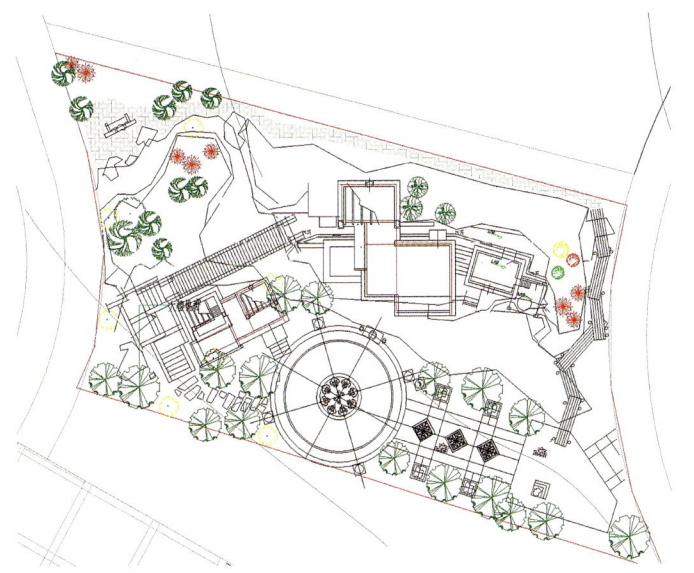

四川成都园 Chengdu Garden, Sichuan

展园以"天府谐园"为名,包括"蜀巷情深"、"塔映花溪"、"景墙艺韵"、"竹篁森森"、"蜀花园艺"等景点。总面积893平方米。各个建筑单体所使用的造型和元素都来源于成都近代建筑形式,利用现代建筑材料和工艺完成,既表达了对历史传统的尊重,也强调了项目所具有的时代性。

Occupying an area of 893 m², Chengdu Garden, Sichuan is also named "Tianfu Harmonious Garden". There are several landscape spots including "Typical Sichuan Lane", "Tower Reflected in Stream", "Landscape Wall Attraction", "Deep Bamboo Forest", "Sichuan Flower Gardening", etc. The design of this Garden is inspired by the style of the modern architecture in Chengdu; modern materials are used in the structure manufacturing. The whole Garden expresses designers' respect to the historical tradition, and shows epochal characters of this project.

甘肃兰州园 Lanzhou Garden, Gansu

总面积1,474平方米,又名百合园,意喻"事事和好",充分体现"和谐"的理念,与本届世园会的主题吻合。设计借用木质水车以及当地植物等特色标志,反映了甘肃兰州地域特色,展示了甘肃深厚的黄河文化传统,同时表现了人民生活蒸蒸日上,人与自然和谐共生。

Lanzhou Garden, Gansu, covering an area of 1,474 m², is also called "Lily (Bai He in Chinese) Garden", with the meaning that "everything goes well", which fully displays the concept of "harmony", coinciding with the theme of Expo 2011 Xi'an. The design features wooden waterwheel and local plants to represent the flavor of Lanzhou and to showcase profound Yellow River civilization of Lanzhou City and improved local residents' quality of life.

黑龙江哈尔滨园 Harbin Garden, Heilongjiang

总面积2,874平方米的展园以"冰雪情怀"作为设计主题,通过"一环、两轴、一中心"的排布形式,将各个富有哈尔滨当地特色的景观小品串联起来,创造出独具北方特色的景观。

Harbin Garden, Heilongjiang occupies an area of 2,874 m². Themed with "sentiment for ice and snow", the Garden is designed in an overall layout as "one circle, two axes and one center" to connect various local-featured landscape decorations. Harbin Garden creates a unique landscape of northern style.

河北唐山园 Tangshan Garden, Hebei

总面积955平方米的河北唐山园,设计主题为"振翅",向游客展示了唐山由工业城市向生态文明城市的发展,从而表达了唐山的建设如"凤凰涅槃、浴火重生"般举世瞩目。

Tangshan Garden, Hebei, covering an area of 955 m², is themed with "to fly" to show visitors the development of Tangshan from an industrial city economy into an ecological civilized city. It also denotes Tangshan's miraculous rebirth as "the Nirvana of the Phoenix through Fire" after the Earthquake 1976.

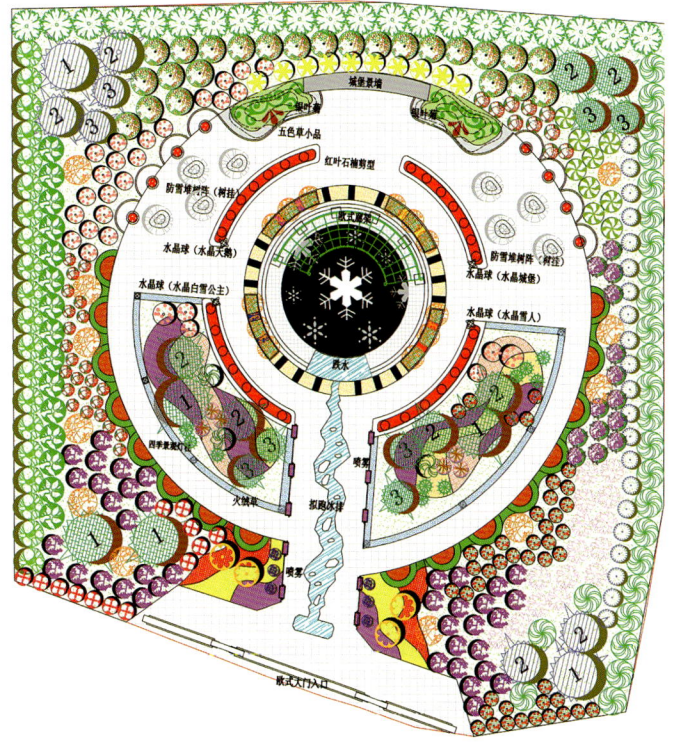

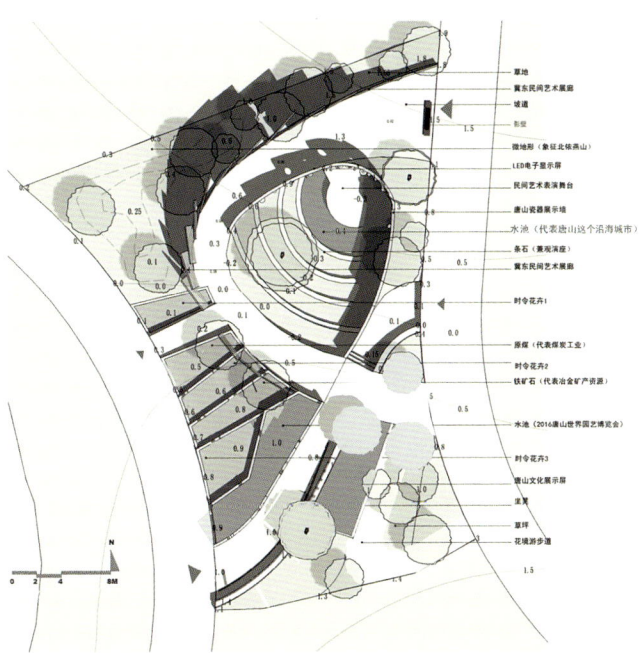

云南园 Yunnan Garden

总面积1,164平方米,设计选用了世界重点保护文物"白塔"、"景真八角亭"作为体现展园主题"有一个美丽的地方"的标志,表现了云南园立足于历史传统,真实呈现云南得天独厚的自然胜景。同时,将园区内的主干道"彩虹道"与其下蜿蜒的溪流结合起来,形成丰富的景观层次。展园除了使用丰茂的竹林元素来体现傣家风情,还会栽种丽江牡丹、贝叶棕、菩提树、洋兰、三七、高山杜鹃等具有云南特色的植物。

Yunnan Garden occupies an area of 1,164m². The main buildings of the Garden are miniatures of the key cultural heritage sites, the "White Tower" and "Jingzhen Octagonal Pavilion" in Yunnan Province. Themed with "A Beautiful Place", the Garden features the history, culture, and nature of Yunnan. The wandering stream is integrated with the "Rainbow Path" above to create rich stratification of landscape. Luxuriant bamboos planted in the Garden bring visitors to the original Dai cultural atmosphere, and other special Yunnan plants such as paeonia suffruticosa in Lijiang, talipot palm, bodhi tree, cattleya, panax notoginseng, and alpine rose are also introduced.

广东广州园 Guangzhou Garden, Guangdong

总面积1,257平方米,以"花城丝路园"为名,设计理念从海陆两条丝绸之路的起源出发,以"丝路—融合与对话"作为主题,让广州与西安这两个在地域上相差很远的城市之间进行对话,整个园区通过景观肌理的变形与抽象,水体沙丘雕塑的运用,体现出海上丝绸之路与陆上丝绸之路的一种形态上的过渡与浓缩。

The 1,257 m² Guangzhou Garden, Guangdong is also called "The Garden of Flower City and Silk Road" mainly themed with "silk road – integration and dialogue", of which the design is inspired by the origin of the overland Silk Road and the maritime Silk Road, making dialogue between the two different cities – Guangzhou and Xi'an. Applying the watering dune sculpture and abstract variant landscape, the Garden gives expression to the modal transition between the marine Silk Road and the overland Silk Road, and to the concentration of the two.

广东深圳园 Shenzhen Garden, Guangdong

总面积1,516平方米的广东深圳展园以"创意深圳"作为主题,通过折纸廊的形式,丰富了展园的内涵,表现了设计者对于城市发展的深层次思考,展园通过这种现代的设计手法,折现出深圳这个城市在过去30年间迅猛发展中,许许多多的人物和故事,同时也强调了深圳作为设计之都所具有的蓬勃的创意力量。

Shenzhen Garden occupies an area of 1,516 m², which is designed with the theme of "Innovative Shenzhen", applying the gallery of paper folding art to enrich the connotation of the garden and to represent the designer's profound thinking of city development. The modern design technique in the Garden reflects 30-year rapid development and the design creativity of Shenzhen.

北京园 Beijing Garden

展园设计结合了北京悠久的历史文化,展现"南襟河济、北枕燕山、西依太行、东临沧海"的地理优势。总面积1,598平方米。在具体布局上,参照了北京燕墩及其上面的乾隆御制碑为主题的园林设计方案,以"台"作为展园的视觉焦点,四周不设围墙,延展了视觉空间,体现了首都的大气豪迈。

Beijing Garden occupies an area of 1,598 m². The design of the Garden combines with the long history and culture of Beijing to present its geographic advantage of "facing Yellow River and Ji River in the south, Yan Mountain in the north, Taihang Mountain in the west, and Bohai Sea in the east". The idea of the layout scheme is borrowed from the Yandun stele in Beijing, which was made upon the order of Qianlong Emperor. The stand, a form of the earliest construction in Chinese garden, is used as the main element of the Garden and the fence-free layout may extend the visual space for visitors, which presents the broadness and generousness of the capital of China.

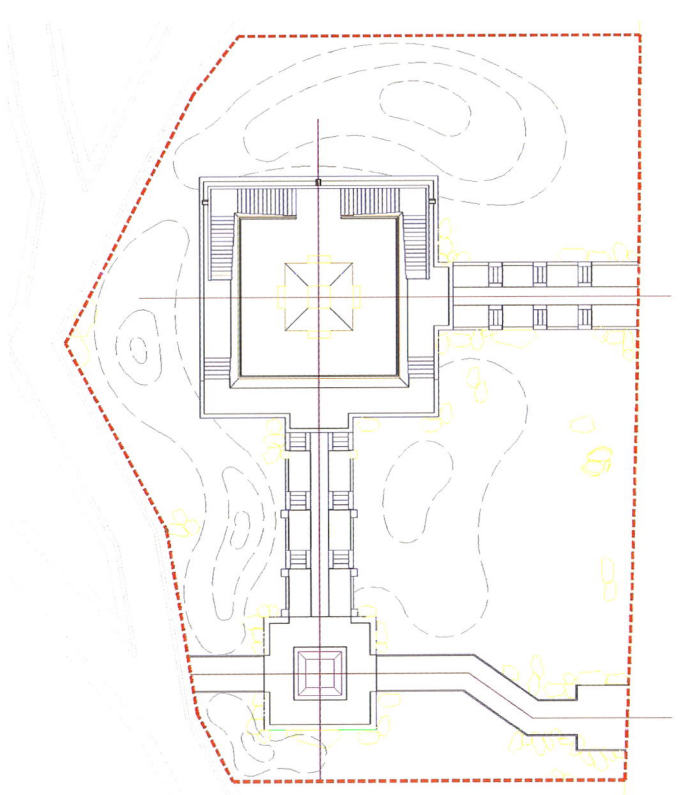

江西南昌园 Nanchang Garden, Jiangxi

总面积945平方米的展园以"打造园林精品工程,树立城市全新形象"为宗旨,整体布局采用自然式园林手法,分别设"落霞云景"、"碧水微澜"和"滕王台"等节点串联参观流线,以沿途的景观小品作为点缀,使得展园的景观层次更为丰富。其中展园的主体建筑"滕王台"采用了传统的修建手法,在装饰上还原了"滕王阁"绚丽的彩绘元素,原汁原味地表达了传统建筑的韵味。

Nanchang Garden, Jiangxi, covering an area of 945 m², aims at "establishing a new city image by building an exquisite garden". The overall layout applies a natural garden mode with the scenery of "sunset and cloud", "green and rippling water" and "Tengwang Tower" as the main axis. Supplemented by natural plants and garden paths as landscape decorations, the Garden creates a magic visual effect for visitors. The main building in the Garden, "Tengwang Tower", designed with Tengwang Pavilion as the prototype, is built with traditional architectural technique and colorful paintings, bringing the unique style of traditional architecture into full play.

河南开封园 Kaifeng Garden, Henan

展园与世园会"天人长安,创意自然"的主题相契合,通过传统的造园手法,突出本园"宋韵、菊香"的特色。展园使用了宋词、景区风貌、古城墙、包公、菊花等要素,并且将它们进行有机结合,从而形成了具有浓郁宋代风情的精致园林庭院。总面积612平方米。

Occupying an area of 612 m², Kaifeng Garden, Henan is designed in accordance with "Eternal Peace & Harmony between Nature & Mankind, Nurturing the Future Earth-a City for Nature, Co-existing in Peace", the theme of Expo 2011 Xi'an. By the traditional garden-constructing technique, the Garden highlights the "Flavor of Song Dynasty and Fragrance of Chrysanthemum" local Kaifeng features. Also, it applies different elements such as Song Prose, landscape, ancient city wall, Bao Zheng (a famous officer in Song Dynasty) and chrysanthemum together to create an exquisite garden space with the profound cultural atmosphere of Song dynasty.

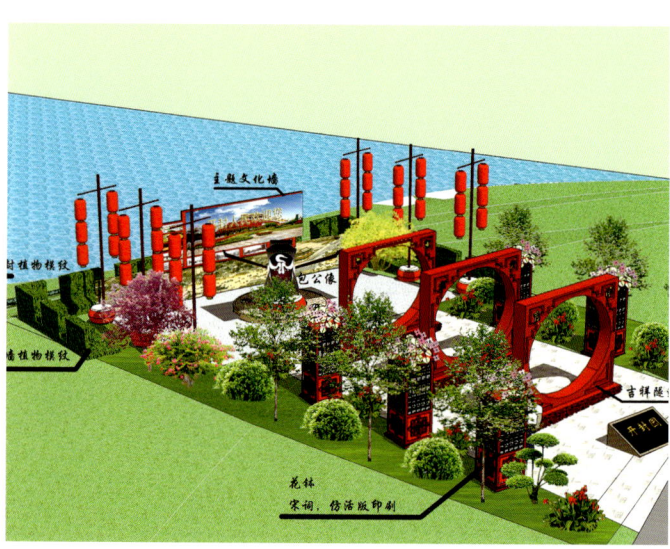

吉林辽源园 Liaoyuan Garden, Jilin

总面积1,028平方米的展园设计以"天人长安,创意自然"为主题,遵循辽源市"盛京围场"、"皇家鹿苑"这一历史脉络,选用当地特有的木质瞭望塔楼建筑形式作为展园入口大门,点出"围场"这一独特主题,使展园显得大气、粗犷。展园布局为近代传统"四合院"空间,外形方正,呼应了辽源的地方气质。在建筑的具体建造手法上,设计师使用了辽源当地特有的材质。不同材质的拼贴,表达出了建筑浑厚的气息,也体现了东北的民俗传统。

Covering an area of 1,028 m², Liaoyuan Garden, Jilin is themed with "Eternal Peace & Harmony between Nature & Mankind, Nurturing the Future Earth-a City for Nature, Co-existing in Peace". Following the historical context of "Shengjing Hunting Park" and "Imperial Deer Garden" in Liaoyuan City, it adopts the form of watchtower made from local birch to build the entrance, focusing on the broadness of "Hunting Park". The layout of the Garden is like a traditional "quadrangle courtyard" with the upright and foursquare appearance to echo with local Liaoyuan ethos. All the architectural materials in the Garden are unique local materials from Liaoyuan to fully present the simple architectural style and folk-custom in Northeast China.

青岛园 Qingdao Garden

总面积1,089平方米的青岛园围绕"一海、一山、一座城"整体布局,通过植物的渐变形成地景的变化,将园内的山景、城景与海景进行有机地联结,从而诠释了青岛的历史传统与现代人文内涵。

Qingdao Garden covers an area of 1,089 m². The overall layout of the Garden is focused on "one sea, one mountain and one city", organically integrating the mountainscape, cityscape and seascape together by the various plant modeling. The whole garden gives full expression to the historical tradition and modern cultural connotation of Qingdao City.

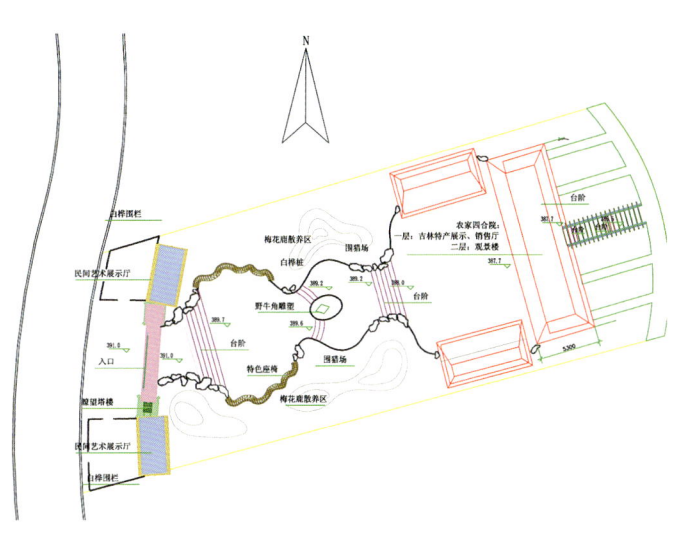

安徽园 Anhui Garden

总面积1,040平方米,以"传承安徽地域和谐文化之德,展示安徽现代和谐发展之音"作为设计主题,将"徽文化"作为主要表现手段。东西向的主体建筑体现了徽派建筑的序列关系以及整洁大气的特点,而"白墙黛瓦"作为传统徽派建筑的代表性语言被保留了下来,简化的细节与当今时代的审美情趣相符合。安徽地区独有的景观材料及地域文化符号,例如马头墙、天井空间、六尺巷空间、牌坊等被引入园中,与主体建筑和谐共生,展现了当代安徽的新气象。

Anhui Garden occupies an area of 1,040 m². It is themed with "inheriting the harmonious culture of Anhui and presenting the harmonious development in modern Anhui" to show the special "Hui Culture". The transmeridional main building in the Garden represents the unique simple serial style and the "white wall with black tile" characteristics of the local Anhui architecture, of which the simplicity concept echoes with modern aesthetic appreciation taste. Exclusive landscape materials and local cultural symbols, such as Matou Wall, patio, courtyard, six-foot wide lane and memorial archway are introduced into the Garden to harmoniously decorate the main building, showing the vitality of new Anhui.

河南园 Henan Garden

展园以"牡丹"之国色天香与中州文明作为其气韵,描绘了一部宏大而绚烂的长卷。总面积1,168平方米。南北向分布的展园,具有清晰的中轴线,在整体上采用了对称的空间序列,而在细节上却多有经营。展园将其空间延伸至水面作为接引入口,并在入口一侧放置箱养牡丹,将参观者自然引至木质平台。园内南北通透,先有"豫园春色"以牡丹元素先声夺人。接着,参观流线步步抬升,道路两侧遍栽河南特色的绿植,以丰富地景呈现出河南深厚的历史底蕴与良好的自然条件。至垂花门而出,点出了展园"盛世美景"的主题。

Henan Garden covers an area of 1,168 m². The north-south Garden is themed with "peony", which is the best symbol of China's national grace. The overall layout of the Garden adopts symmetrical spatial series with the grand and splendid long lane as the axis. The detailing decorations are also well-designed. The exhibition space extends to the water surface as the entrance, along one side of which set a line of box-planted peonies, leading visitors to the wooden terrace. Walking ahead along the long lane, visitors will see the local green plants fully planted on both sides. The design of this garden gives expression to the profound historical connotation and favorable natural conditions of Henan.

湖北园 Hubei Garden

展园以弘扬三峡大坝工程为主线，总面积1,091平方米，以"水"为题材，力求展现湖北的植物特色，努力营造景观的多变，以植物造景为主，体现绿色时尚的现代园林。

Hubei Garden occupies an area of 1,091 m². With the promotion of "Three Gorges Dam Project" as the theme and "water" as the subject, the Garden is committed to presenting the plant characteristics of Hubei and to creating a modern garden filled with varied garden landscapes that reflect a green philosophy by focusing on the plant landscaping.

广西园 Guangxi Garden

总面积1,011平方米，以广西母亲河、红水河、壮族特色木楼、对歌亭等为设计要素，营造出一个自然的壮乡小景，阐述广西人与自然和谐发展的生态理念，突出"城市与自然和谐共生"的世园会主题，展现广西绿意盎然的山水风景以及淳朴的民俗风情。

With an area of 1,011 m², Guangxi Garden creates a natural landscape of Zhuang Region with Hongshuihe River, distinguished wooden houses of Zhuang ethnic minority, Duige Pavilion (pavilion for singing in antiphonal style) as the design elements. It elaborates the ecological philosophy of harmonious development between Guangxi People and the nature, highlighting the theme "a City for Nature, Co-existing in Peace" of Expo 2011 Xi'an, and showcasing the green landscapes and simple folk customs in Guangxi.

杨凌园 Yangling Garden

总面积约831平方米。展园的主题为"现代园艺",强调现代农业科技示范效应,是现代农业中新技术与新成果为主体的展示性场地。景观形式突出现代气息,注重新技术、新材料、新工艺。

Yangling Garden covers an area of 831 m². The Garden is designed with the theme of "modern horticulture", highlighting the demonstration effect of modern agricultural science and technology, which can be a stage to show new achievements on modern agriculture. Modern landscapes in the Garden also reflect modern technology and new materials applied in agriculture.

商洛园 Shang Luo Garden

总面积约1,026平方米。"商法自然,落水青山"的自然形态传达了寄情山水的设计意念。通过抽象山水的写意形式诠释出"商山洛水",重点突出了自然生态的地域特色及文化、人文、特产等内容,满足游客观赏、休憩、交流的需求。

Shang Luo Garden occupies an area of 1,026 m². The Garden gives expression to "Shang Shan Luo Shui" (the local natural scenery of Shaanxi) by abstract landscape design technique and highlights cultural and regional features, creating an ideal place for visitors to enjoy the sight, take a rest and communicate with one another.

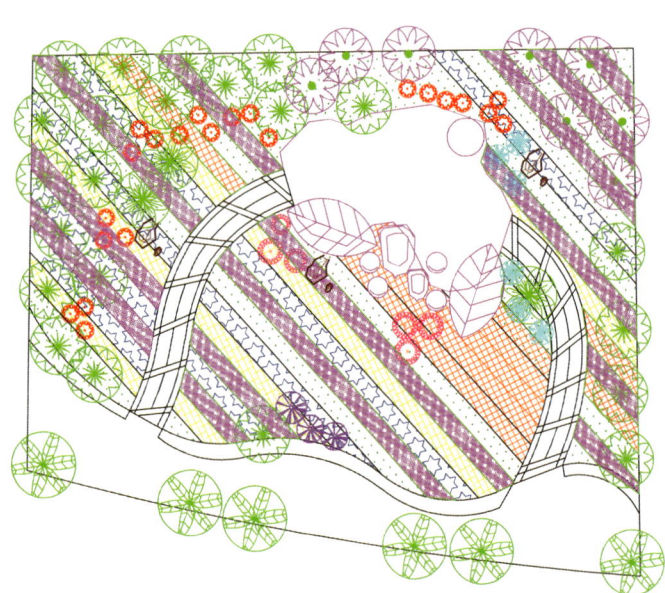
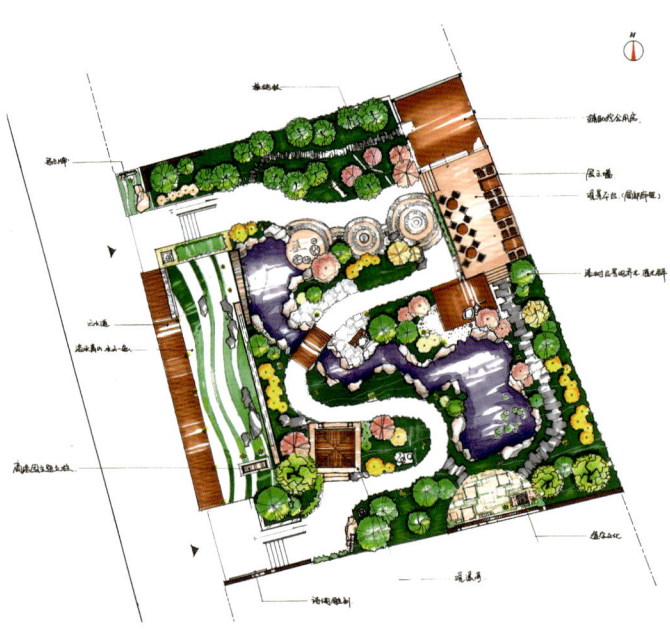

安康园 Ankang Garden

总面积约1,013平方米。为了表达出"秦巴山水写意",展园以安康典型的秦巴山水景观、农耕景观和人文景观为主题,抒发出安康人民在这片土地上与生态和谐共处的美好愿望。

Occupying an area of 1,013 m², Ankang Garden is designed with the theme of "Enjoyable Mountains and Waters of Qinling-Bashan Region" to show the fascinating local natural landscapes, expressing best wishes of Ankang people to coexist harmoniously with nature.

汉中园 Hanzhong Garden

总面积约975平方米。汉中,简称"汉",素有"汉家发祥地、中华聚宝盆"的美誉。通过汉江两岸秦岭山麓下村庄与现代城市形式上的冲突,在园林意境中展现汉中秀美的自然风光、深厚的历史文化底蕴与城市发展的传承、和谐、统一。

Hanzhong Garden covers an area of 975 m². Hanzhong City is called "Han" for short, which always gains a good reputation of "Hanjiayuan (Birthplace of Han)". An ancient pavilion of Han style is built as the main scenery of the Garden in the junction of Hanjiang River and Qinling Mountains; the characteristic landscape plant thriving around the main scenery to display the beautiful natural scenery of the south region of Shaanxi, which is near the mountain and by the river with evergreen plants throughout the year.

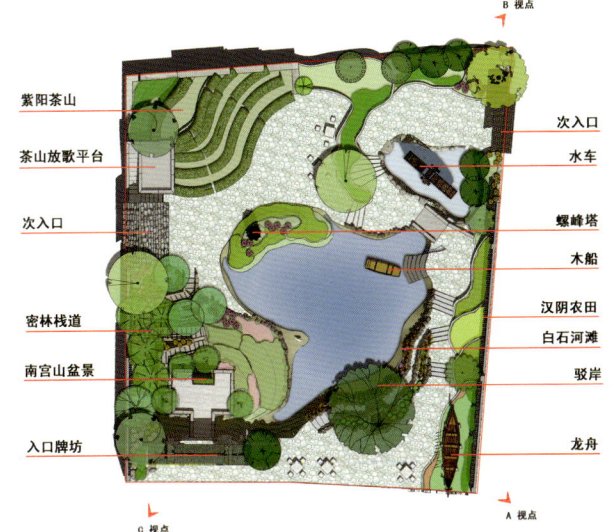

渭南园 Weinan Garden

总面积约1,040平方米。展园主题是"人文渭南、山水渭南",突出渭南地区华夏民族文明发源地的重要地位,宣传渭南改革开放30年以来取得的巨大成就,展现当地历史文化、地域文化和民俗文化的特色。

Weinan Garden covers an area of 1,040 m². The design of the Garden focuses on the theme of "Culture and Landscape of Weinan" to highlight Weinan as the birthplace of Chinese Civilization, and to show that Weinan has got gigantic achievements since implementing reform and opening-up policy 30 years ago.

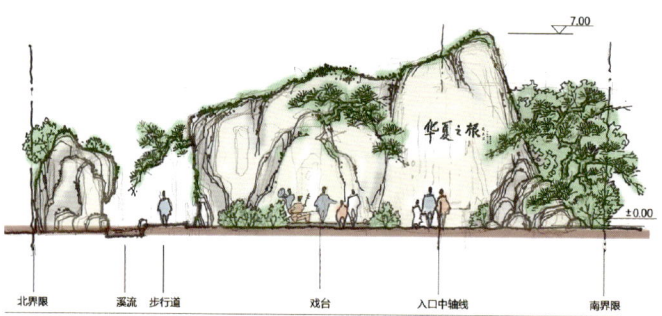

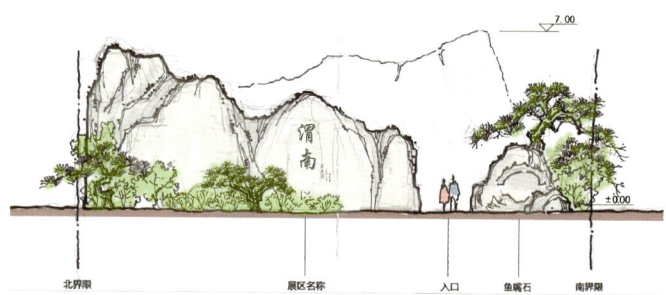

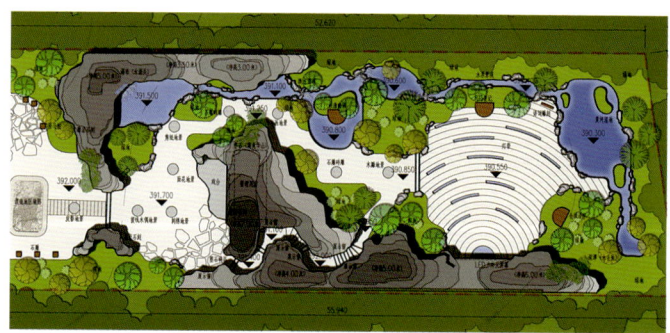

铜川园 Tongchuan Garden

总面积约1,015平方米。展园主题是"铜川故事",以铜川陈炉陶瓷和耀州青瓷为中心,结合玉华宫、铜川民间故事和植物景观,彰显铜川深厚的历史风韵以及铜川人千年来"创意自然"的智慧,这种智慧将泥土变为具有艺术生命力的陶瓷。

Tongchuan Garden covers an area of 1,015 m². The Garden is themed with "Tongchuan Story", integrating the folktales of Tongchuan and local flora landscape with the Yaozhou kiln culture to show the profound historical connotation and people's wisdom of making artistic ceramics from clay in Tongchuan.

咸阳园 Xianyang Garden

总面积约1,295平方米。展园主题为"秦人、秦风、生态咸阳",以秦文化为主基调,侧重于自然的回归、文化的传承和人性的体验感受,宣扬生态和谐与人文情怀。

Occupying an area of 1,295 m², Xianyang Garden is themed with "Qin people, Qin culture, and green Xianyang". Focusing on the inheritage of nature, culture and customs, the Garden presents an ecological and harmonious Xianyang.

宝鸡园 Baoji Garden

总面积约1,350平方米。根据宝鸡的现实地貌,通过园林景观中周鼎流瀑、青铜流殇和陈仓栈廊等景点的设计,突出展示炎帝故里和周秦文化的历史标志,世界佛都法门寺的文化标志及中华石鼓园的地理标志,表现了一个具有绿色、生态理念的森林城市,既反映了周秦文化,也表现了宝鸡的历史和现代和谐发展。

Baoji Garden covers an area of 1,350 m². The design of this Garden is inspired by the physical landform of Baoji to show the local famous scenery spots, such as Zhongzhouding Waterfall, Chenchang Corridor, Famen Temple and so on. It highlights that Baoji is the homeland of Yan Emperor in the Garden and reflects the city's long history from Zhou and Qin Dynasties. The Garden also gives expression to a new and modern image of ecological and harmonious Baoji.

榆林园 Yulin Garden

总面积约992平方米。榆林沉淀着悠久厚重的文化之宝,传承着不畏艰难的精神之宝,蕴藏着世界罕见的矿产之宝。通过园区景观的标志建筑"镇北台"、樟子松林、能源之都、塞外驼城,把榆林的地域特色全部展现出来,可谓和谐、生态的宝藏榆林。

Yulin Garden occupies an area of 992 m². Yulin is a city with a long history, rich culture and rare mineral resources. The garden sets a lot of local landmarks and landscapes such as miniature of Zhenbei tower and pinus sylvestris woods, and desert landscape to show a harmonious and ecological Yulin with plenty of mineral treasures.

延安园 Yan' an Garden

总面积约1,015平方米,体现延安文化绚丽、生态和谐、山川秀丽、社会发展的地域特色。设计弘扬城市历史与文化,突出红色文化主线的现代景观体验,表现黄土高原的自然景观——山、水、城、园融为一体的空间格局。

With an area of 1,015 m², Yan'an Garden displays the gorgeous culture, harmonious ecological system and beautiful natural scenery of Yan'an, highlighting the theme of holy land of revolution with modern elements. Applying micro-scenes of mountain, river and tower, the Garden also shows the natural landscape of on the Loess Plateau.

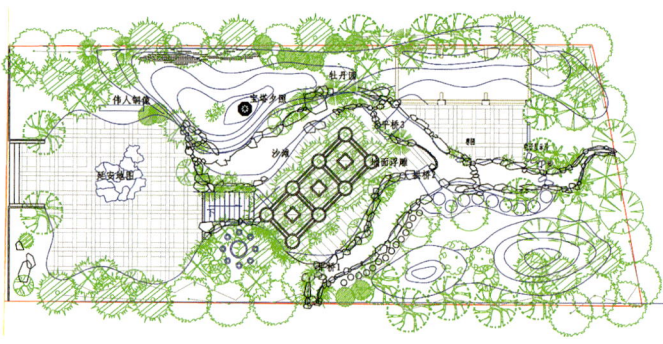

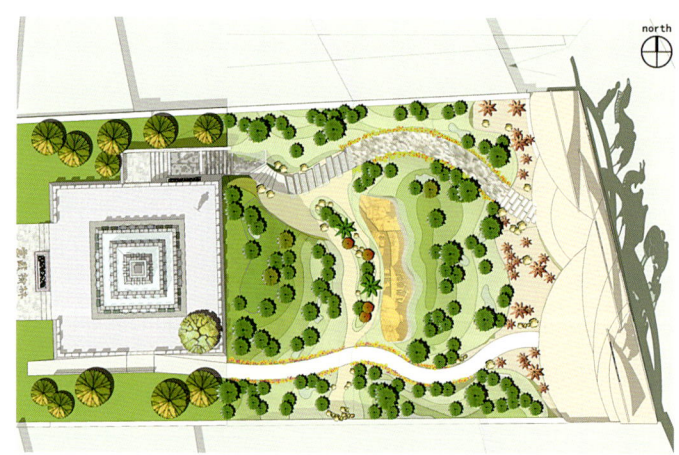

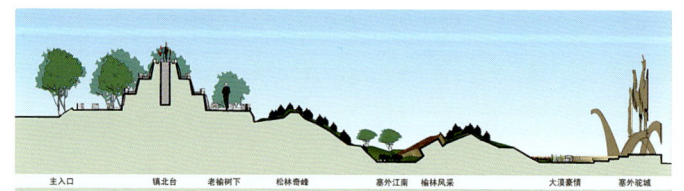

Bonsai Garden

专类园·盆景园

展园位于世界园艺博览会小终南山之上，紧邻秦岭园与长安塔，总面积约17,000平方米。因盆景起源于中国，始于唐代，今盆景园在发源地展出，可谓千年古都长安之盛世。在世界艺术之林中，盆景艺术是富有自然情趣的东方艺术精品之一，也是我国独特的传统园林艺术之一。园中展出各式盆景，展示出盆景艺术在其发源地和世界范围内的发展历程和独特魅力。

Occupying an area of 17,000 m², Bonsai Garden is located on the Young Mount Zhongnan, and adjacent to Qinling Garden and Chang'an Tower. Since bonsai derives from Tang Dynasty in China, the exhibition of bonsai at Expo 2011 Xi'an implicates the profound historical significance. Visitors could experience thousand-year history and development of Chinese Bonsai art in the Garden, which is one of the art treasures in the Orient with natural implications.

Qinling Garden

专类园·秦岭园

展园位于世界园艺博览会小终南山南麓,园区规划集中体现秦岭的生态特性,展示具有秦岭特色的生物系统,内容涉及珍稀植物和秦岭四宝馆,其中馆内大熊猫、金丝猴、朱鹮、羚牛四种珍稀动物的展示,在国内世园会中属于首创,对体现本次世园会"天人长安,创意自然"的主题具有非常重要的意义。

Qinling Garden is built at the south foot of the Young Mount Zhongnan, focusing on the ecological feature of Qinling Mountain. Aiming at displaying the bio-diversity of Qinling Mountain, the Expo exhibits the "Four Treasures of Qinling", the unique local animals including panda, golden monkey, crested ibis, and takin for the first time, which reflects the theme of Expo 2011 Xi'an - "Eternal Peace & Harmony between Nature & Mankind, Nurturing the Future Earth-a City for Nature, Co-existing in Peace".

Poetic Landscape in Chang'an Garden

专类园·人文山水 诗意长安园

展园借用《诗经·国风》中秦地风光诗句，以及唐朝王维《辋川别业》诗句，用中国写意式的造园手法，表现中国古代文人诗词中的山水诗画及中国古代文人的生活场景，展示《诗经》中为人熟知的植物与古代文人偏爱的植物及其蕴含的文化寓意，用中国传统园林构景手法，以时代文化主旋律——"爱"与"和谐"造林思想为纲要，勾勒出一个突出东方文化特色，应对时代"生态危机"、"文化挑战"的21世纪时代新园林。

The design of Poetic Landscape in Chang'an Garden is inspired by artistic conception from the verses of *The Book of Songs and of Tang poem*. It adopts traditional Chinese landscaping technique with enjoyable poetic style to show the profound implicit cultural sense in *The Book of Songs*, also to show a modern and fashionable society filled with love and harmony with special landscape decorations.

桑园
文杏馆
次入口
竹里馆
主入口
临湖亭

Chang'an Garden

专类园·长安园

居小终南山之上，傍长安塔，隐青翠竹林之中，依湖水而建，望东南亚风情，于山水妙境之中，隐现几间茅舍，谓之长安园。展园位于世界园艺博览会浐灞生态区终南山西北侧，紧邻长安塔，总面积约56,000平方米。以彰显"隐"的自然美为主题，以竹为基底竭力营造了清新飘逸、诗意典雅的景观氛围，表达了人们对于平静、闲适、幽远的精神境界的追求，宁可食无肉，不可居无竹，竹与中国人的灵魂已经融为一体。草药园、文化水岸、牡丹园、芍药园是对传统文人山水园林的时代表达。瓦片、竹子也可以歌唱现代，古朴中流露大气。航天园与其他展园迥然不同，彰显了现代设计的张力与疯狂，其层次、序列、界面均沿袭了传统园林的精神内涵。

With an area of 56,000 m², Chang'an Garden is located on the Young Mount Zhongnan along the lake, close to the Chang'an Tower, and surrounded by the verdant bamboo forest. The main element of the Garden is bamboo, by which it aims to create a fresh, poetic and elegant landscape atmosphere to give expression to human's pursuit of tranquil and leisurely life. There are Herb Culture Exposition Zone, Cultural Water Front Exposition Zone and Peonies Exposition Zone in the Garden to represent the traditional landscape in modern times. Space Plants & Space Science Education Exposition Zone is different from other zones in the Garden, which displays space plants and space horticultural varieties to interpret the history of human's understanding and exploring of the universe.

VIP酒店
观东南亚风情街
观vip酒店
观长安花谷
观长安塔
航天园
竹园
草药园
水文化岸
牡丹芍药园
长安塔

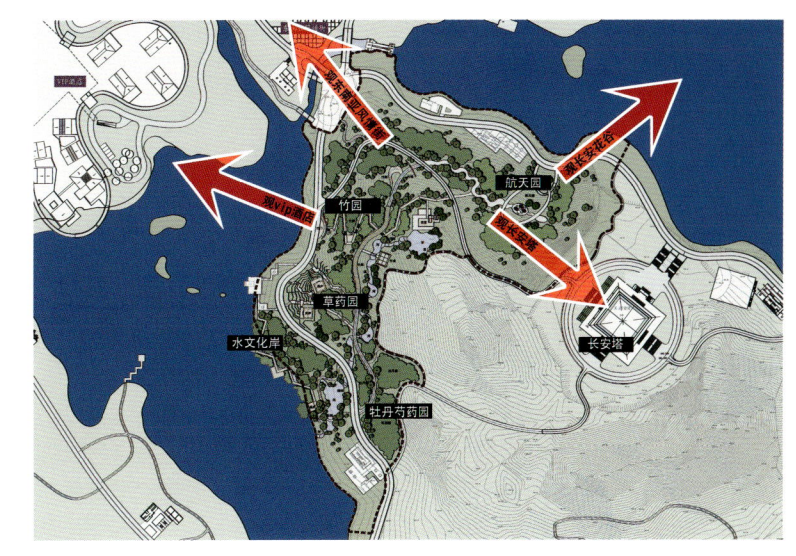

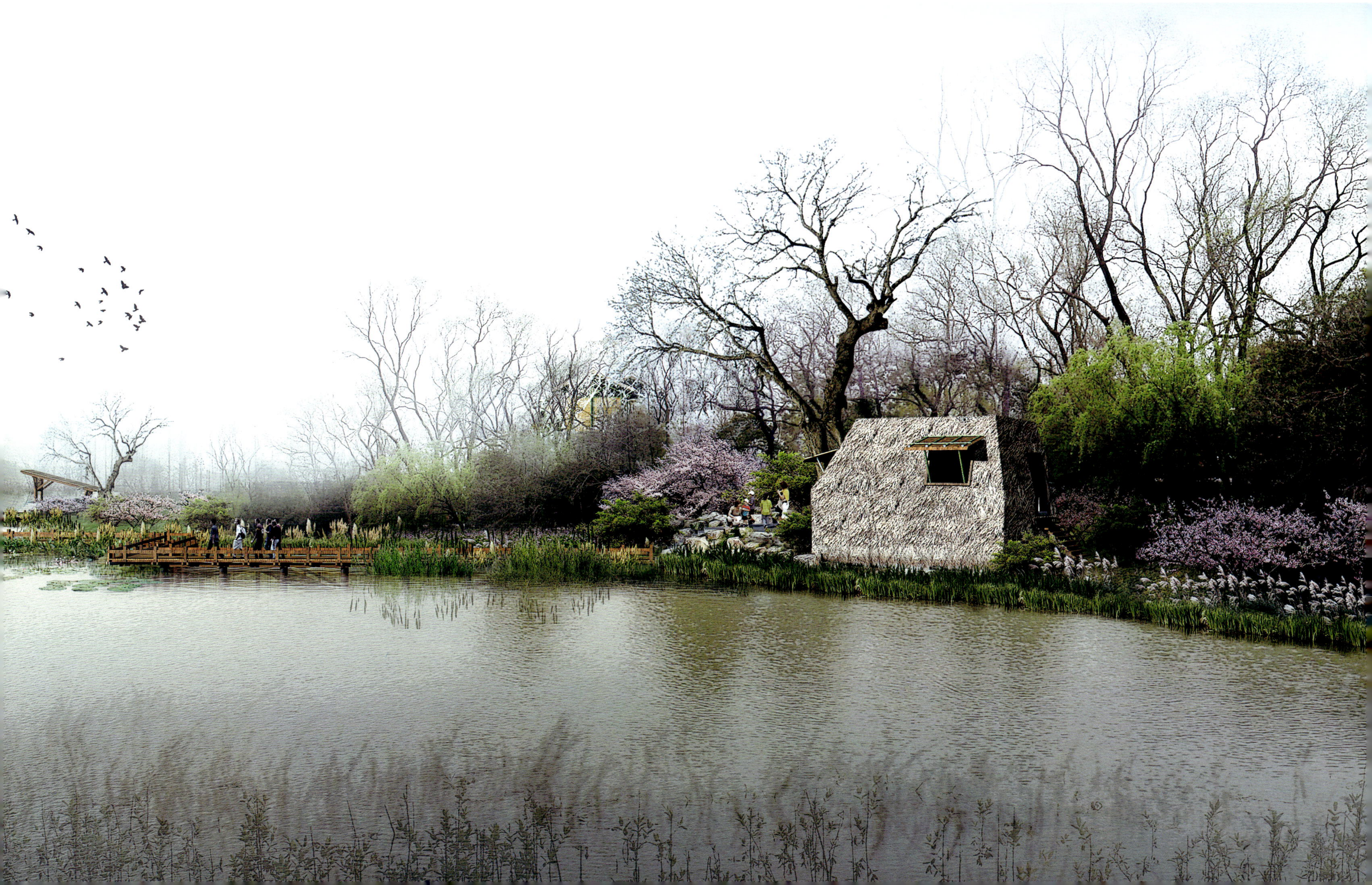

西农科技生态园
Northwest Agriculture & Forestry University Sci-tech Ecological Garden

本展园是西北农林科技大学对外展示的窗口，在园林建筑、水体和植物配置上体现"农、林、水"的特色。总面积约817平方米。主体建筑为一百多平米的温室（内培育无土栽培的花卉等产品），并分为3个景观区：入口处的景观设计围绕着"农"的主题进行设置，特色的景观柱以麦穗的形象为原型进行艺术的提炼；中心景观区由水系景观、园林小品景观和丰富变化的植物景观构成；温室景观区室外是以"燃烧绿色的火焰"为主题的艺术喷绘，节能温室内展示区主要展示一些利用农业科技手段培育的植物。

The Garden is located in the Enterprises Gardens of Expo 2011 Xi'an with an area of 817 m². It is a demonstration window for Northwest A & F University. Its arrangement of garden architecture, water, and plant reflects the characters of "agriculture, forestry, and water". The main building of the Garden is a greenhouse (in which planted flowers with soilless cultivation) occupying an area of more than 100 m², and it is divided into three landscape sections. The entrance landscape section is designed with the theme of "agriculture", while the unique landscape columns are inspired by the image of wheat. The central landscape section consists of water system landscape, garden ornament landscape and various plant landscapes. Outside the greenhouse landscape section is an art painting with the theme of "green flames" and the indoor exposition section of the energy-saving greenhouse displays some plants cultivated by using agricultural science and technology.

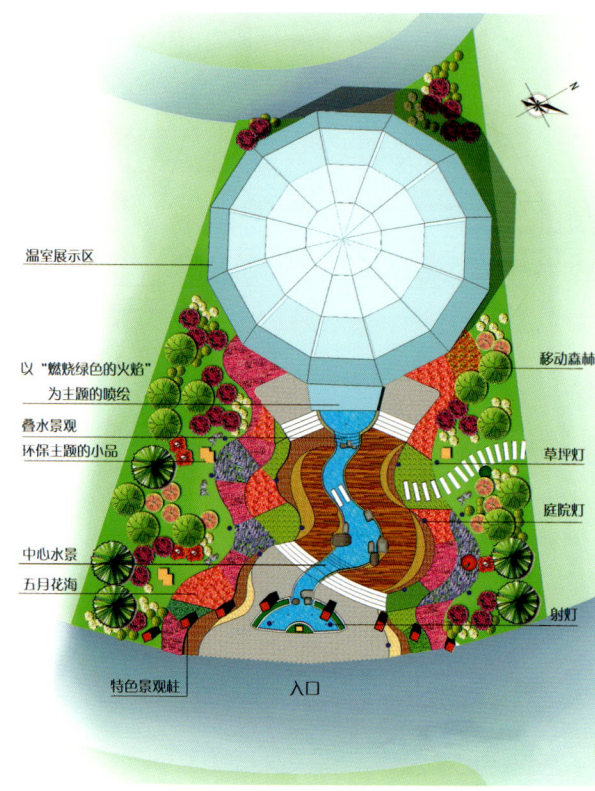

Enterprises Gardens
企业园

无痕的"建筑" Traceless "Building"

无痕的"建筑"是一个概念型的空间艺术作品,意图通过环境、建筑与人之间的本质关系在视觉过程中的相互换位,自然地产生出错综复杂的、微妙的情境环境意识。对当代人类与未来生存过程中的关系进行心理的、意识的、精神的相关研究。建筑主体以镜面不锈钢、玻璃为主,整体设计以水为创意元素,建筑形态与建筑材质形成统一协调的变化关系,并通过周边环境、建筑和人之间时间与空间概念的转化,完美地诠释出未来人类生存环境的和谐关系。总面积约782平方米。展馆主要分为两个围合区域,主体建筑之外有6条自然通道将整个建筑错落串联在一体。

The Garden occupies an area of 782 m² in the Enterprises Gardens. The "Traceless Building" is a conceptual space artwork that explores the fundamental relationship among environment, architecture, and people to inspire corresponding complicated and subtle ideas, and also explores the psychology, consciousness and vitality of human beings nowadays in the process of future survival. Mirror stainless steel and glass are main materials of the main body of the building. Water is a creative element of its overall design. The building materials match the building form perfectly. Combining with the change of surrounding environment, architecture, and people in terms of time and space, the building perfectly reveals the relation between man and our future living environment. The exhibition hall consists of two enclosed zones. There are six natural channels scattered outside the main building, which integrate the whole building.

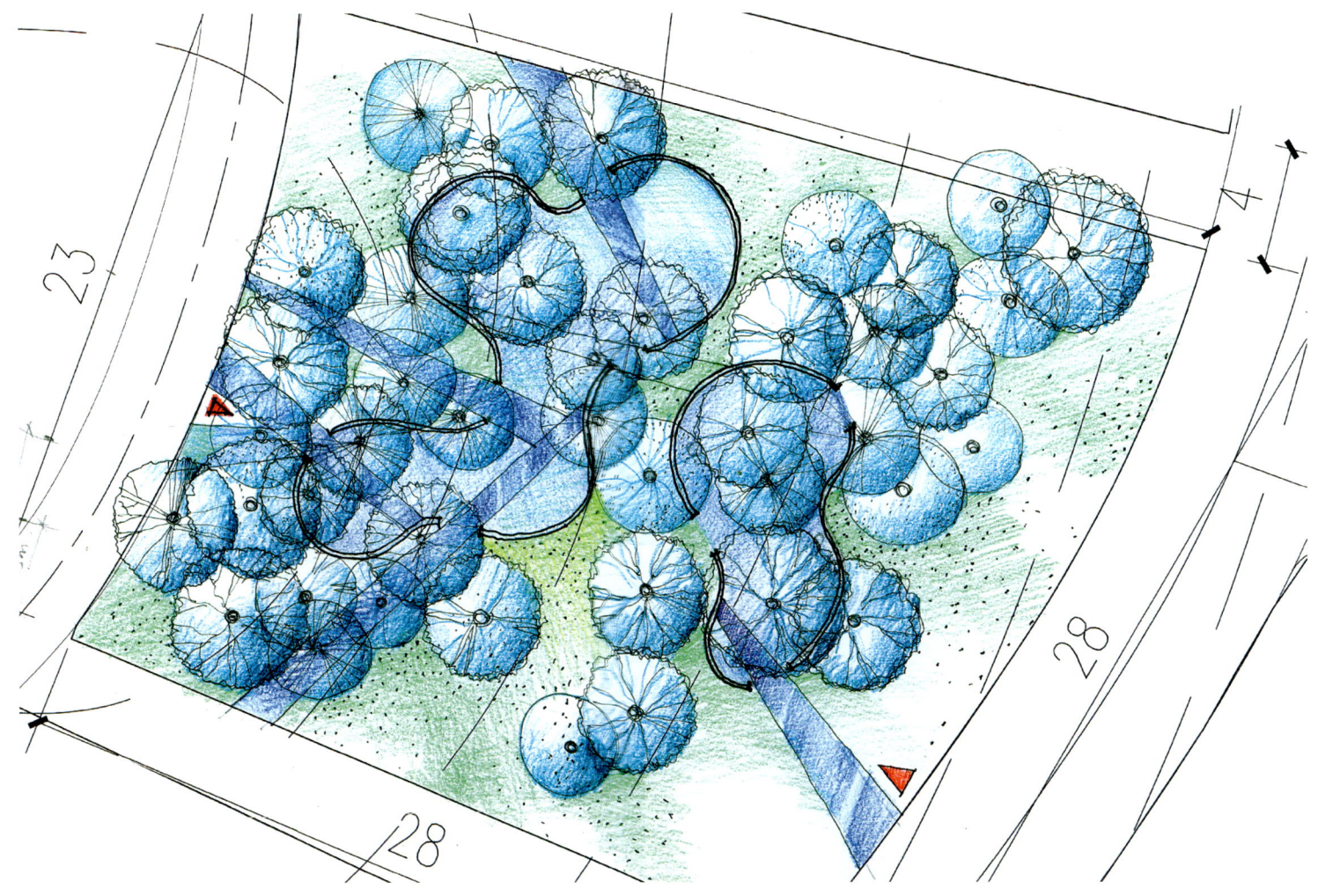

豪景创意园 Magnificent & Creative Garden

总面积约802平方米,豪景创意园以营造生态城市节能减碳为中心,保护地球自然生态为主题;重点展示"新统麟活性合成泥土"等高科技产品。该活性合成泥土可造性极强,可随意造型,组合成山水缩影等一系列小品,并可养殖植物。该展馆利用现代媒介方式使室内室外空间互为一体;在植物配置上利用多种植培方式,使人们有赏心悦目之感。

The Garden is built with an area of 802 m² in the Enterprises Gardens. With the theme of protecting the earth's natural environment, the garden focuses on creating energy-saving and low-carbon ecological cities and mainly displays "Xintonglin Active Synthesis Soil" among other high-tech products. The active synthetic soil has a strong plasticity and can be made into different shapes. It can be used to make a series of ornaments such as miniature mountain and water landscape and to breed plants. The Garden integrates the outdoor space with indoor space by using modern media; it adopts multiple plant cultivation ways in the plant configuration and brings pleasant scenery for people.

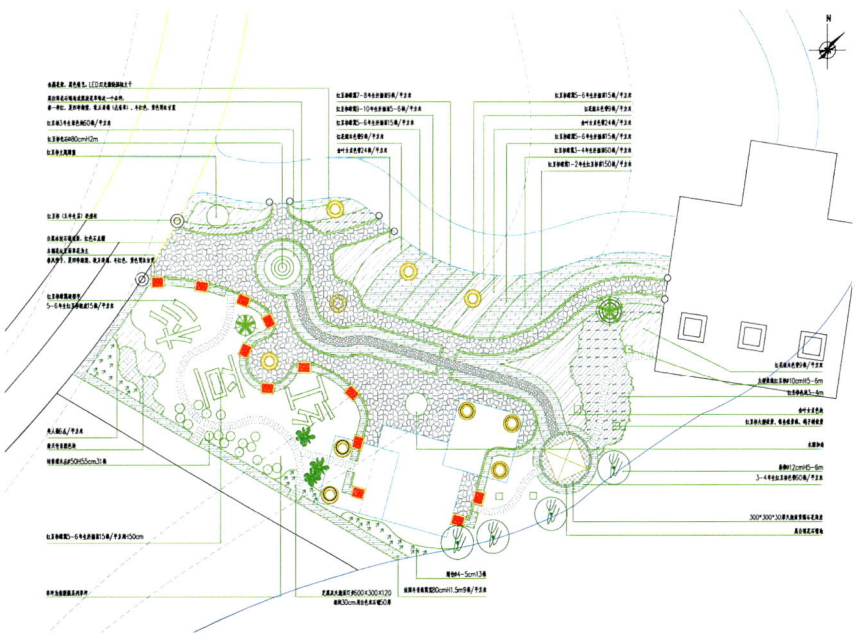

红豆杉主题园 Taxus Theme Garden

总面积约1,075平方米。园内展示区分为室内、外两部分，室外部分充分地将红豆杉以各种形式穿插其中。包括红豆杉造型绿塔、红豆杉灌木色带等。园区内建有玻璃温室，并设展厅详细介绍红豆杉的育苗、培育、生长的过程。另一方面介绍红豆杉的药用价值及其提取"紫杉醇"的工艺流程。通过两个方面全面深入地介绍红豆杉这一濒危珍贵植物。

Occupying an area of 1,075 m², the Taxus Theme Garden consists of outdoor and indoor exposition sections. The outdoor section displays various Chinese yews including shape-based Chinese yew green towers and Chinese yew shrub belts. The glass greenhouse section demonstrates Chinese yew, an endangered rare plant, in an all-around and in-depth manner, including the process of its seedling raising, cultivation and growth, its medicinal value, and "taxol" extraction process.

电信科技园 China Telecom Sci-tech Garden

总面积约1,816平方米。考虑到可持续利用,建筑主体选材以钢结构、玻璃幕墙为主,玻璃的通透象征了中国电信的开放和自由。屋顶采用了玻璃幕墙和张拉膜的造型,这个造型像一艘扬帆起航的大船,象征了中国电信不断超越的理念和追求。

The Garden covers an area of 1,816 m². With the consideration of sustainable utilization, the main building in the Garden adopts steel structures and glass curtain wall, of which the modern and transparent façade represents the openness and freedom of China Telecom so as to make visitors get much understanding of China Telecom. The tensile-membrane-pattern roof is made of glass curtain wall to symbolize China Telecom as a sail-setting vessel to make constant progress.

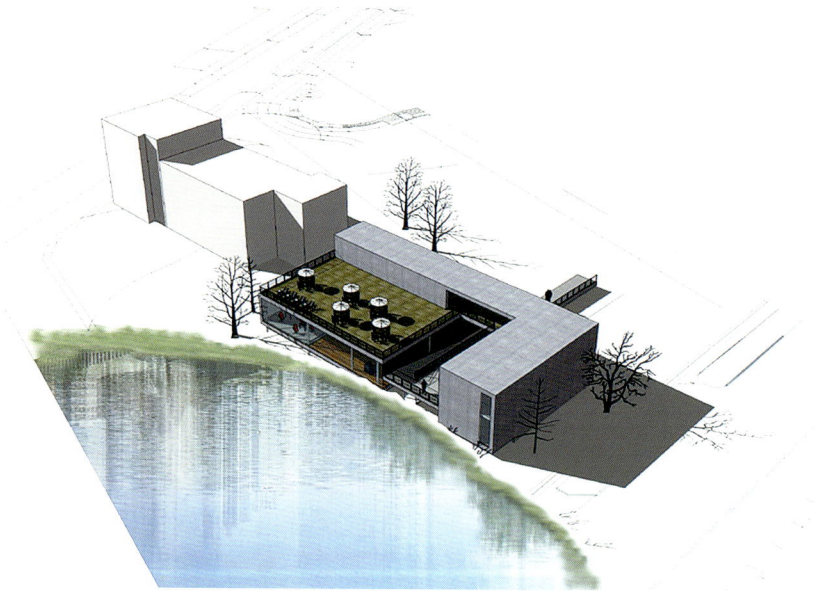

绿地生态科技馆
Greenland Ecological Sci-tech Garden

该馆具象地将建筑设计成一段像被人类砍伐而残留的树根，用超现代的设计表达手法契合世界园艺博览会的主题。总面积约1,034平方米，建筑强调创新与美感，形态似"0"，以体现建筑与环境最流畅的对话，通过展示未来可持续发展技术，传递"0"能耗、"0"排放、对环境"0"介入的理念。展馆内大量应用了声、光、电等多样的展示手法来展示生态建筑的自身优势，重点展示了零能耗的九大系统，四十项技术可以使建筑能耗接近零，并采用新型装饰材料体现低碳经济特性。

The Garden, built in the Enterprises Gardens of Expo 2011 Xi'an with an area of 1,034 m², is designed to be a residual root of a tree cut down by man. This ultra-modern design and presentation method fits the theme of the Expo. The 0-shaped architecture highlights innovation and aesthetic sense and embodies the harmony between architecture and environment. It conveys the concepts of zero energy consumption, zero emission, and zero environmental interference by demonstrating future sustainable development technology. It applies a large number of presentation means such as sound, light, and electricity to display its own advantages as an ecological architecture. It mainly displays nine zero energy-saving systems. Forty technologies adopted by the pavilion allow its energy consumption close to zero, and its new decorative material reflects the character of low carbon economy.

节能馆
Energy-saving & Environmental Protection Pavilion

全新的节能理念结合绿色园林、景观建筑特色，体现出人类追求绿色、环保、节能、时尚的新风尚。总面积约1,976平方米。建筑主体由最低层向上逐渐收拢，其剖面形式呈等腰梯形，采用中空玻璃，可调控百叶舢板，整个建筑每日所消耗的能量大部分来自于舢板所吸收的能量，是一种完全的自给自足的绿色循环模式。同时馆内利用保温、防火、防震材料以及地暖设施节能；采用LED照明充分节电及其它节能新材料、新技术。节能环保馆的设计是传统文化与现代技术的紧密结合，从而达到天人合一，城市与自然和谐共生。

The pavilion occupies an area of 1,976 m² in the Enterprises Gardens. The brand-new energy conservation concept is integrated with the distinct landscape architecture in the pavilion to show human's up-to-date pursuit of ecology and environmental protection. The pavilion's main building converges gradually from bottom to up with an isosceles-trapezoid section. It applies insulating glass and adjustable blinds plates. Most energy consumed daily by the pavilion comes from the energy absorbed by these plates. This is a completely self-sufficient green circular pattern. In addition, the pavilion adopts thermal insulation, fireproof, and shockproof materials as well as under-floor heating facilities to save energy and LED lighting to save electricity. It also introduces other new energy-saving materials and technologies. Combining the traditional culture with modern technologies, the pavilion's design shows the peaceful coexistence between city and nature.

天鹅湖园 Swan Lake Garden

展馆主题外形酷似一台DV摄像机，将整个天鹅湖的百鸟飞翔的动态美尽录其中。总面积约1,054平方米。建筑用钢结构框架，是集绿色环保、健康、舒适、设计风格独具个性等诸多优势于一身的健康型展馆，同时辅以合理的人流环线、亲水平台、休息平台等人性化设计。天鹅湖展园将完美展现人与自然和谐相处的自然建筑艺术理念。

Occupying an area of 1,054 m², Swan Lake Garden is located in the Enterprises Gardens. The Pavilion is shaped like a DV camera, which can fully record the dynamic beauty of birds flying above the lake. It adopts steel structure frame and boasts environmental protection, health, comfort, unique design style and other features. Moreover, it combines with such visitor-friendly design as the pedestrian flow ring road, the waterfront platform, and the rest platform. The tranquil lakescape looks like sapphire blue brocade, on which swimming swans sing loudly. The Garden perfectly displays the architectural concept to harmonize the relationship between man and nature.

Planning for Plants

植物种植规划

乔灌木种植现状

广运潭景区内原已栽植乔木5万多株，灌木6万平方米左右，品种以国槐、刺槐、杨树、油松、大叶女贞、柿树、板栗、丁香、红瑞木等乡土树种为主。
本次规划根据实际情况将变动部分的乔灌木进行移栽调整，约需要新增加乔木1万株。

乔灌木变更方案

园区内大部分乔木将保持现状，在主要特色园区、服务区等局部区域进行优化调整。主入口和长安园区区域的乔木因竖向地形的变化，需全部进行移植。园艺重点展示区域如：入口、广场、重点展园等重要景观节点则配合景观需要做重点配置设计，适当增加观赏树种及特色树种。
园区灌木原则上保持现状，路边和林下部分区域根据景观的需要进行梳理或移植，并结合乔木，在道路两边、密林和疏林下栽植草本花卉，形成疏密有致、浓淡相宜的植物空间格局。

水生植物种植规划

原水生植物设计面积为5.3万平方米左右，品种以日本芦苇Phragmites japonica、香蒲Typha minima Funk、花叶芦竹Arundo donax var versicolor、荷花Paeonia Suffruticosa、千屈菜Herba Lythri Salicariae等为主，由于部分岸线有所变动，所以本次规划根据实际情况，在原有水生植物的基础上进行补充和调整设计。

Existing (Trees & Shrubs) Plants

There are over 50,000 arbors and 60,000 m^2 of shrubs in the Guangyuntan Ecological Scenic Spot, which are mainly native tree species such as Chinese scholar tree, silver chain, poplar, Chinese pine, Ligustrum lucidum Ait, persimmon, Chinese chestnut, clove tree and Coruns alba L. etc.
The planning includes the transplant of some arbors and shrubs according to the actual condition, and another 10,000 arbors will be added.

Planting Design Incorporate with Existing Plants

Most arbors in the Expo Site will maintain the current situation, but those in major special zones and service zones will experience optimization. Due to the vertical changes in the terrain, arbors in the main entrance and Chang'an Garden all need to be transplanted. Key landscape nodes such as the entrances, plazas and core gardens will be designed in accordance with the entire landscape. More ornamental trees and featured trees will be planted appropriately.
In principle, shrubs will maintain the existing state. However, those on roadsides and under trees need to go through trim and transplant according to landscaping demand. The shrubs, combined with arbors and herbaceous flowers on roadsides or in dense groves and sparse woodlands, form a spatial framework of plants in appropriate density & color mix.

Planting Design for Aquatic Plants

The area for aquatic plants was designed to be about 53,000 m^2. Aquatic plants mainly include phragmites japonica, Typha minima Funk, Arundo donax var. versicolor, Paeonia Suffruticosa, Herba Lythri Salicariae etc. Due to the alteration of part of the waterfront, supplementation and adjustment will be made on the basis of existing aquatic plants according to the actual situation.

花卉组合及选用品种概览

• 景观花卉选用组合

根据园艺规划，园区分为重点园艺景观区和扩展园艺景观区。采用不同的设计风格，外环区域花卉组合采用自然、野趣的设计风格；内环区域以精致、华丽为设计风格。

(1) 外环（景观扩展区）花卉组合设计

林下花卉多采用耐荫品种组合，如大滨菊Chrysanthemum maximum、小滨菊Leucanthemella linearis (Matsum.) Tzvel.、钓钟柳Penstemon campanulatus、抱茎金光菊Rudbeckia amplexicaulis组合；花葱Cobaea scandens、剪秋罗Lychnis senno Sieb. et Zucc.、楼斗菜Aquilegia vulgaris组合；勿忘草Myosotis sylvatica、一串蓝Salvia farinacea Benth等等。

开阔地带结合草坪采用混播种植野花，拟选用品种有：紫松果菊Echinacea purpurea Moench.、虞美人Papaver rhoeas L.、紫花地丁Viola ycdoensis Mak.、蓝箭菊Catananche caerulea、宿根亚麻Linum perenne L.等。

道路两边花卉组合以颜色鲜艳、花期持久、抗旱抗热的野花品种为主，组成不同色块，如红色组合可选用虞美人Papaver rhoeas L.、大金鸡菊Coreopsis lanceolata、百日草Zinnia elegans、天人菊Gaillardia pulchella、翠菊Callistephus chinensis、串红Salvia splendens、射干Belamcanda chinensis (Linn.)DC.、鸡冠花Celosiae Cristatae、福寿花Adonis aestivalis L.；黄橙色组合可选用蛇目菊Coreopsis tinctoria、黑心菊Rudbeckia hirta、硫华菊Cosmos sulphureus、野菊花Dendranthema indicum (L) Des Moul. [Chrysanthemum indicum L.]、月见草Oenothera biennis L.、万寿菊Gerbera jamesonii Bolus、孔雀草Tagetes patula L.、赛菊芋Heliopsis scabra；粉白色组合可选用百日草Zinnia elegans、翠菊Callistephus chinensis、波斯菊Cosmos bipinnatus Cav.、满天星Cupheahyssopifolia、矢车菊Centaurea ruthenica Lam.、醉蝶花Cleome spinosa、常夏石竹Dianthus plumarius、蓝亚麻Linum perenne L.、麦仙翁A-grostemma githagol等。

(2) 重点园艺景观区花卉组合设计

重点景观区的花卉组合主要分布在道路两边、广场、建筑周边；造型灵活多变，花色搭配丰富多样，如郁金香Tulipa、水仙Narcissus tazeta var chinensis、葡萄风信子MuscaribotryoidesMill.、三色堇Viola tricolor L.组合；四季海棠Bedding begonia, Wax begonia、鸡冠花Celosiae Cristatae、彩叶草Coleus blumei Benth.、矮牵牛Petunia hybrida Vilm、一串红Salvia splendens Ker-Gawler组合；长春花Herba Catharanthi rosei、丰花百日草Zinnia hybrida、观赏辣椒capsicum frutescens、一串蓝Salvia farinacea Benth组合；孔雀草Tagetes patula L.、雏菊Bellis perennis Linn.、太阳花Portulaca grandiflora、夏堇Torenia fournieri组合；凤仙花Impatiens balsamina组合等等。

• 拟选用花卉品种

长春花Herba Catharanthi rosei、四季秋海棠Bedding begonia, Wax begonia、矮牵牛Petunia hybrida Vilm、虞美人Papaver rhoeas L.、白晶菊Chrysanthemum paludosum、黄晶菊Chrysanthemum multicaule、一串红Salvia splendens Ker-Gawler、孔雀草TagetespatulaL.、万寿菊Tagetes erecta L.、彩叶草coleus blumei benth、金鱼草Antirrhinum majus L.、蓝雏菊Bellis perennis、香雪球Lobularia maritima、何氏凤仙Impatiens holstii Engler et Warb.、银叶菊Centaurea Cineraria、石竹Dianthus plumarius、藿香蓟Ageratum conyzoides、南非万寿菊Tagetes erecta L.、太阳花Portulaca grandiflora、百日草Zinnia elegans、夏堇Hyacinthus orientalis L.、一串蓝Salvia farinacea Benth、一串白Salvia splendens var.alba、美兰菊Melampodium Lemon Delight、美女樱Verbenahybrida、麦秆菊Helichrysum bracteatum Andr.、鸡冠花Celosiae Cristatae、丰花百日草Zinnia hybrida、鸢尾花Iris tectorum、香彩雀Angelonia angustifloia、红花文藤Mandevilla sanderi、波斯菊Cosmos bipinnatus Cav.、观赏辣椒Capsicum frutescans、萱草Hemerocallis esculenta Koidz.、紫松果菊Echinaceaperperea、金光菊Rudbeckia laciniata、月见草Oenothera biennis、醉蝶花Cleome spinosa、花荵Polemonium coerulcum Linn.等等。

Planting Design

• Ground Cover

According to the planning, the Expo Site is divided into the core horticultural landscape area and extended horticultural landscape area. Different design styles are adopted for the two areas: the flower mixes in the extended area bear the style of naturalness and rustic charm, while in the core area there are mainly delicate and gorgeous flower mixes.

(1) Flower mixes in extended horticultural landscape area

Flower mixes under trees mostly adopt shade-tolerance varieties, such as the mix of Chrysanthemum maximum, Leucanthemella linearis (Matsum.) Tzvel., Penstemon campanulatus and Rudbeckia amplexicaulis, the mix of Cobaea scandens, Lychnis senno Sieb. et Zucc. and Aquilegia vulgaris, Myosotis sylvatica and Salvia farinacea Benth, etc.

In open areas, lawns will be mixed with wild flowers like Echinacea purpurea Moench., Papaver rhoeas L., Viola ycdoensis Mak., Catananche caerulea and Linum perenne L., etc.

The flower mixes on roadsides are mainly wild flowers with bright colors, long florescence, and drought and heat resistance, which constitute patches of different colors. The red mix can choose Papaver rhoeas L., Coreopsis lanceolata, Zinnia elegans, Gaillardia pulchella, Callistephus chinensis, Salvia splendens, Belamcanda chinensis (Linn.)DC., Celosiae Cristatae and Adonis aestivalis L.; the yellowish-orange mix can choose Coreopsis tinctoria, Rudbeckia hirta, Cosmos sulphureus, Dendranthema indicum (L) Des Moul. [Chrysanthemum indicum L.], Oenothera biennis L., Gerbera jamesonii Bolus, Tagetes patula L. and Heliopsis scabra; the pinkish-white mix can choose Zinnia elegans, Callistephus chinensis, Cosmos bipinnatus Cav., Cupheahyssopifolia, Centaurea ruthenica Lam., Cleome spinosa, Dianthus plumarius, Linum perenne L. and A-grostemma githagol, etc.

(2) Flower mixes in core horticultural landscape area

Flower mixes in the core horticultural landscape area are mainly distributed on roadsides, plazas and surrounding areas of buildings. They are arranged in flexible patterns and colorful mixture, such as the mix of Tilipa, Narcissus tazeta var chinensis, Muscaribotryoides Mill. and Viola tricolor L., the mix of Bedding begonia, Wax begonia, Celosiae Cristatae, Coleus blumei Benth, Petunia hybrida Vilm and Salvia splendens Ker-Gawler, the mix of Herba Catharanthi rosei, Zinnia hybrida, capsicum frutescens and Salvia farinacea Benth, the mix of Tagetes patula L., Bellis perennis Linn., Portulaca grandiflora and Torenia fournieri, the mix of Impatiens balsamina, etc.

• **Species**

Herba Catharanthi rosei, Bedding begonia, Wax begonia, Petunia hybrida Vilm, Papaver rhoeas L., Chrysanthemum paludosum, Chrysanthemum multicaule, Salvia splendens Ker-Gawler, TagetespatulaL., Tagetes erecta L., coleus blumei benth, Antirrhinum majus L., Bellis perennis, Lobularia maritime, Impatiens holstii Engler et Warb., Centaurea Cineraria, Dianthus plumarius, Ageratum conyzoides, Tagetes erecta L., Portulaca grandiflora, Zinnia elegans, Hyacinthus orientalis L., Salvia farinacea Benth, Salvia splendens var.alba, Melampodium Lemon Delight, Verbenahybrida, Helichrysum bracteatum Andr., Celosiae Cristatae, Zinnia hybrid, Iris tectorum, Angelonia angustifloia, Mandevilla sanderi, Cosmos bipinnatus Cav., Capsicum frutescans, Hemerocallis esculenta Koidz., Echinaceaperperea, Rudbeckia laciniata, Oenothera biennis, Cleome spinosa, and Polemonium coeruleum Linn., etc.

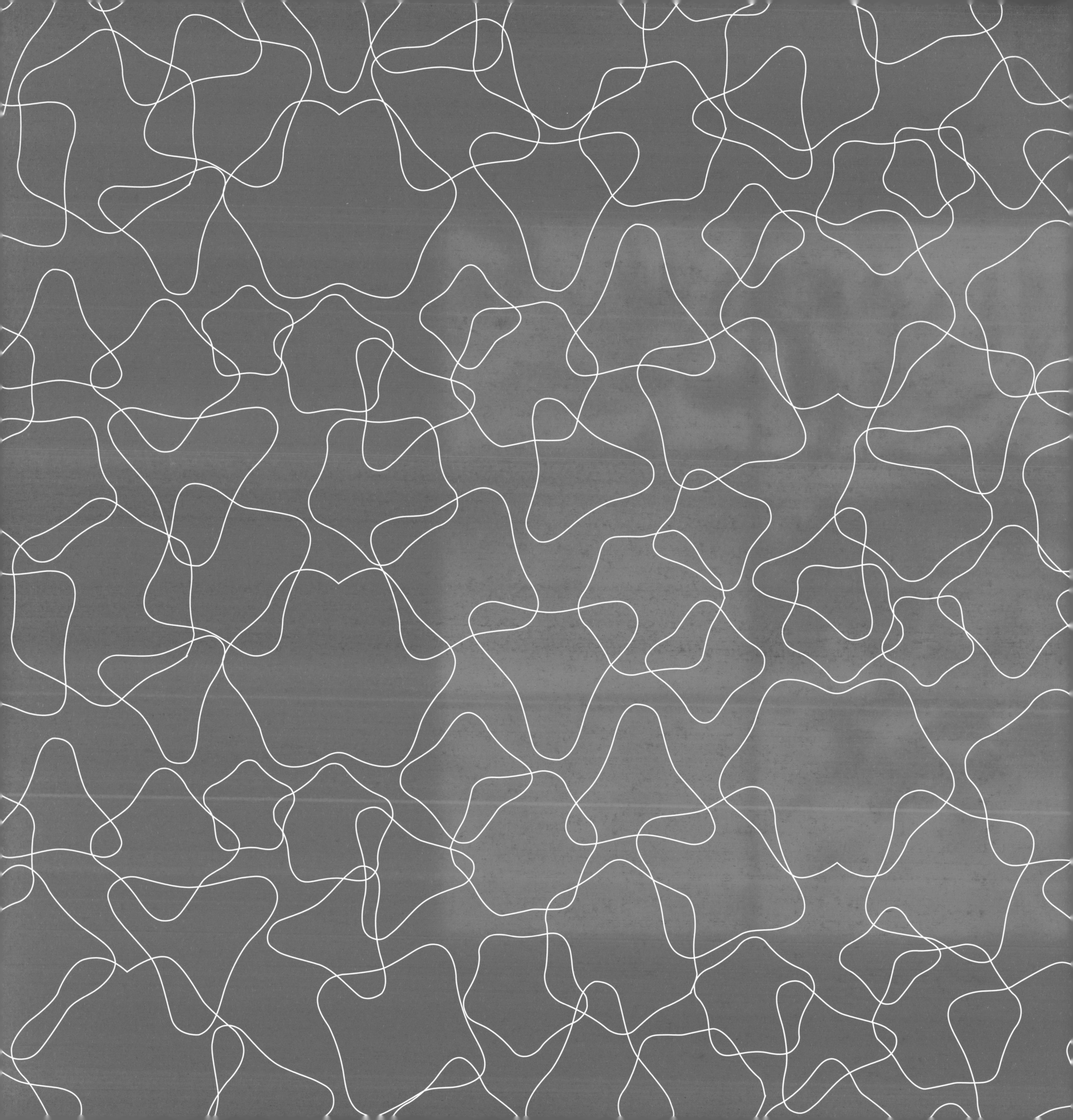

SUPPORTING
支撑体系

如果把西安世界园艺博览会看做是一席盛宴,多彩展园和魅力园区显然是其中必备的佳肴,而园区的一系列支撑体系可谓盛宴的幕后功臣。它们由服务设施、基础设施、安全保障、消防设施、控制指挥中心系统以及景观照明等共同构成一套紧密有序的背景体系。虽然不是主角,但同样充满着精心的构思和精巧的设计。本章主要介绍这些配套设施的设计理念和建造效果。

If the Expo is a feast, the colorful Expo Site and charming Expo Zones are apparently indispensable delicacies, whereas a series of supporting facilities in the Expo Site are contributors behind the scenes. The service facilities, infrastructure, safety insurance and fire-control facilities, control center system and landscape lighting system constitute a tightly organized background system. The supporting facilities do not take the leading role, but thoughtful ideas and elaborate design are exercised. This chapter mainly focuses on the design concept and construction effects of the supporting facilities.

Traffic Planning
交通规划

园区外部交通组织

在到达世园会园区的各种车辆中，由于出租车使用灵活，且不受停车场地因素的限制，很难对其路径进行控制或引导，故对出租车流线不作过多考虑。而后勤车辆与VIP车辆总量比较少，且有专门的出入口进出园区，其路径选择有一定局限性，在交通组织时可以对其进行控制；公交专线车、旅游巴士及常规公交车，都是有固定线路的，因此其路径控制也很容易实现；对于社会车辆，本方案则主要通过停车场容量及其布设的位置来实现对其到达及离开世园会园区路径的控制引导。

园区内部交通组织

园区内部人流分析

根据《2011年西安世园会游客市场基础预测》，西安世园会游客量保守估计为960多万人次；加上诸多有力、有效的营销推广行动，以及重大节日的吸引力和票务引导方案，得出1,200万人次，作为适量数据，并作为交通规划、基础设施配套等一系列问题的基础数据；将1,300万人次作为目标数据。

根据相关数据，世园会期间，游客规模按照平时7万人次/天、一般高峰日9-10万人/天、极端高峰日12万人考虑。西安世园会每天的开园时间为9:00-22:00，而在一天中，游客的入园时间又具有很强的不平均性。9:00-12:00入园游客占50%，12:00-14:00入园游客占40%，14:00-16:00占10%。

园区人流在空间分布上也将呈现出不平均性。主入口广场在入场高峰时段人流最为密集；长安花谷预计在9:30-12:30出现人流高峰期；内地园区的高峰期预计出现在10:30-13:30；椰风水岸及创意园的高峰期预计出现在12:00-14:30；国际园、欧陆风情的高峰期预计出现在15:00-16:00，但人流已大幅减少。

The Traffic Organization outside the Expo Site

Among the vehicles arriving at the Expo Site, taxis may be the most difficult to control, for they are elastic to operate, and are not restricted by the parking lot. The traffic controllers will not consider much about taxi routes. Special entrances are offered for vehicles of rear-service and VIPs, and the amount is not large, which are easy to control with the limitation of driving routes. For special lines, tour buses and normal buses, certain routes are designated. They are much easier to control. For social cars, the controllers may direct their entering and exiting according to the capacity of parking lot and parking positions.

The Traffic Organization inside the Expo Site

The Analysis of the Stream of Tourists in Expo Site

According to *The Basic Market Prediction of Tourists in International Horticultural Exposition 2011 Xi'an China*, it is conservatively estimated that the number of tourists of this Expo will surpass 9.6 million. But various kinds of effective sales marketing methods, gala days and tickets directing plans may do a lot in revising the data. That is 12 million, which is a basic data for reference for traffic planning and facility arrangement. 13 million will finally be the target number.

According to relevant data, the amount of tourists is supposed to be 70,000 per day at ordinary times during Expo 2011 Xi'an, 90,000 to 100,000 at peak times, and 120,000 will be reached during peak holidays. The opening hours are from 9:00 to 22:00. While the number of tourists at different time differs greatly with 50% of the visitors arriving between 9:00 and 12:00; 40% between 12:00 and 14:00 and the other 10% between 14:00 and 16:00.

At the same time, the attribution of tourists is also out of balance. The number of tourists will reach its high point during the peak hours at the main entrance square. The peak hours of the Chang'an Flowery Valley is expected to be from 9:30 to 12:30; Domestic Gardens from 10:30 to 13:30, while the peak hours of Southeast Asian Street and Creativity Gardens are predicted between 12:00 and 14:30. The International Gardens and the European Avenue may meet their summit during 15:00 and 16:00. However, the number of tourists in these areas has begun to decline.

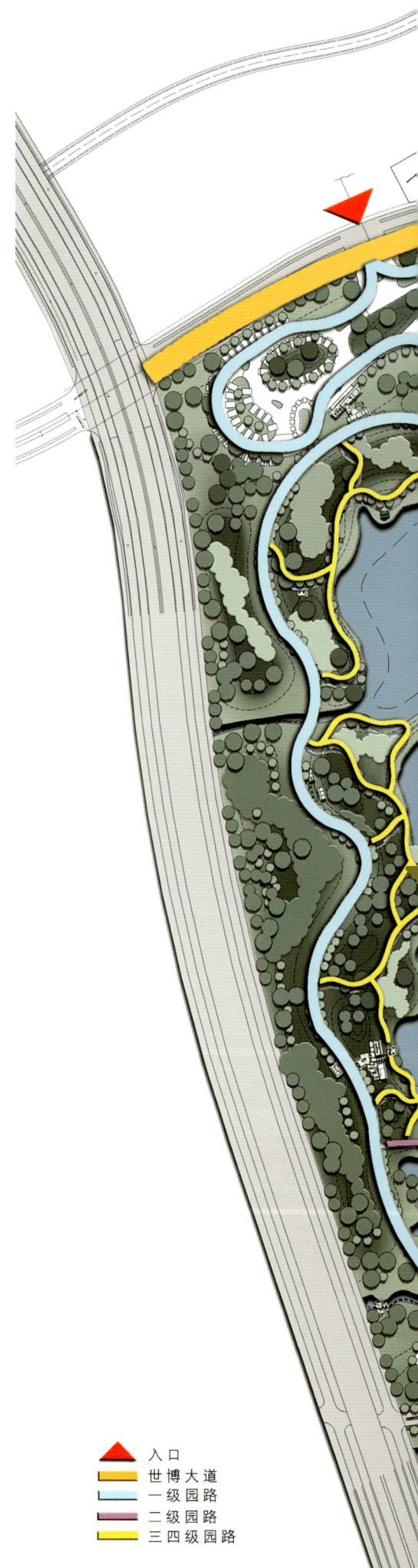

入口
世博大道
一级园路
二级园路
三四级园路

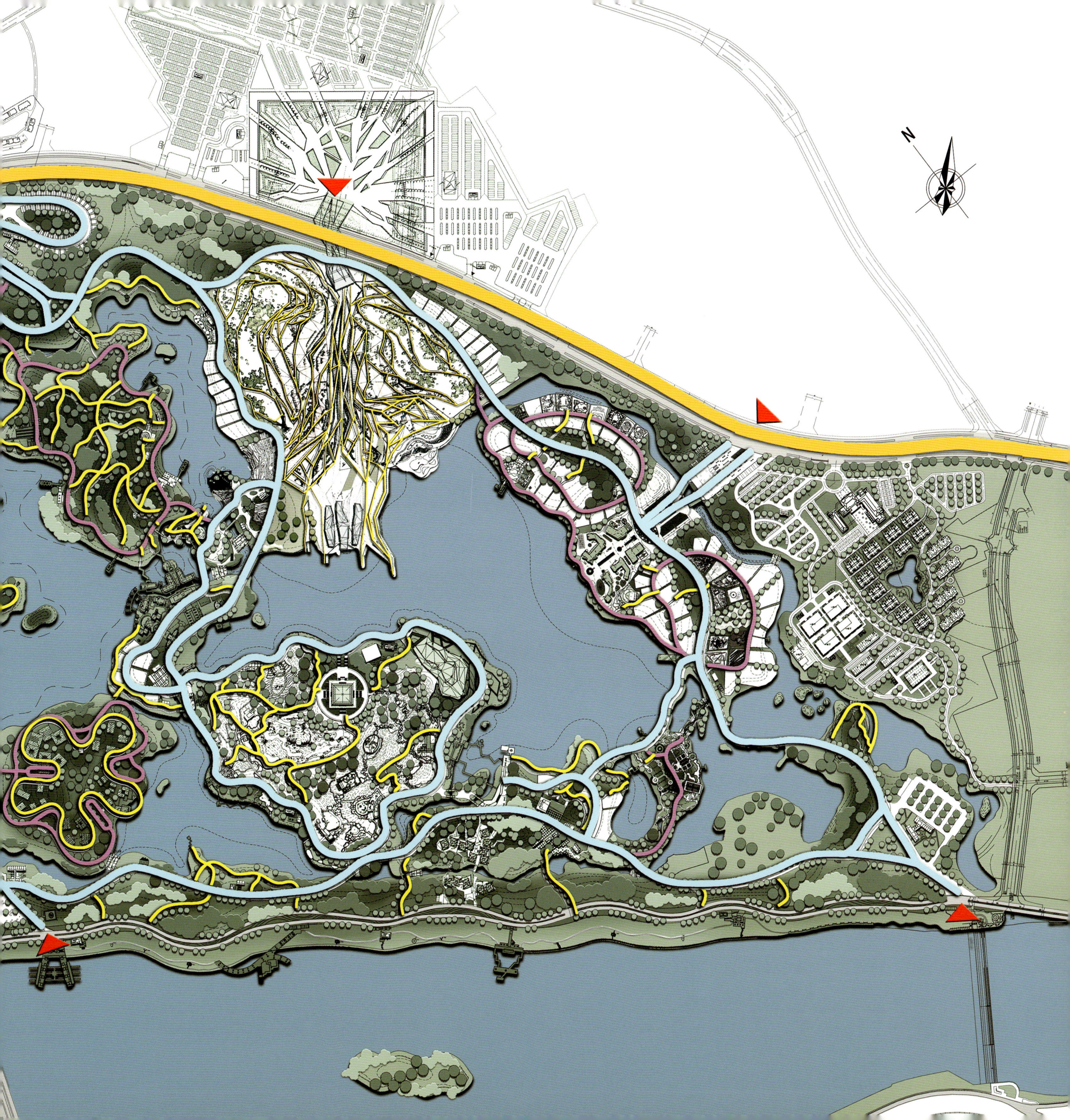

园路

2011西安世界园艺博览会园路总体分为一二级园路和三四级园路。一二级园路分别为7米、5米与4米园路，以沥青路面为主；三四级园路分别为3.5米、2米、0.9米园路，主要以石材为主。整体园路主要以曲线形式贯穿于园区内，其中：7米园路共 11,100.6 米长；5米园路共 3,717 米长；4米园路共 615 米长；2米园路共 6,330.2 米长；0.9米园路共 369.5 米长。整个园区的道路总长约28公里，大约是西安古城的周长。

Roads

The roads of the Expo Site are generally divided into main roads and secondary roads. The main roads are basically 7 meters, 5 meters and 4 meters wide, with asphalt paved on the ground; while the secondary roads are separately 3.5 meters, 2 meters and 0.9 meters wide, with stones as the main materials. The whole design of roads through the Expo Site is mainly shaped curved lines: the 7-meter-wide roads are totally 11,100.6 meters long; the 5-meter-wide roads, 3,717 meters long; the 4-meter-wide roads, 615 meters long; the 2- meter-wide roads, 6,330.2 meters long; the 0.9-meter-wide roads, 369.5 meters long. The total length of the entire roads is about 28 kilometers, approximately equal to the circumference of the ancient Xi'an city.

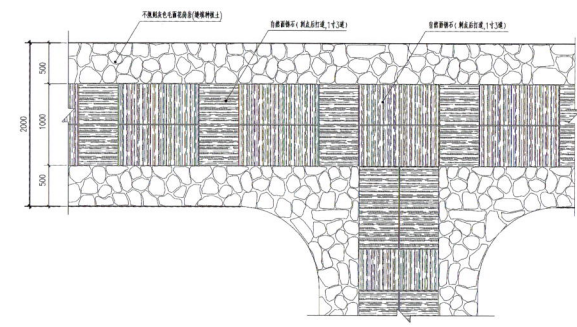

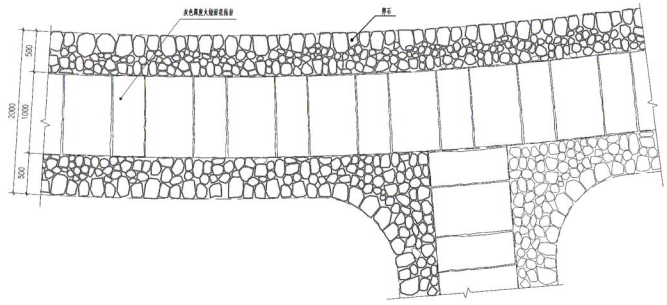

园桥

本届世界园艺博览选址富含水系，桥梁由此成为移步换景的重要工具。整个园区共建设园桥42座，以尺寸规格分为12种类型，风格跨越了中西各个时代，可谓锦上添花的"桥"艺博览。园区中的桥或宏伟壮丽，或轻盈质朴，或平桥贴水，游客凌波信步，将水与岸的景致尽览无遗。园林中的桥，联系风景点的水陆交通，组织游览线路，变换观赏视线，点缀水景，增加水面层次，兼有交通和艺术欣赏的双重作用。

各款园桥之中，以欧式桥和中式七孔桥最具备代表性。欧式桥长30米，宽8.1米；桥面以花岗岩铺装构成为主，别致的罗马洞石装饰柱坐落于桥面两侧，每根装饰柱上装有花钵，使整个桥体绿色、生动；铁艺栏杆与之连接。桥身两侧为罗马洞石装饰面层，面部配有极具欧式风情的水舌造型装饰。中式七孔桥，长60米，宽8米，位于长安花谷与椰风水岸相接处，园区的中心区域。桥面铺装以沥青为主，栏杆为石材，端头设计抱鼓花岗岩，桥身两侧配有水舌造型装饰，使其别具一格。

Bridges

Due to the vast water area of the Expo Site, bridges are surely the important tool for tourists to go through within the Expo Site. Divided into 12 types according to measurements, the bridges in the Site, totally 42, show people with modern styles in either China or western countries. It can be simply called a Bridge Expo, which increases the delight of the originally perfect Expo. Some are grand, some graceful, and some are quite close to water. Tourists may feel as if walking on the water when drinking themselves into the scenery all around. The bridges, embellish water, and link up water and land together. However, the water seems much better arranged against bridges. Whenever you pass by a bridge, you will see different pictures. All in all, the bridges not only function as transportation, but also stand as art.

The European bridge and Chinese seven-arch bridge stand out among various bridges in the garden. The former is 30 meters long and 8.1 meters wide. The deck is mainly made of granite. Pillars decorated with unique Roman travertines stand on each side. Each pillar has flowerpots on it, making the bridge lively. Iron railings connect with pillars. The surfaces on both sides, embellished with Roman travertines, are decorated with the appearance of European nappe. Chinese seven-arch bridge is 60 meters long and 8 meters wide. Located at the juncture of Chang'an Flower Valley and Southeast Asian Street, it is in the center of the Expo Site. The pavement is asphalt, and the railings are built of stones. The bridge looks much more unique with drum stones granite at either end and nappes on either side.

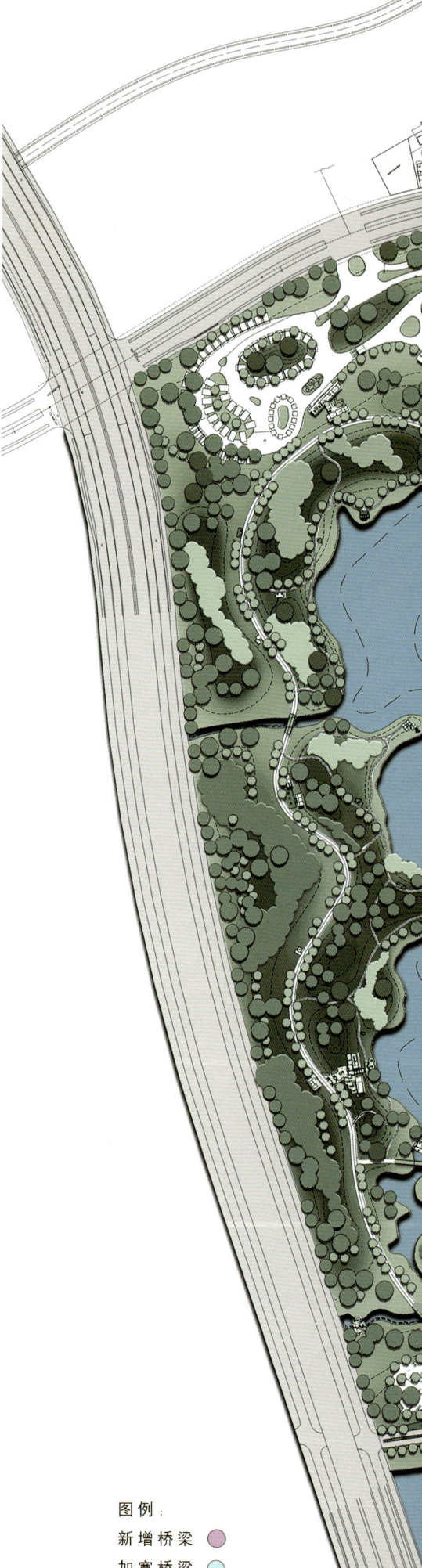

图例：
新增桥梁 ●
加宽桥梁 ●

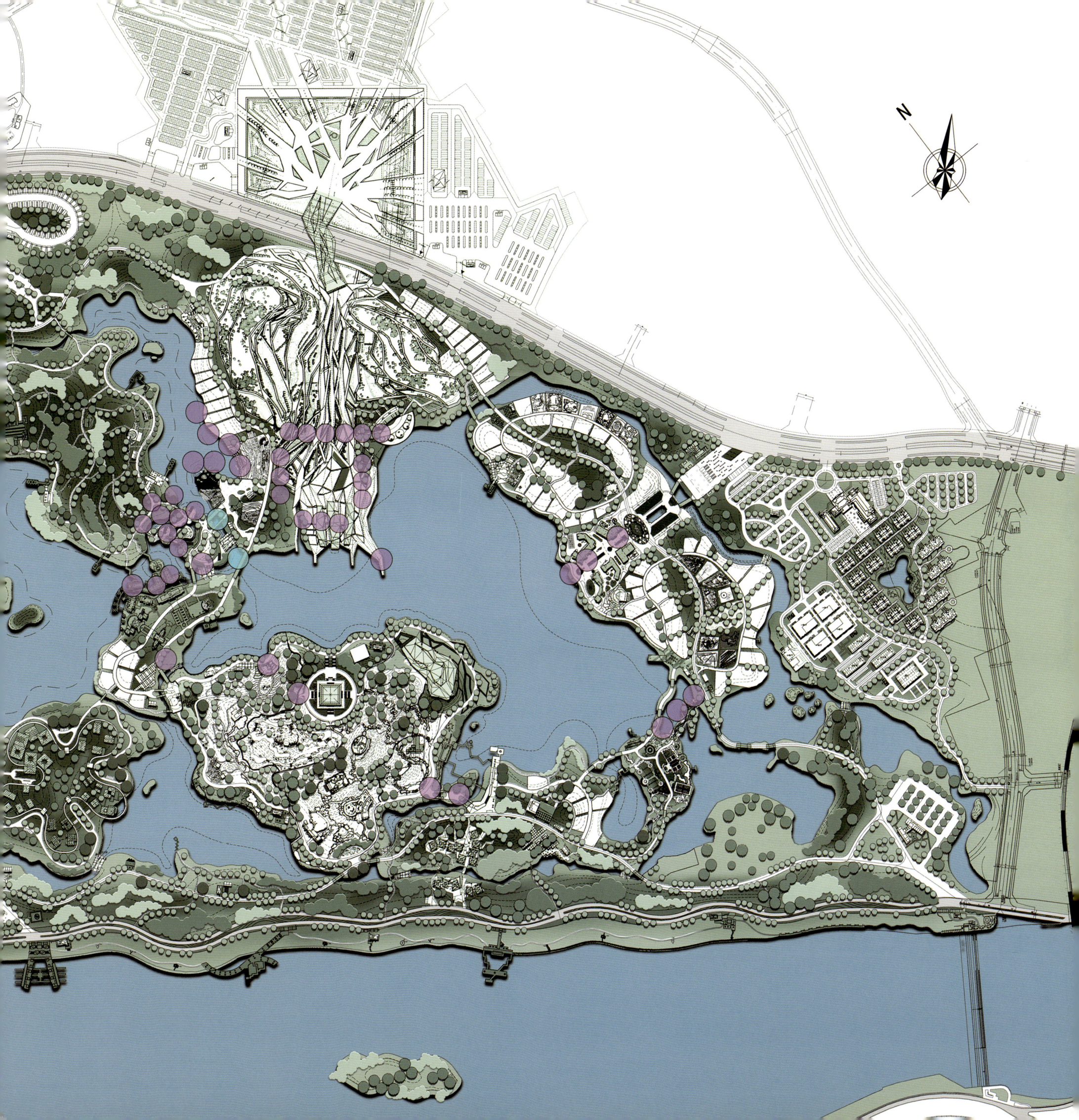

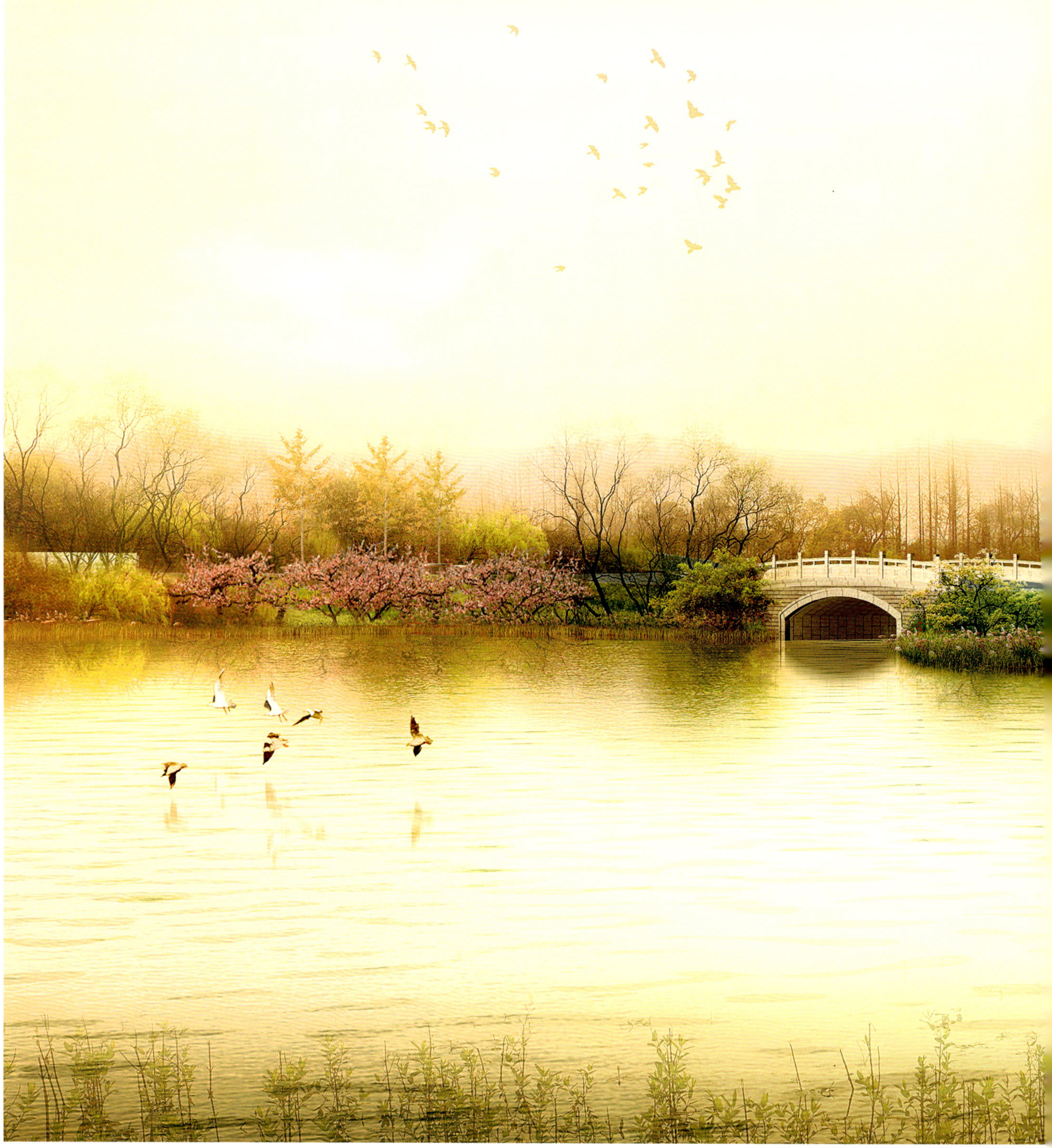

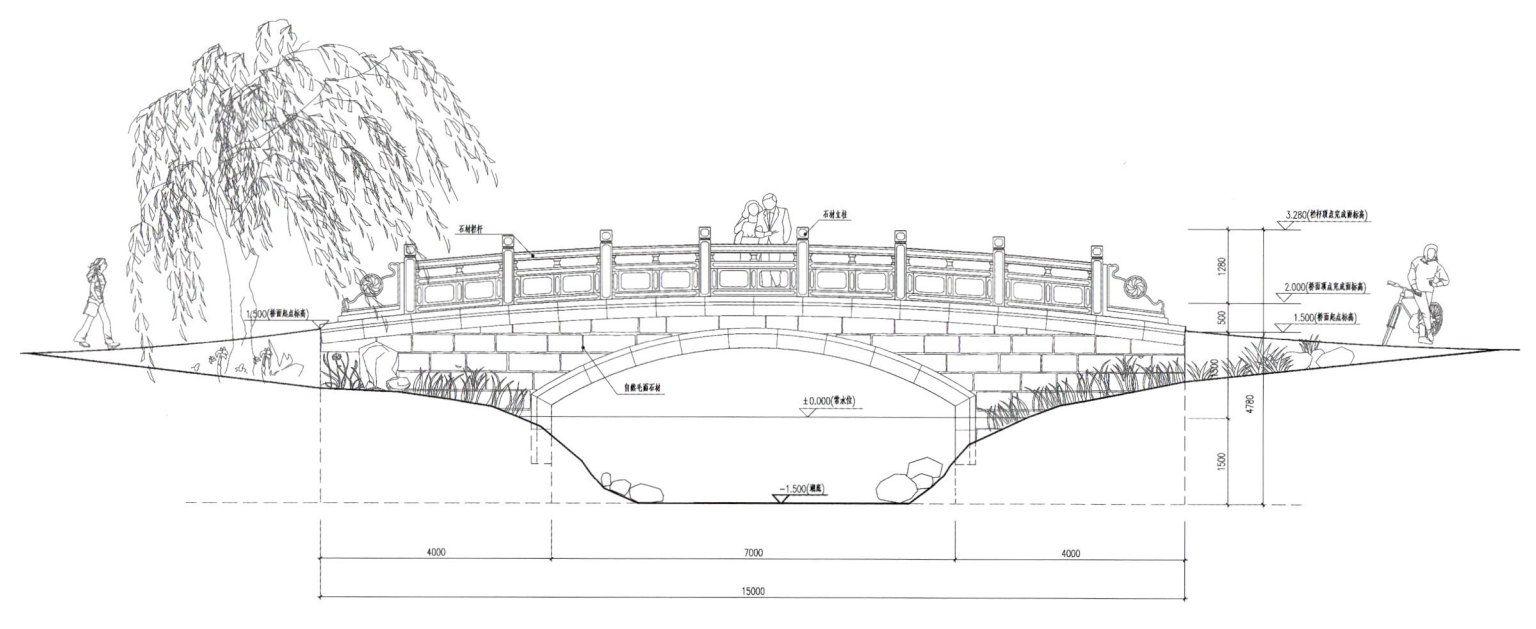

单孔桥

单孔桥，长15米，宽8.2米；桥面铺装以沥青为主，栏杆为石材，端头设计抱鼓石装饰，桥身两侧为自然毛面石材装饰面层，整体精巧、自然。

The Single-arch Bridge

The 15-metre-long and 8.2-metre-wide Single-arch Bridge is mainly paved by asphalt. The railings of the bridge are built of stones. The top part of the main body is decorated with drum stones, and the two sides of it are covered by natural rough stones. The whole bridge is exquisite and natural.

三孔桥

三孔桥，长30米，宽7.9米；桥面铺装以沥青为主，栏杆为石材，桥身两侧为石材装饰面层，整体造型简洁、明快。

The 3-arch Bridge

The 30-metre-long and 7.9-metre-wide 3-arch Bridge is mainly paved by asphalt, either side of which is constructed by stones to represent a simple, lucid and lively look.

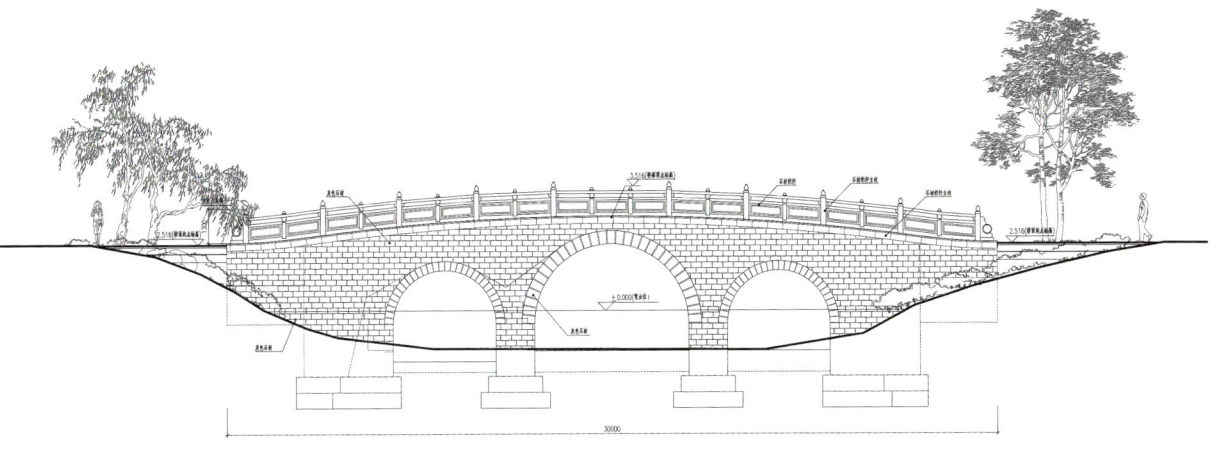

七孔桥

七孔桥,长60米,宽8米,位于长安花谷与东南亚风情区相接处,处于园区的中心区域。桥面铺装以沥青为主,栏杆为石材,端头设计抱鼓花岗岩,桥身两侧配有水舌造型装饰,使其别具一格。

The 7-arch Bridge

The 60-metre-long and 8-metre-wide 7-arch Bridge is located at the joint of the Chang'an Flower Valley and Southeast Asian Street, where is the centre of the Expo Site. The bridge is paved by asphalt, and the railings are built of stones. The top part of the bridge is decorated with drum granite, and either side of it is set nappe-shaped decorations to make the bridge peculiar.

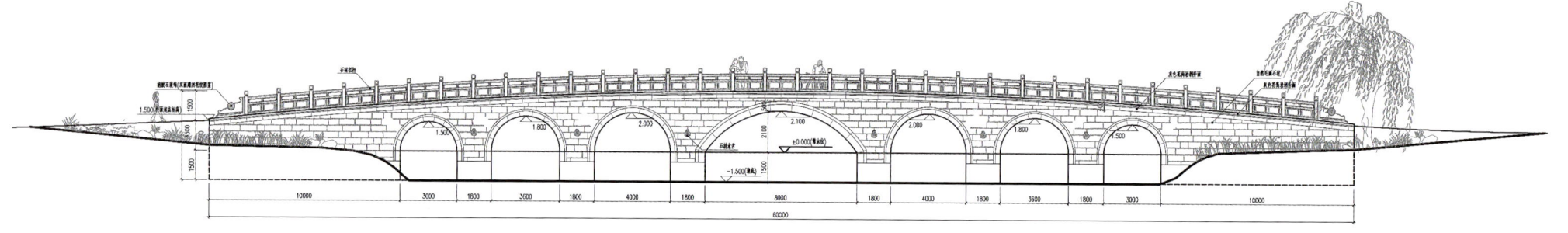

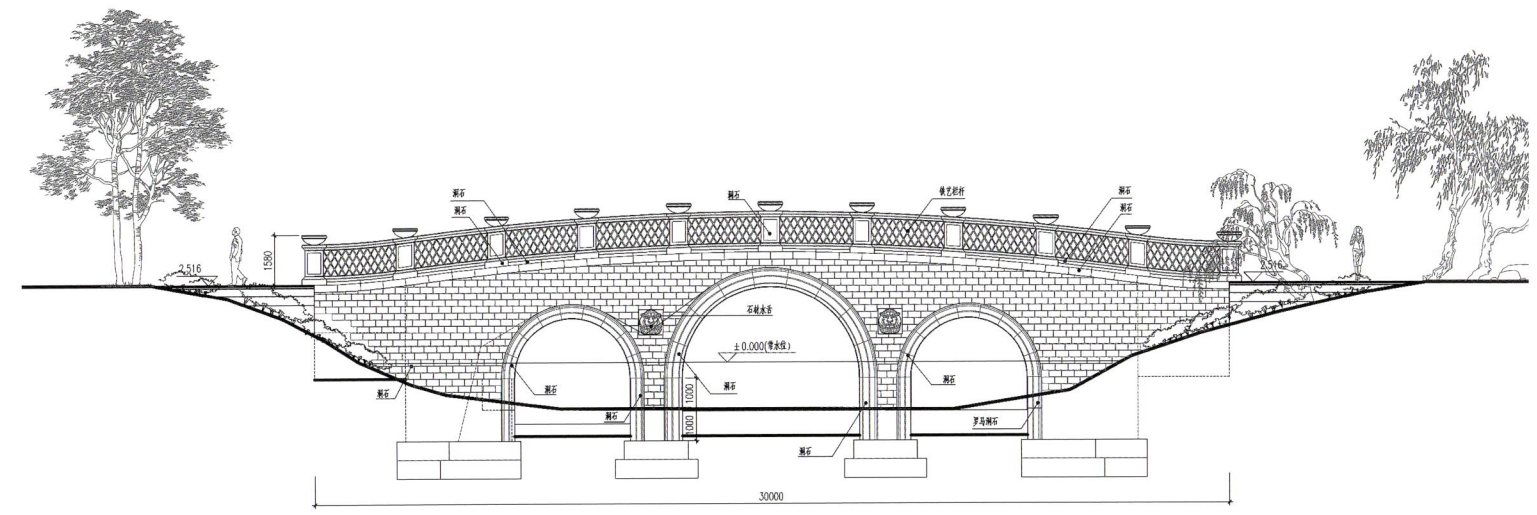

欧式桥

欧式桥,长30米,宽8.1米。桥面以花岗岩铺装构成为主,别致的罗马洞石装饰柱坐落于桥面两侧,每根装饰柱上装有花钵,使整个桥体绿色,生动;铁艺栏杆于之连接。桥身两侧为罗马洞石装饰面层,面部配有极具欧式风情的水舌造型装饰。

European-style Bridge

The 30-metre-long and 8.1-metre-wide European-style Bridge is mainly paved by granite. The decorative Roman travertine pillars are relatively set on either side of the bridge which is also paved by travertine and decorated with European-style nappe-sculptures. Each pillar is decorated with the flowerpots connected by iron railings to echo with the ecological and green concept.

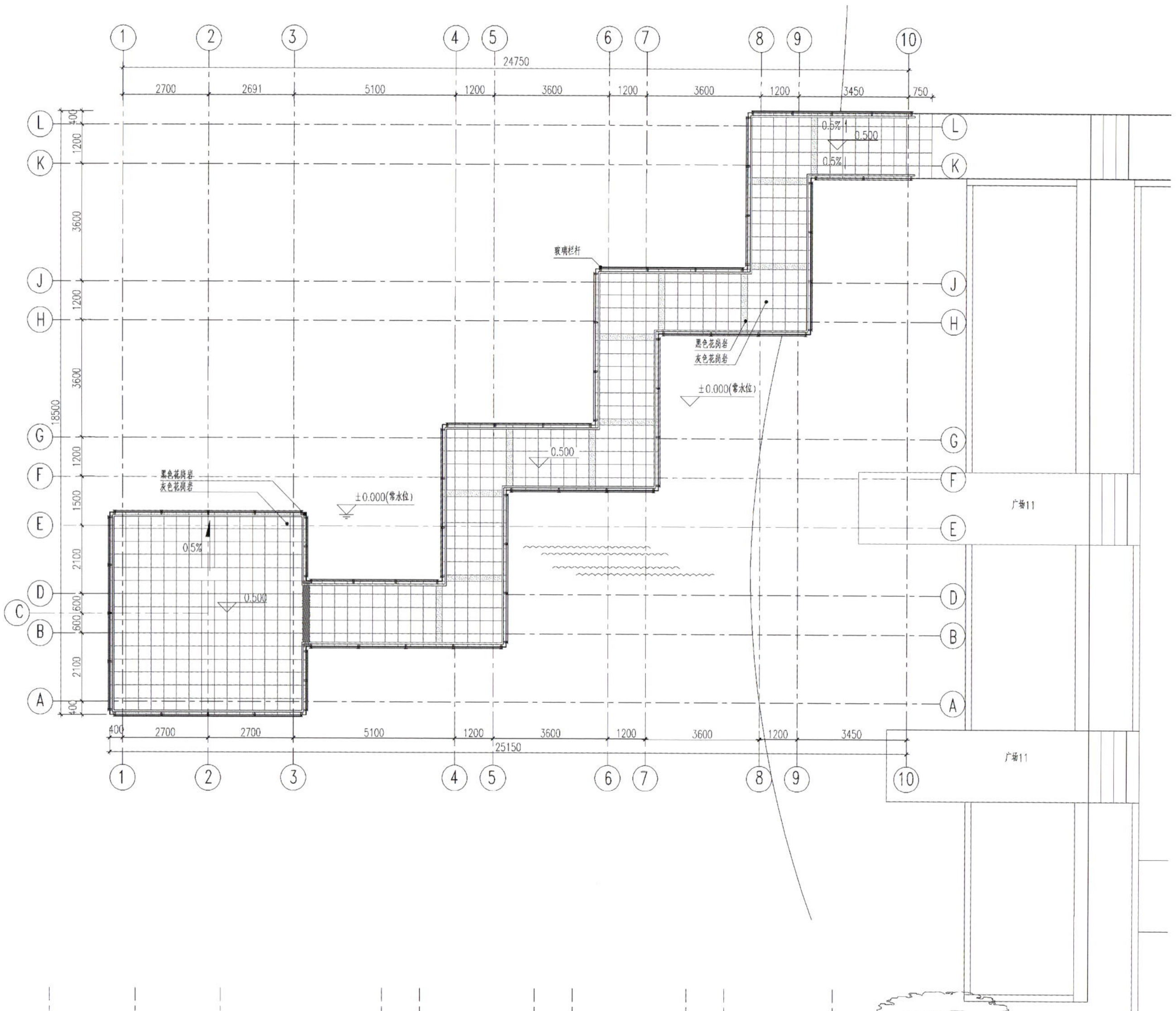
24750
25150
0.500
0.5%
±0.000(常水位)
玻璃栏杆

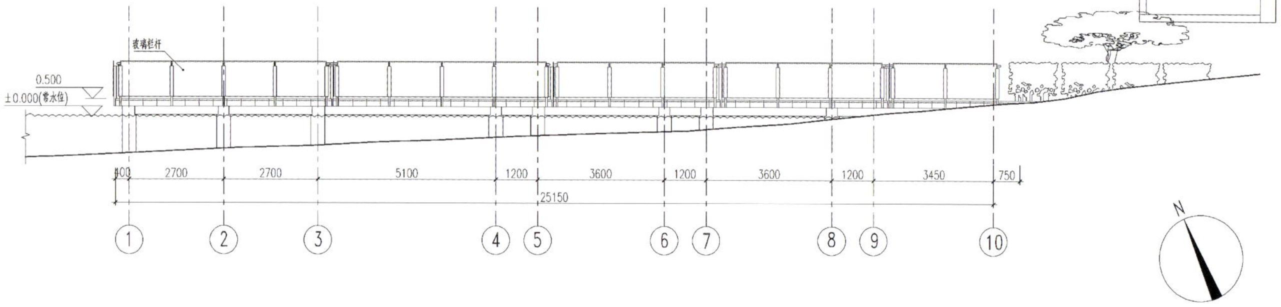
玻璃栏杆
0.500
±0.000(常水位)
25150

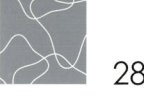

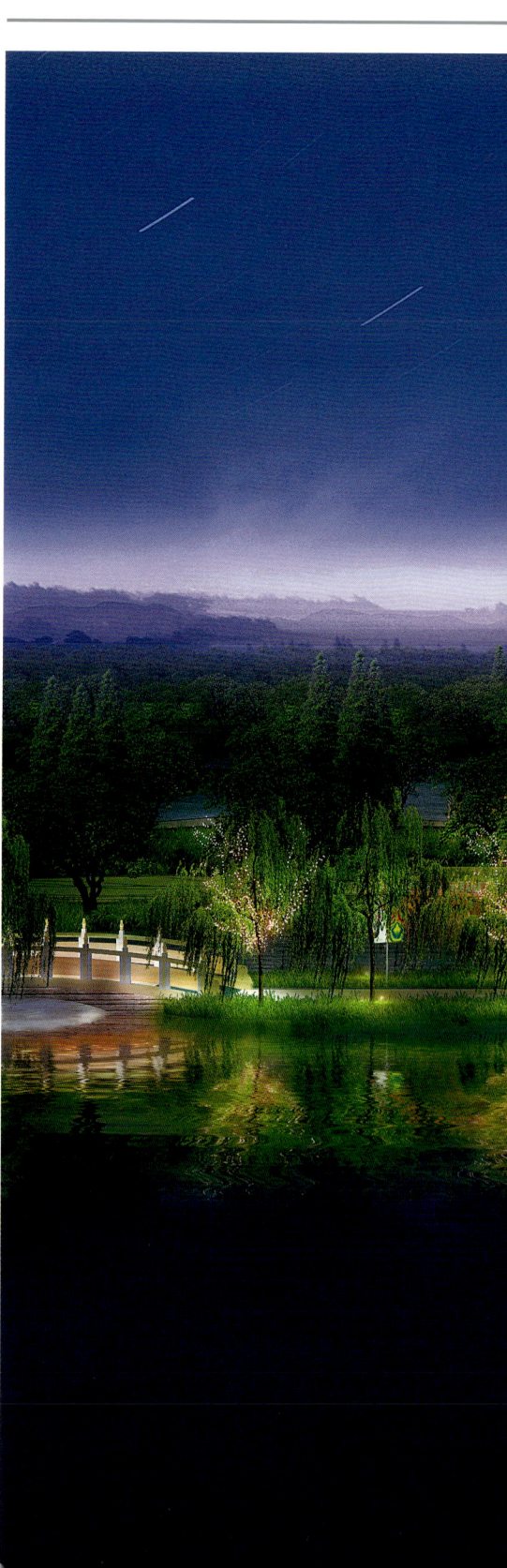

亲水平台、码头

2011世界园艺博览会园区内亲水平台和码头共计12个，其中亲水平台为8个总面积约为7,346.5平方米，码头为4个，总面积约为1,231平方米。亲水平台和码头造型简洁大方，为世园游客提供了戏水，赏水，驻足远眺的开敞空间，满足人们休憩、观景及通行的需求。

Waterfront Decks and Docks

Waterfront Decks and Docks are built in the Expo Site, with a total number of 12. Among them, waterfront decks take 8, covering an area of 7,346.5 m²; docks take 4, covering an area of 1,231 m². Both of them appear simple and modern, meeting the needs of lounging, sightseeing and passing through. People may enjoy their trips with paddling, sightseeing and overlooking with enough open space.

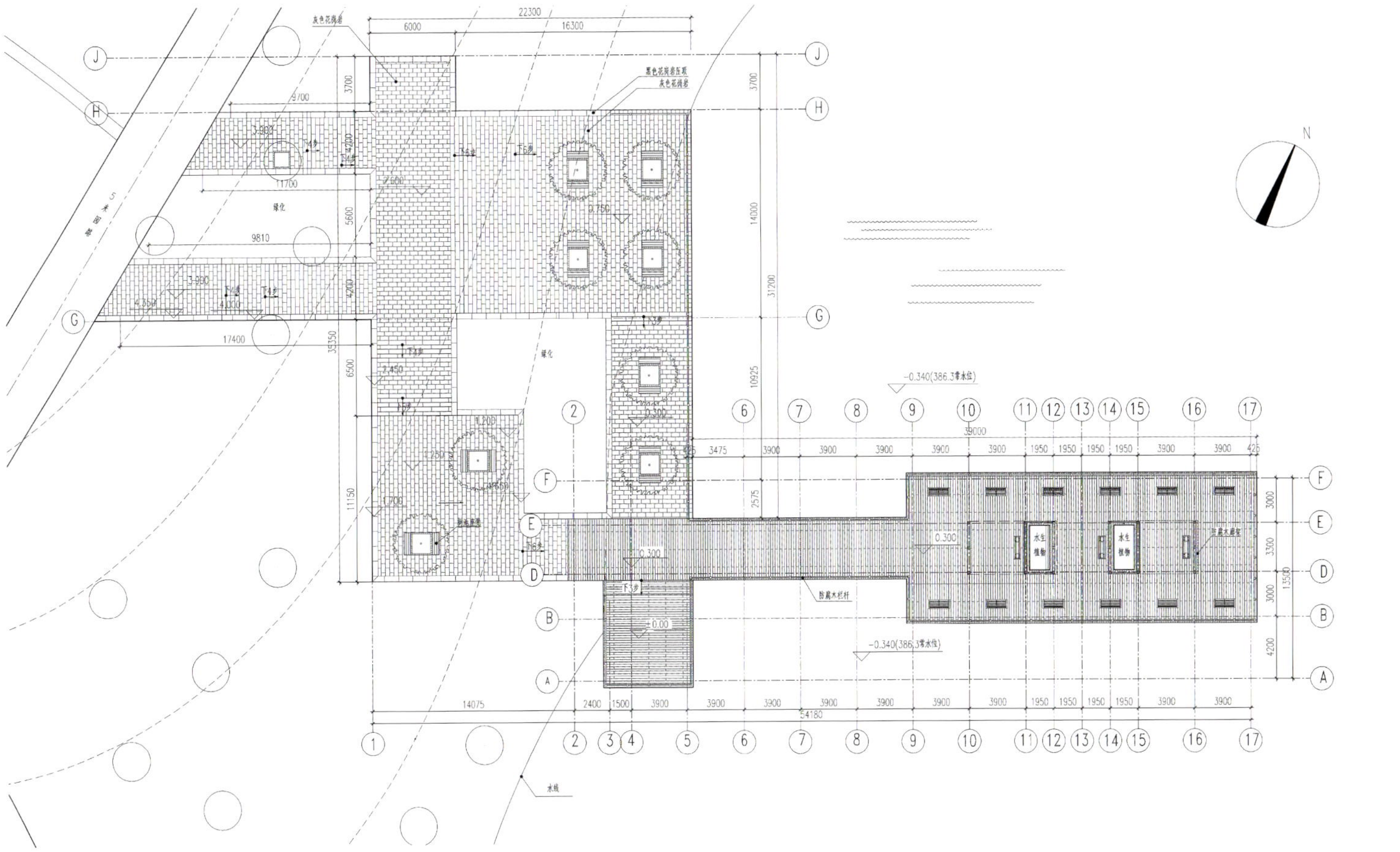

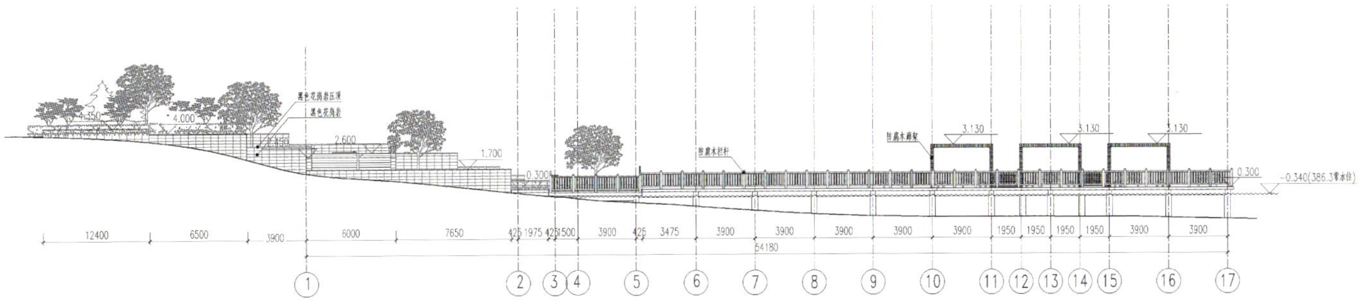

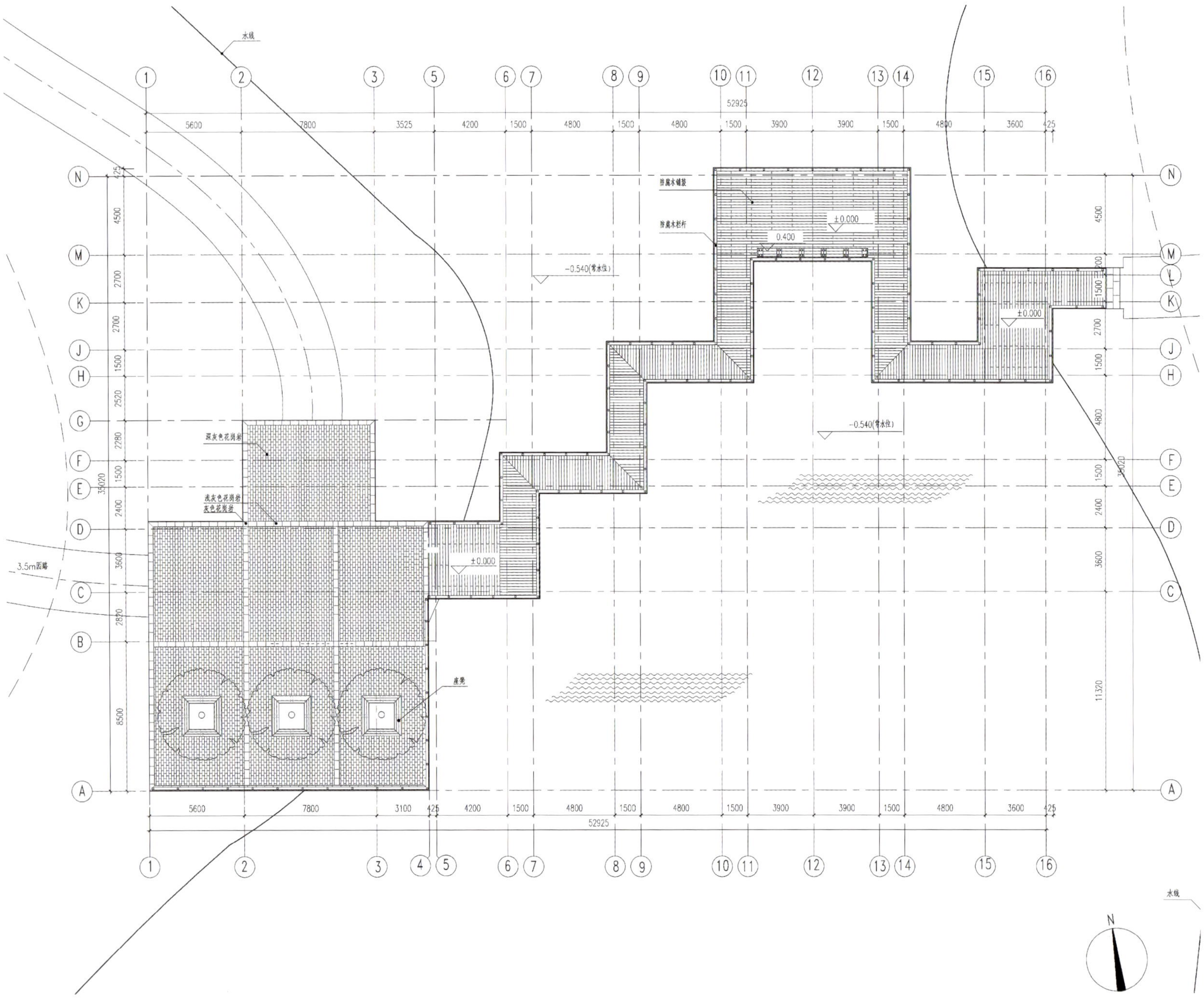
水线
52925
5600
7800
3525
4200
1500
4800
1500
4800
1500
3900
3900
1500
4800
3600
425
防腐木铺装
防腐木栏杆
±0.000
0.400
-0.540(常水位)
-0.540(常水位)
深灰色花岗岩
浅灰色花岗岩
灰色花岗岩
±0.000
3.5m园路
座凳
35020
4500
2700
2700
1500
2520
2280
1500
2400
3600
2820
8500
11320
5600
7800
3100
425
4200
1500
4800
1500
4800
1500
3900
3900
1500
4800
3600
425
52925
水线
N

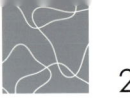

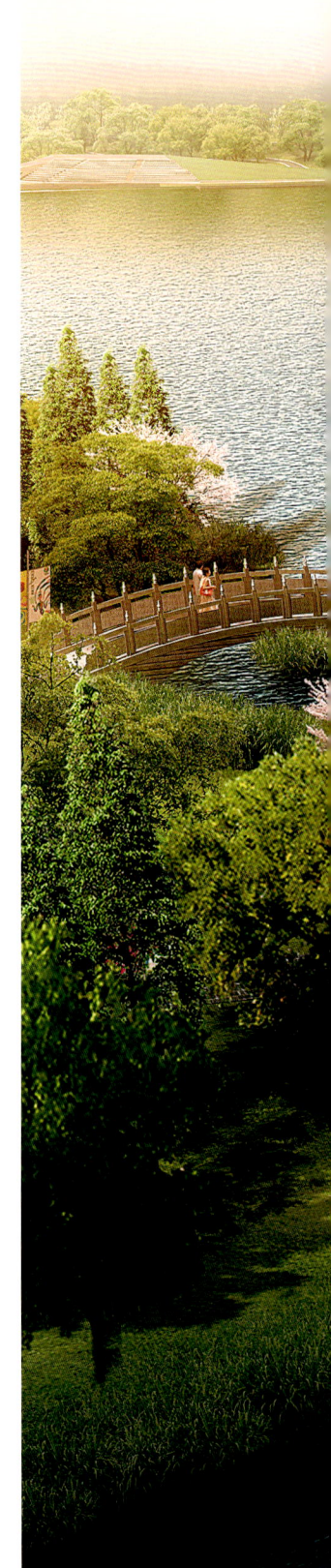

售卖亭
防腐木栏杆
防腐木坐凳
1.040
0.640
0.340
0.040
0.040
-0.300(常水位)
0.600(广场面)
防腐木坐凳
425
1500
1800
3000
425
2150
425
3600
2400
850
2750
3000
3000
2950
28275
1
2
3
5
8
10
11
12
13
14
15
16

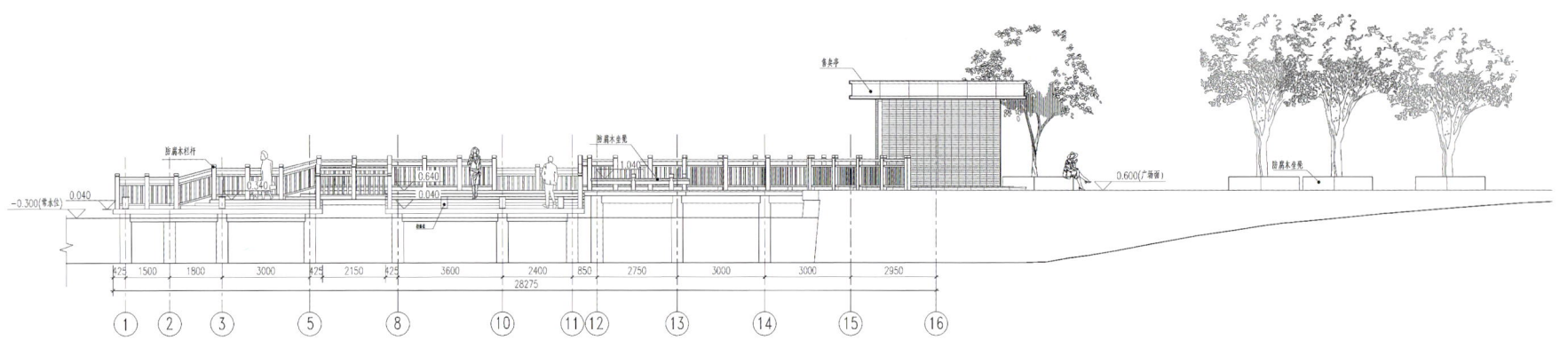

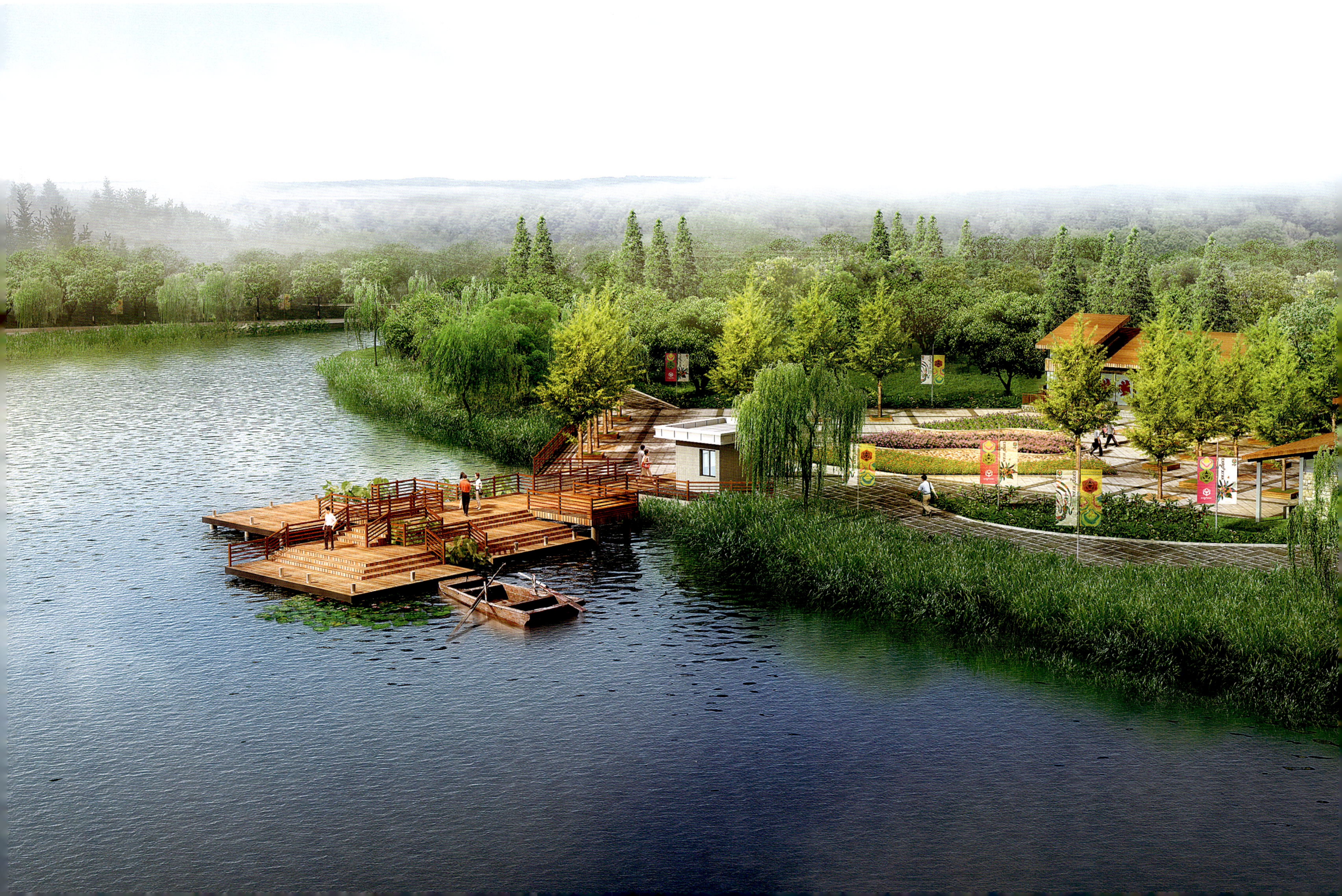

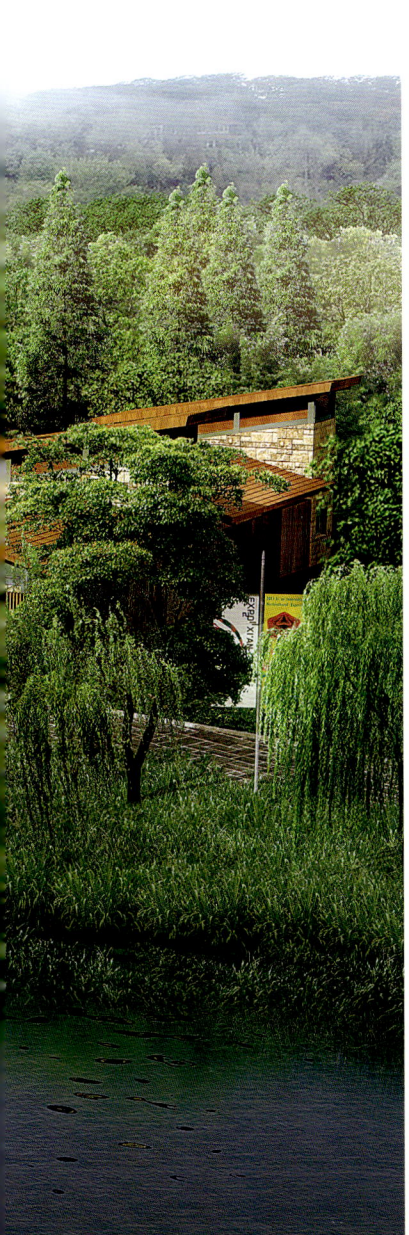

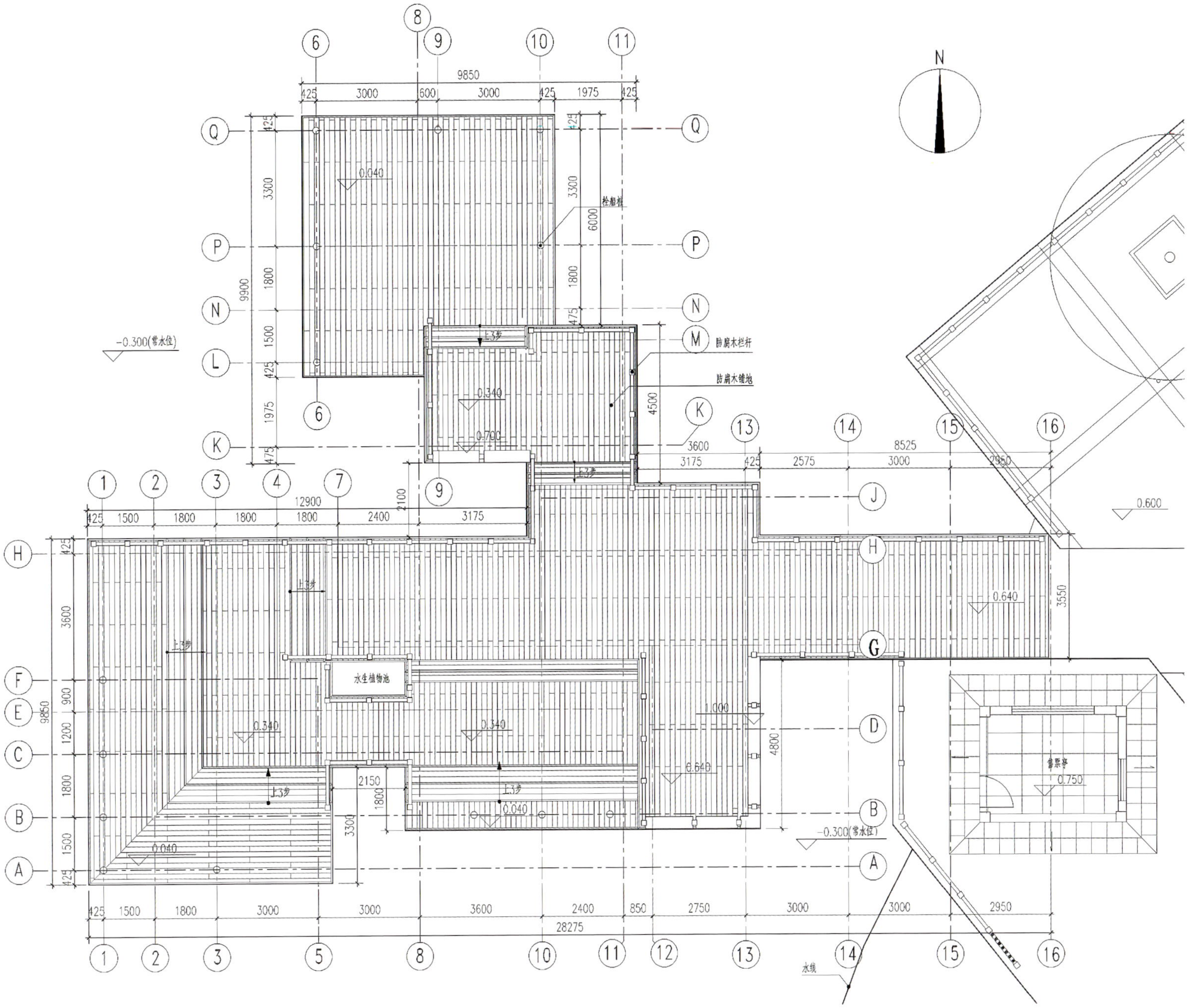

N
栓船桩
防腐木栏杆
防腐木铺地
-0.300(常水位)
水生植物池
售票亭
0.750
0.600
0.640
0.340
0.040
1.000
0.700
上3步
水线
9850
12900
28275
8525
9900

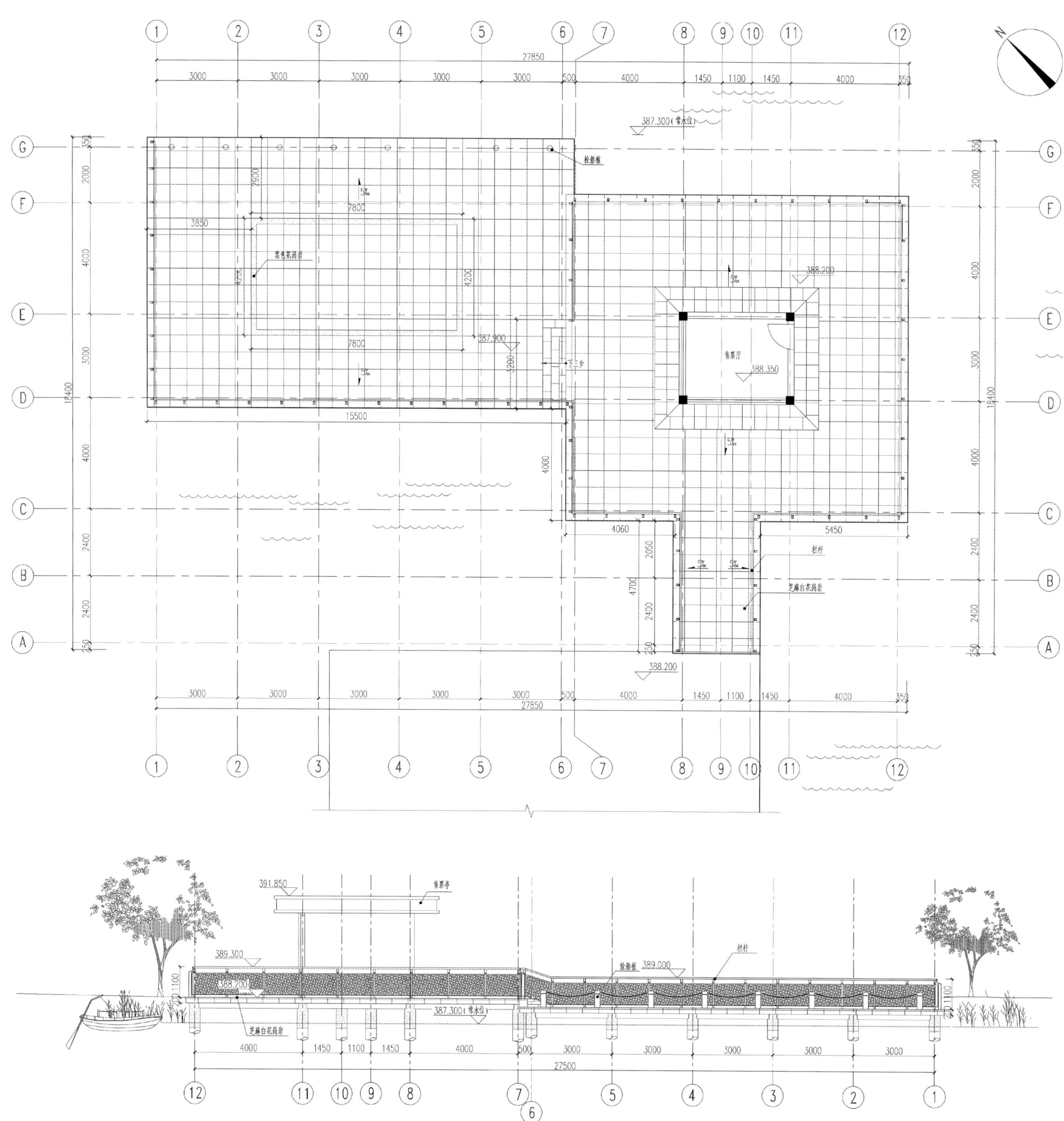

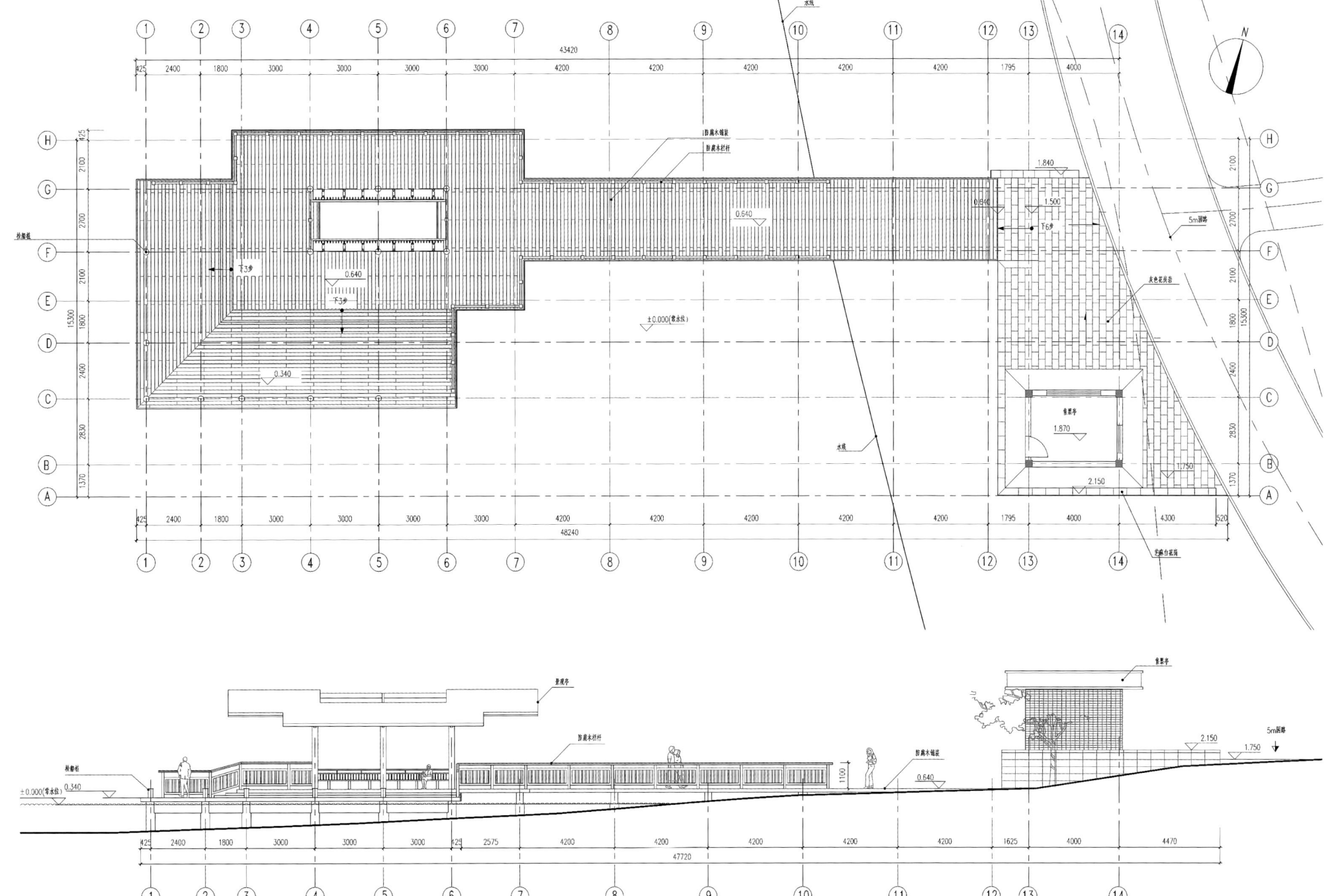

Planning of Service Facilities

服务设施规划

1．服务设施规划

园区内景观服务建筑包含卫生间、商店、服务用房等。服务建筑与周边景观环境完美结合，不突兀、不怪异，采用后现代建筑设计手法和理念，加入传统的建筑样式，形成优美、和谐的单体和群体建筑；建筑材料的选用以绿色、环保为主要原则，传统材料与现代材料相互结合，使建筑生态化、自然化。

根据建筑样式及风格可将服务建筑划分为4种类型，每种类型的建筑都分别包含商店、卫生间及服务用房。

第一类建筑其风格朴实、淳厚，带有浓郁的乡土气息，建筑材料以木材和青砖等传统材料为主，搭配少量的玻璃等现代材料，以一种极为自然生态的形式隐于绿树花草间；

第二类建筑其风格延续了第一类建筑的淳朴，又增添了更多的现代构成元素；材料的运用上仍以木材为主，结合片石与现代钢结构，使整个建筑在保持古朴与传统的同时又增添了几分现代气息；其中值得一提的是卫生间的设计，打破了传统的完全私密性，降低了隔墙的高度，以木百叶作为阻挡外界视线的屏障，但从建筑的内部却可以看见周围的景色，使游客即使在卫生间也能欣赏到户外的美景；

第三类建筑形式上更加现代，并结合陕西"房子半边盖"的地域文化特点，使整个建筑拥有独特的魅力；卫生间的做法依然采用半开敞式，给人以新奇的感觉；

第四类建筑与前三类建筑最大的区别在于其屋顶的形式为带挑檐的平屋顶，而其他的三类建筑均为坡屋顶；建筑的选材更多样化，石材、木材、金属和玻璃有机地结合在一起，建筑形式也更加灵活、现代。

2．餐饮设施布置规划及方案设计

展会期间，园区游客数量大，停留时间长。大约60%至70%的游客选择在园区内进餐，预计将有10%的游客选择餐厅就餐。50%-60%的游客可在分散式快餐供应点购买快餐。

规划集中餐饮服务区三处：灞上人家、椰风水岸以及欧陆风情是主要服务区。建筑面积15,000平方米，可满足每天12,000人就餐，23处广场服务用房可满足快餐需求。指挥中心餐厅主要为内部工作人员服务。

1. Service Facilities

Landscape service buildings, include washrooms, shops, and service rooms etc. The designers adopt a combination of modern artchitectural skills and concepts with traditional styles, forming elegant and harmonious single buildings and building complex, which are perfectly in harmony with surroundings. And in terms of the architectural materials, environmental protection is the rule. With a combination of traditional materials and modern elements, the buildings are ecological and natural.

According to the style, the service buildings are divided into four types, and each of the four includes shops, washrooms and service rooms.

The first type presents visitors with a plain and pure country style. Wood and grey bricks are the main materials; while a few modern materials, such as glass, embellishing among the grass and flowers, forming a natural and ecological life pattern.

On the basis of reserving the pureness of the first type, the second type is added to some modern elements. Still adopting wood as the main materials, slabstones and modern steel strucure are new elements, which lend modern flavor to the traditional building. What is worth to mention is the design of the washroom. The designer lowers the partition, instead, the woodern louvres are chosen as the shelter. Of course, tourists can see scenery outside. Getting rid of the traditional total secret character, it's a new vision!

The third type is much more modern with the regional culture, reserving the custom of Xi'an,"Half Built". The washrooms are also half open, freshening our eyes. This "HALF" feature leads the buildings charming!

The big difference of the fourth type is the roof. Unlike the above three types adopting abat-vent, it's built flat roofed with corbel table. And the materials are much more elastic and modern, with a integration of stones, woods, metal and glass.

2. Catering Facilities

During the Expo, there will be a large number of tourists staying in the Expo Site for a long time. It is predicted that 60%~70% of the tourists will take meals in the Expo Site, and 10% will choose to have meals in restaurants. 50%~60% of the tourists can buy fast food at the scattered fast-food points.

There are three concentrative catering service zones: Romance by the Ba River, Southeast Asian Street and European Avenue. The building area of the catering zones is 15,000 m², which can accommodate 12,000 people per day. The 23 service rooms can meet tourists' fast-food demand, and the command center restaurant mainly serves internal staff.

图例：
餐饮服务区 ●

3. 商业设施布置规划及方案设计

为满足游客的购买需求,园区设有大量的购物设施,在布局上,可以单独布置,也可与餐饮设施及其他建筑、广场相结合。

共设商业设施17处,其中3处规模较为集中,分布在灞上人家、椰风水岸以及欧陆风情内;其他为小型购物点,主要售卖纪念品。

3. Commercial Facilities

To meet the shopping demand of tourists, commercial facilities are established in the Expo Site, which can be set independently or combined with catering facilities or other buildings and plazas.

17 commercial facilities have been established, and three of them concentrate on Southeast Asian Street, Romance by the Ba River and European Avenue, while others are small shopping points selling souvenirs.

图例:

商业服务设施分布点 ●

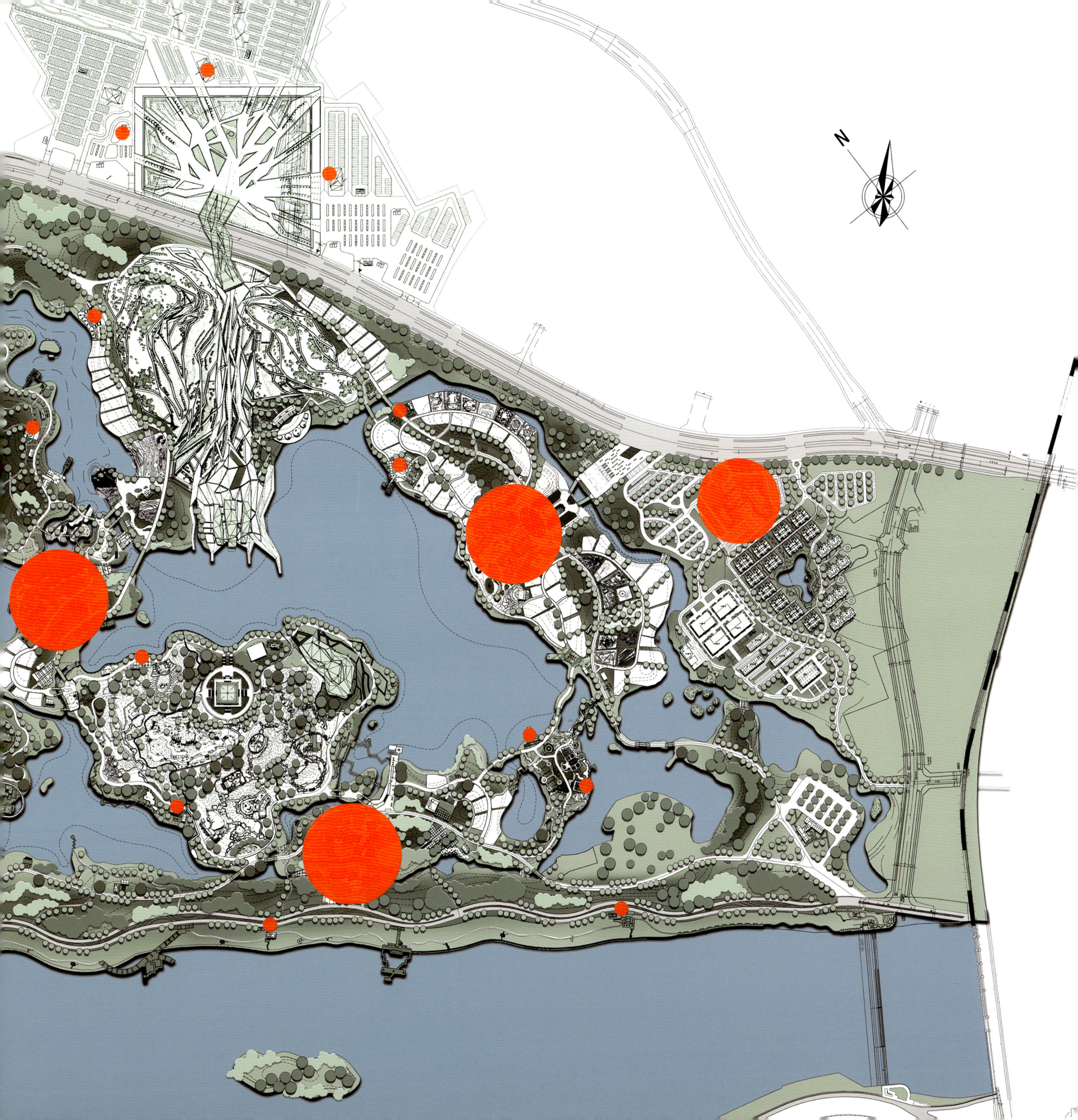

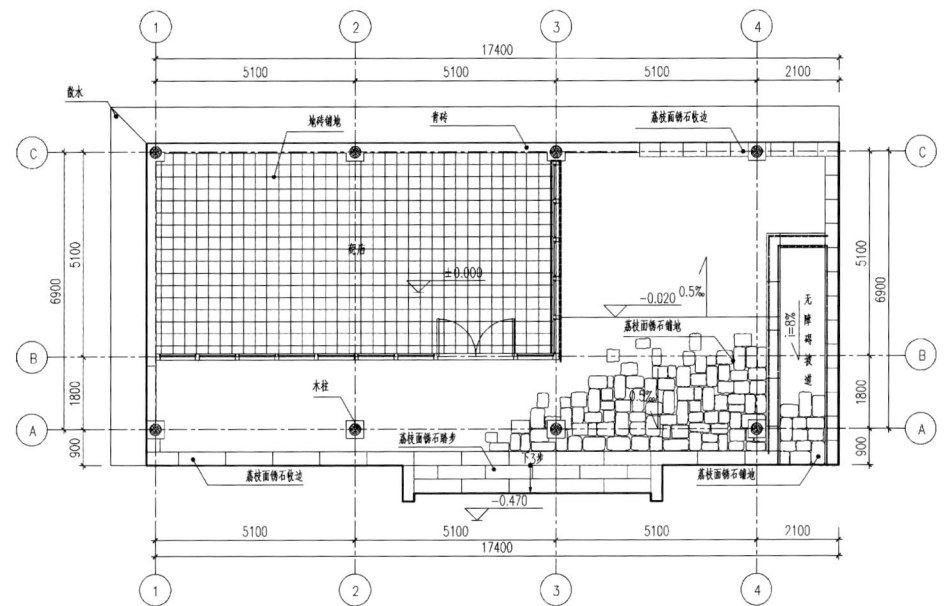
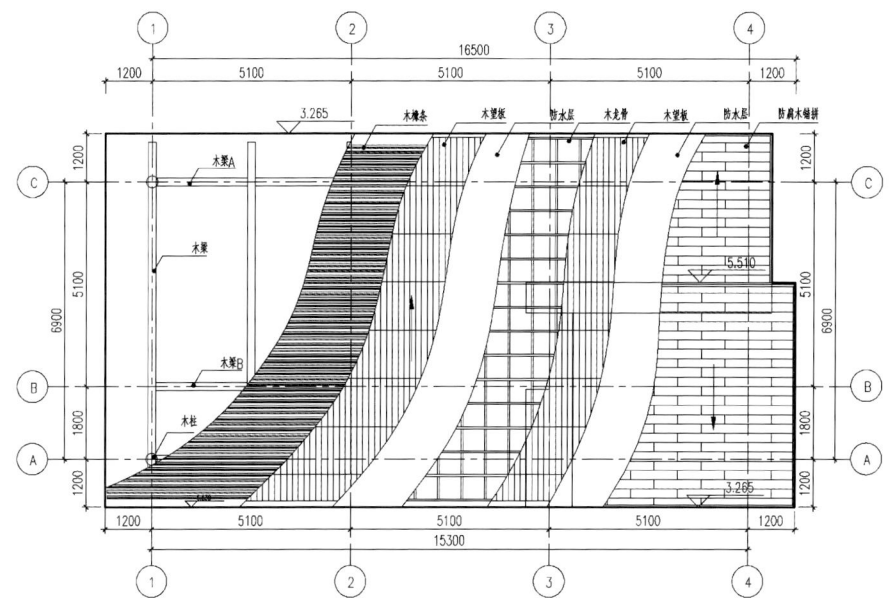
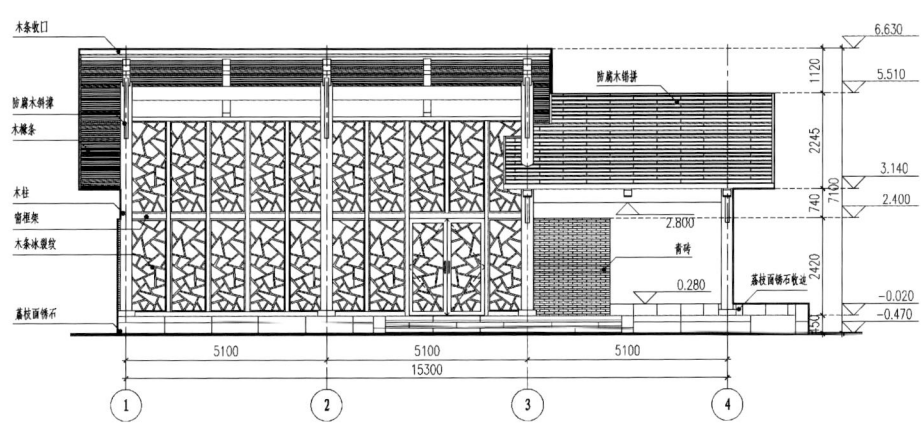
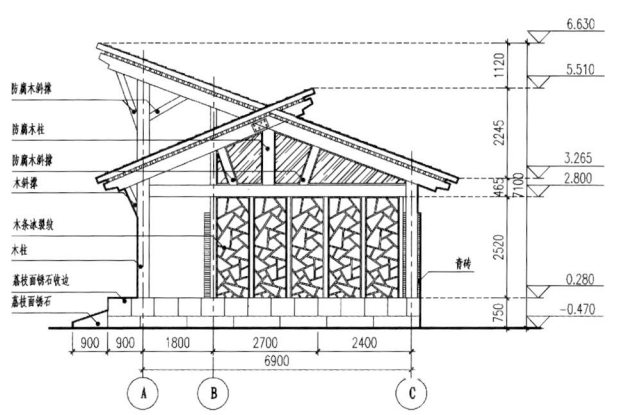

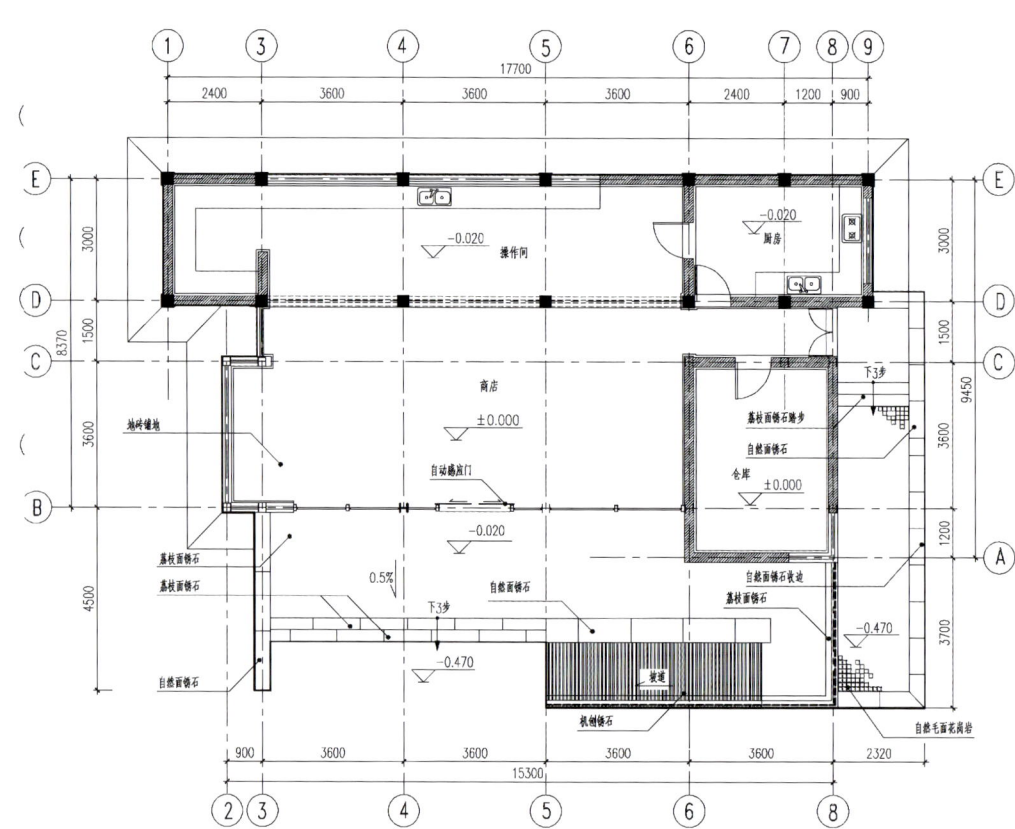
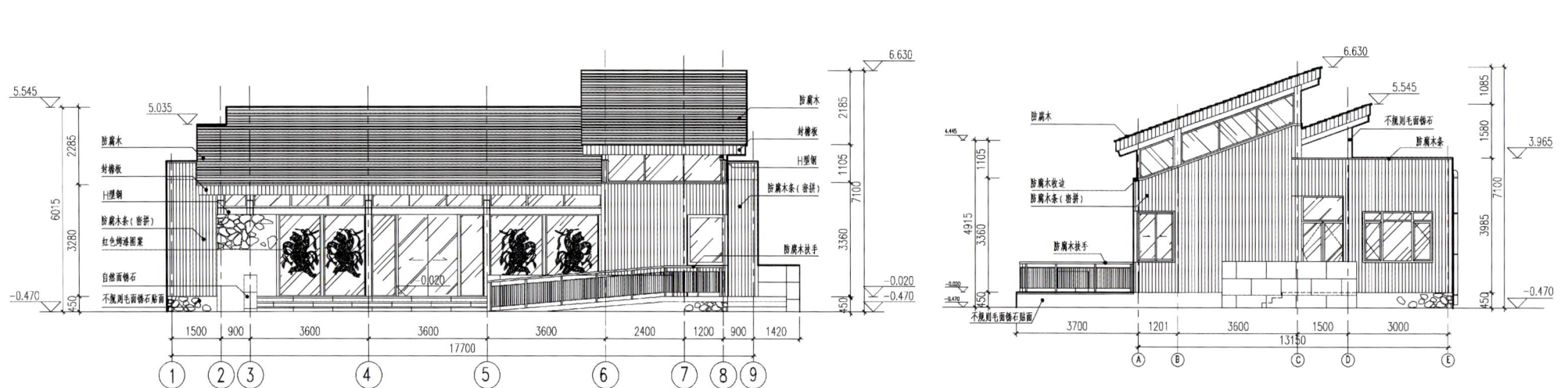

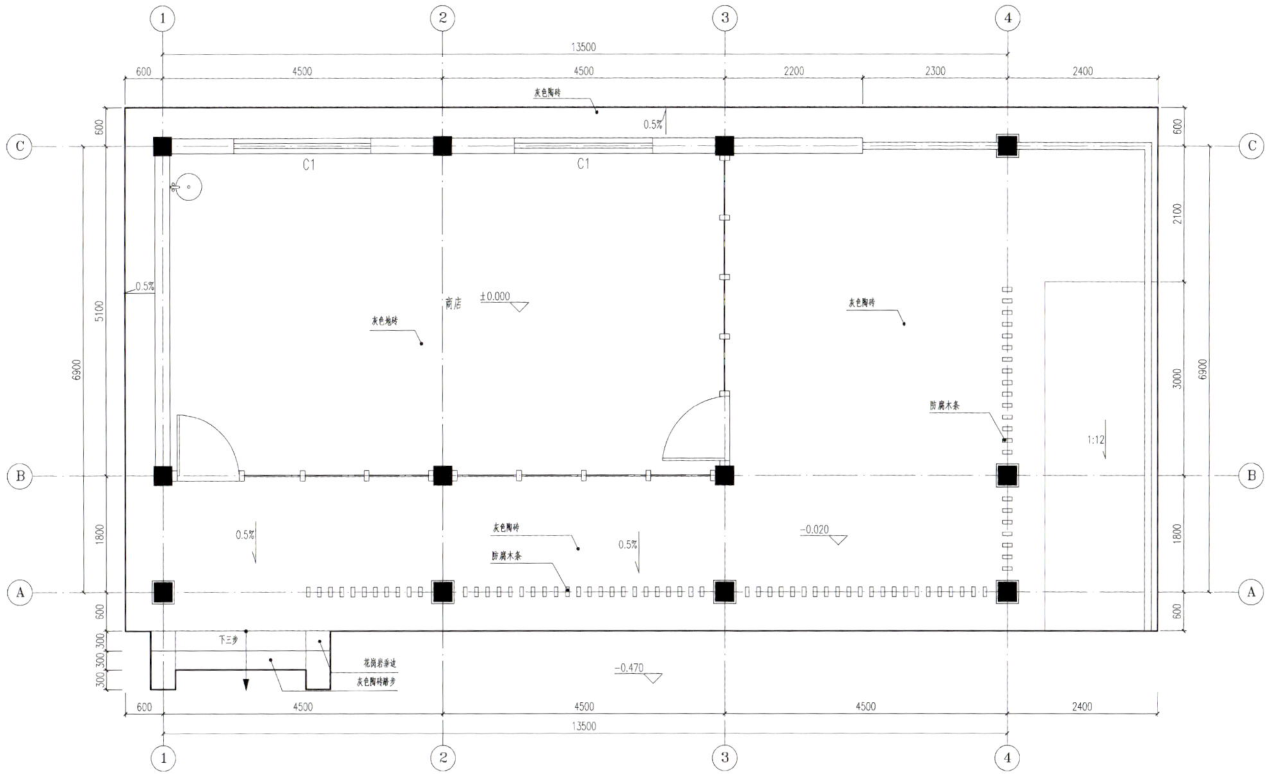
13500
600
4500
2200
2300
2400
C1
C1
0.5%
±0.000
-0.020
-0.470
1:12
6900
5100
1800
2100
3000

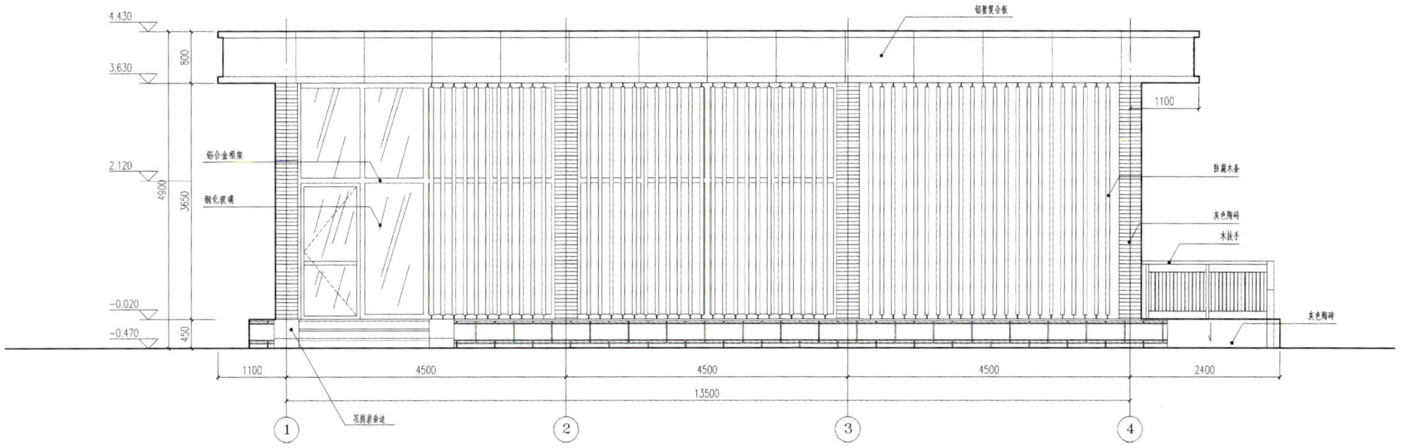
4.430
3.630
2.120
-0.020
-0.470
1100
4500
2400
13500
4900
3650

shopping

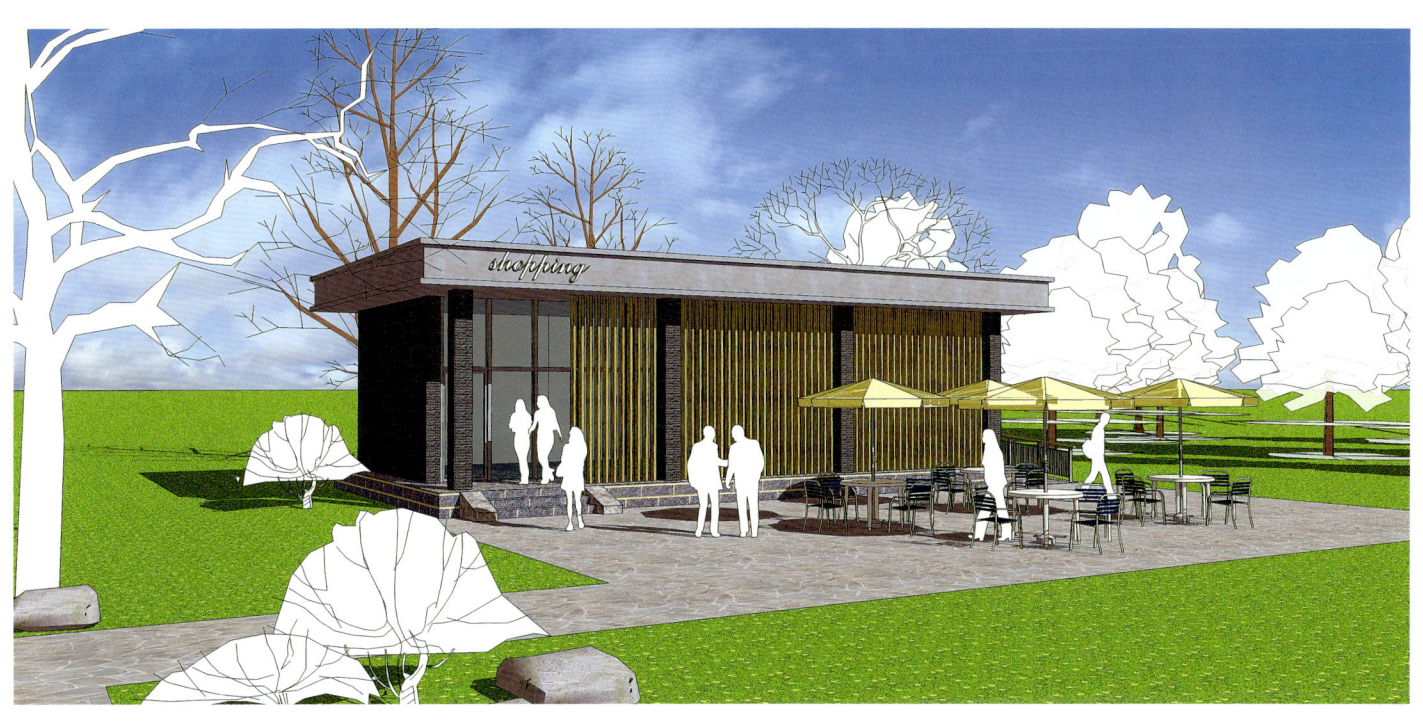

4．厕所布置规划及方案设计

厕所是园区重要的服务设施之一，男女蹲位比例设置为1:2.5左右。按照相关国家标准，考虑到极端日人流状况，总蹲位为1,400个，其中男厕蹲位400个，女厕蹲位1,000个。

厕所布置根据总体布局特点，按照服务半径100-300米布置，人流集中区域相应增加布设。建造方式采用临时永久相结合模式，便于会后调整厕所的数量。

4. Toilet Facilities

Toilets are important service facilities in the Expo Site, and the proportion of squatting seats for men and women is 1/2.5. According to relative national standard, the total number of squatting seats is designed to be 1,400 (400 for men and 1,000 for women) on account of extremely large people flow.

According to the characteristics of the general layout, toilets are arranged with a service radius of 100-300 meters. Population-concentrating areas should be equipped with more toilets. Temporary toilets and permanent ones are both constructed to facilitate the adjustment after the Expo.

图例：
卫生间分布点

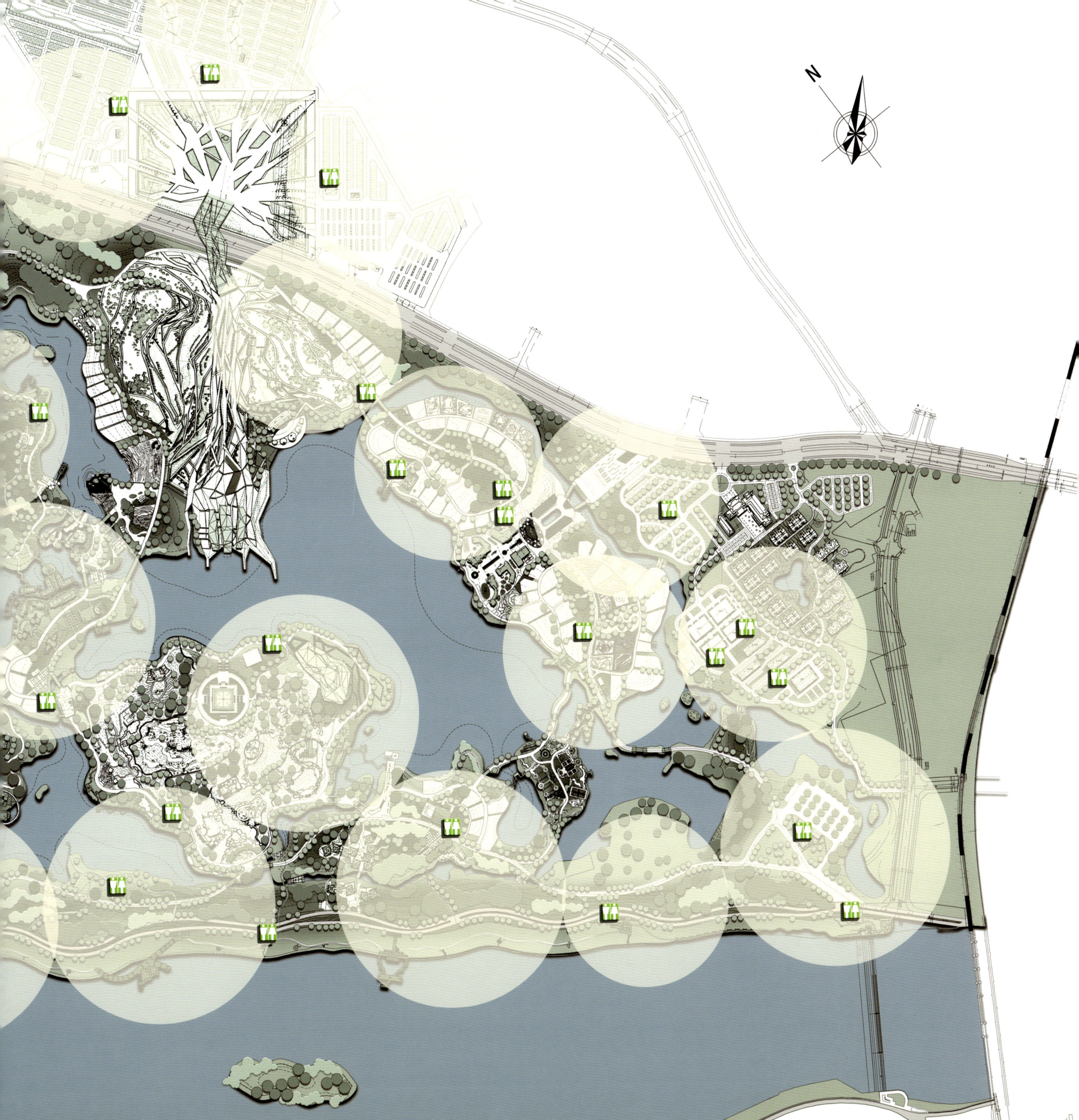

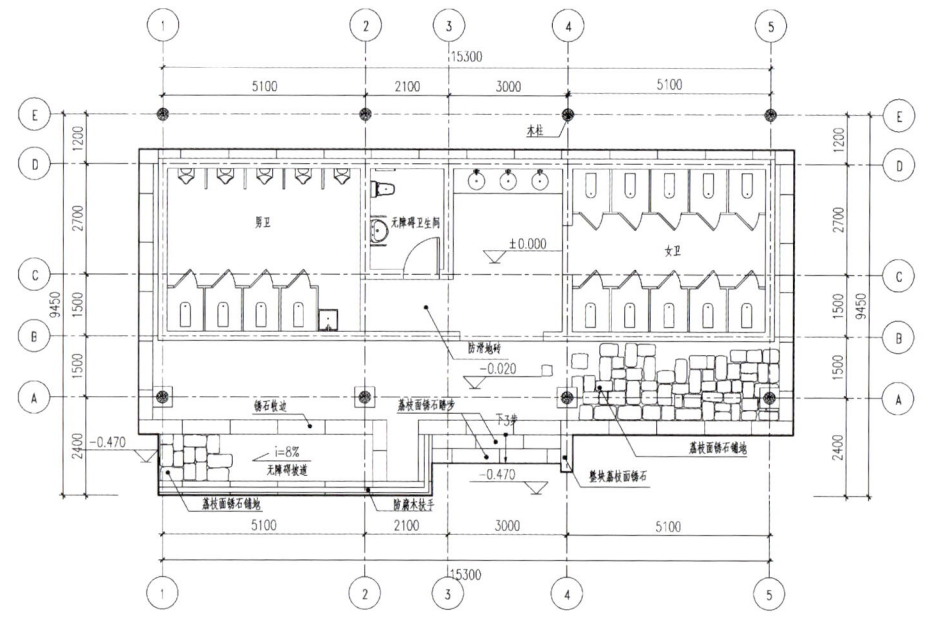

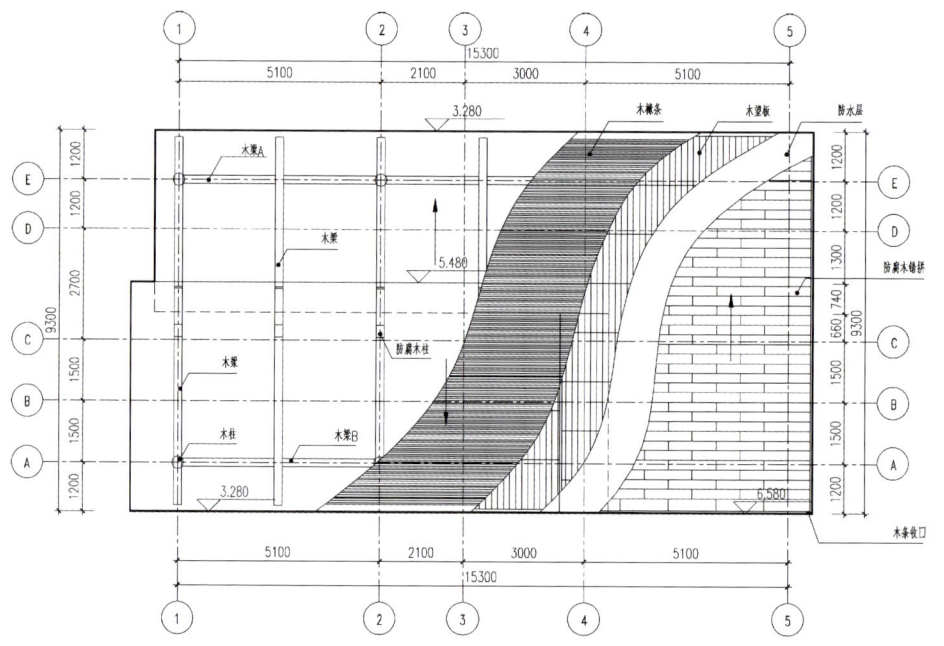

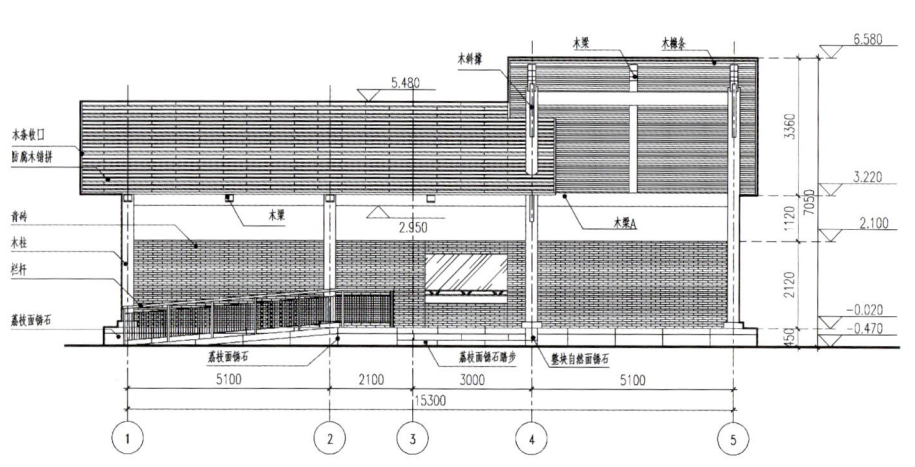

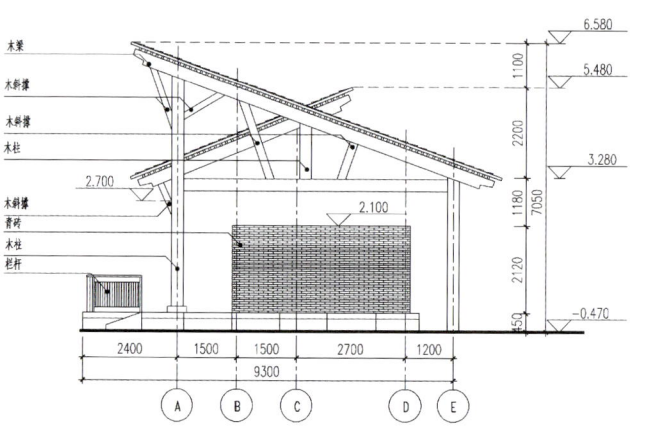

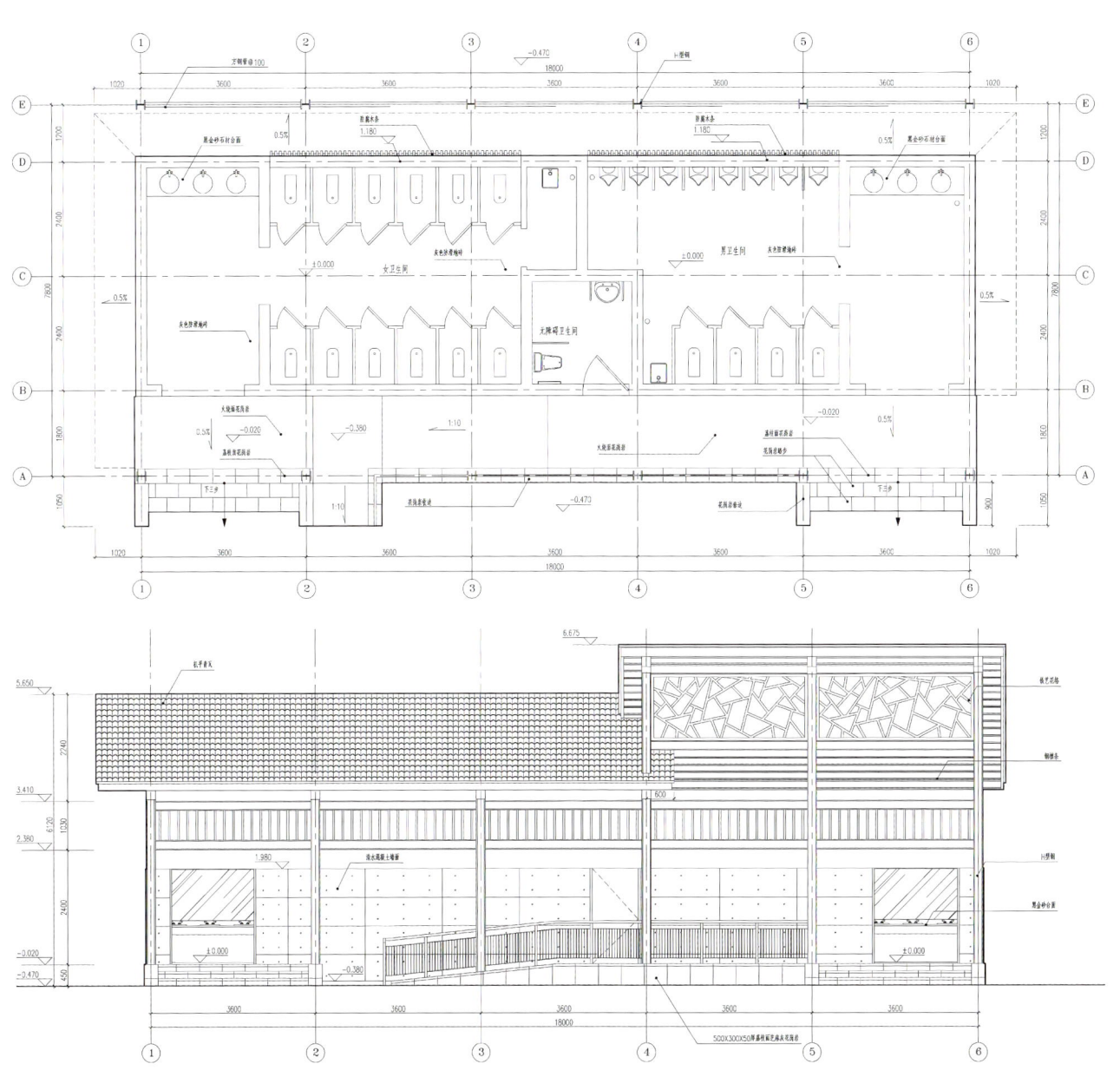

5. 标识系统规划

- **标识系统构成概述**

  外部标识系统：国际、国内、省内、市内

  内部标识系统：园内

- **标识系统要素构成**

  ①户外广告

  ②交通指示牌

  ③花卉摆放及绿化雕塑

  ④服务台

  ⑤景区指示牌

  ⑥工作人员服装

  ⑦专用车辆

  ⑧其他

- **外部标识系统规划**

  ①国际范围内，在参展国和主要客源国适当地点布置户外广告。

  ②国内主要在各个参展城市以及重点客源城市适当地点设置户外广告。

  ③省内应在各个地市甚至县城布置户外广告。

  ④市域范围内是重点区域，除布置大量的户外广告外，还应在机场、火车站（西安站及新建成的北客站）、各个高速公路出入口、市区主要出入口、市区主要商业区等地布设大型主题花卉造型或植物雕塑，营造展会气氛；在快速干道、主干道主要路口设置交通指示牌，方便游客前往园区。

- **内部标识系统规划**

  ①园区导游图

  ②景点名称标示牌

  ③出入口指示牌

  ④内外部交通指示牌

  ⑤紧急疏散方向指示牌

  ⑥医疗指示牌

  ⑦服务设施指示牌

  ⑧安全警示牌

  ⑨温馨提示牌

  ⑩工作人员及专用车辆的明显标识

  ⑪多媒体信息牌

5. Signage System Planning

- **Overview of the constitution of signage system**
  External signage system: international, domestic, provincial and civic
  Internal signage system: within Expo Site
- **Elements of signage system**
  ① Outdoor advertisements
  ② Traffic signs
  ③ Flowers and green sculptures
  ④ Service platforms
  ⑤ Signs in scenic spots
  ⑥ Internal staff's clothes
  ⑦ Special purpose vehicles
  ⑧ Others
- **Planning of external signage system**
  ① Post outdoor advertisements in appropriate locations of participating countries and major guest source countries.
  ② Post outdoor advertisements in appropriate locations of participating cities and major guest source cities of China.
  ③ Post outdoor advertisements in every city and even county of Shaanxi.
  ④ The urban Xi'an is the key area. Besides large quantities of outdoor advertisements, large theme flowers and plant sculptures have to be placed in airports, railway stations (Xi'an Railway Station and Xi'an North Railway Station), expressway accesses, main entrances to the urban district, major business districts, etc. to create an exposition atmosphere. Traffic signs should be established in major crossings of expressways and trunk roads to facilitate tourists' heading for the Expo Site.
- **Planning of internal signage system**
  ① Tourist map of the Expo Site
  ② Spot signs
  ③ Gateway signs
  ④ Internal and external traffic signs
  ⑤ Emergency evacuation direction signs
  ⑥ Medical treatment signs
  ⑦ Service facility signs
  ⑧ Safety warning signs
  ⑨ Signboards of friendly tips
  ⑩ Clear marks for staff and special purpose vehicles
  ⑪ Multimedia information signs

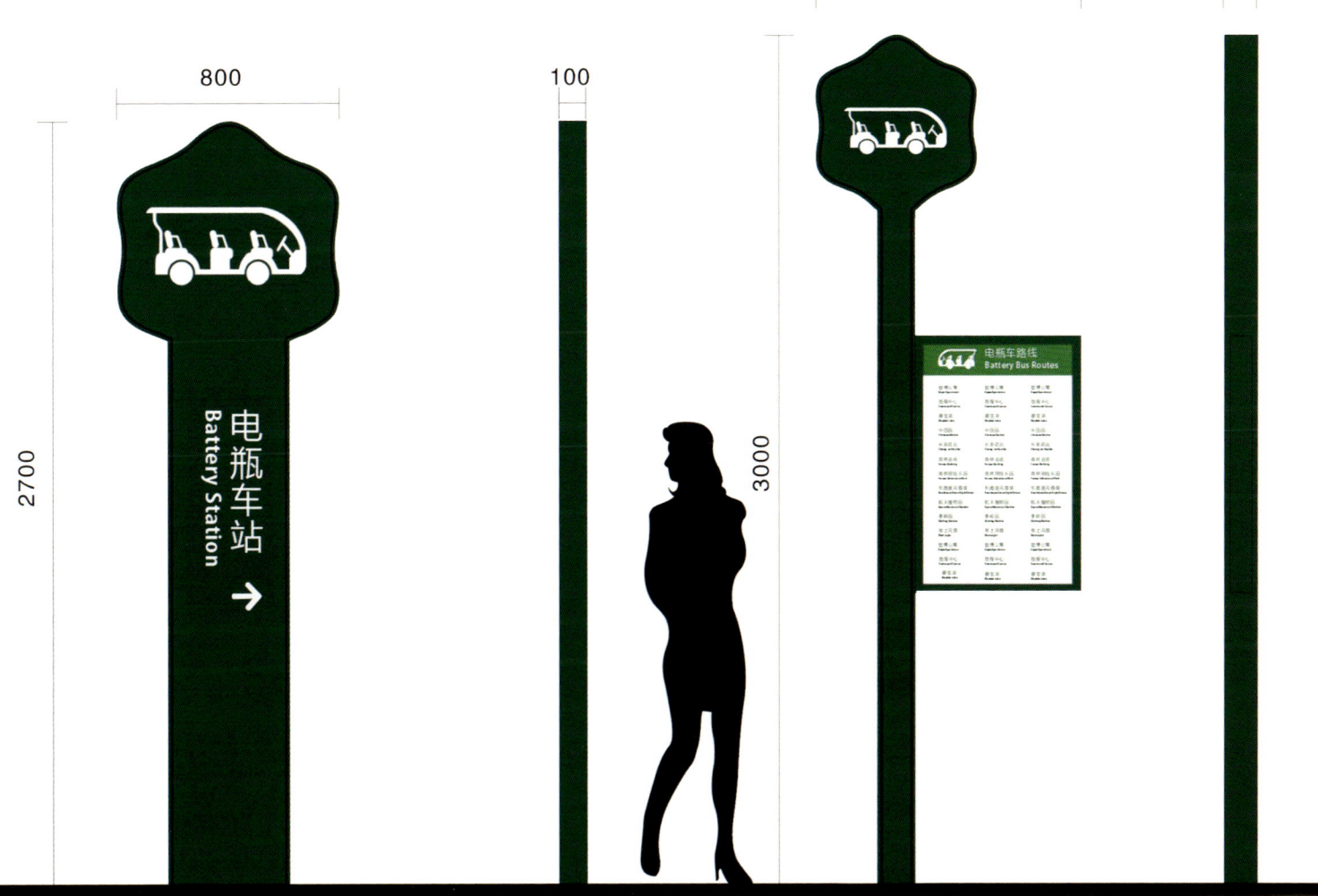

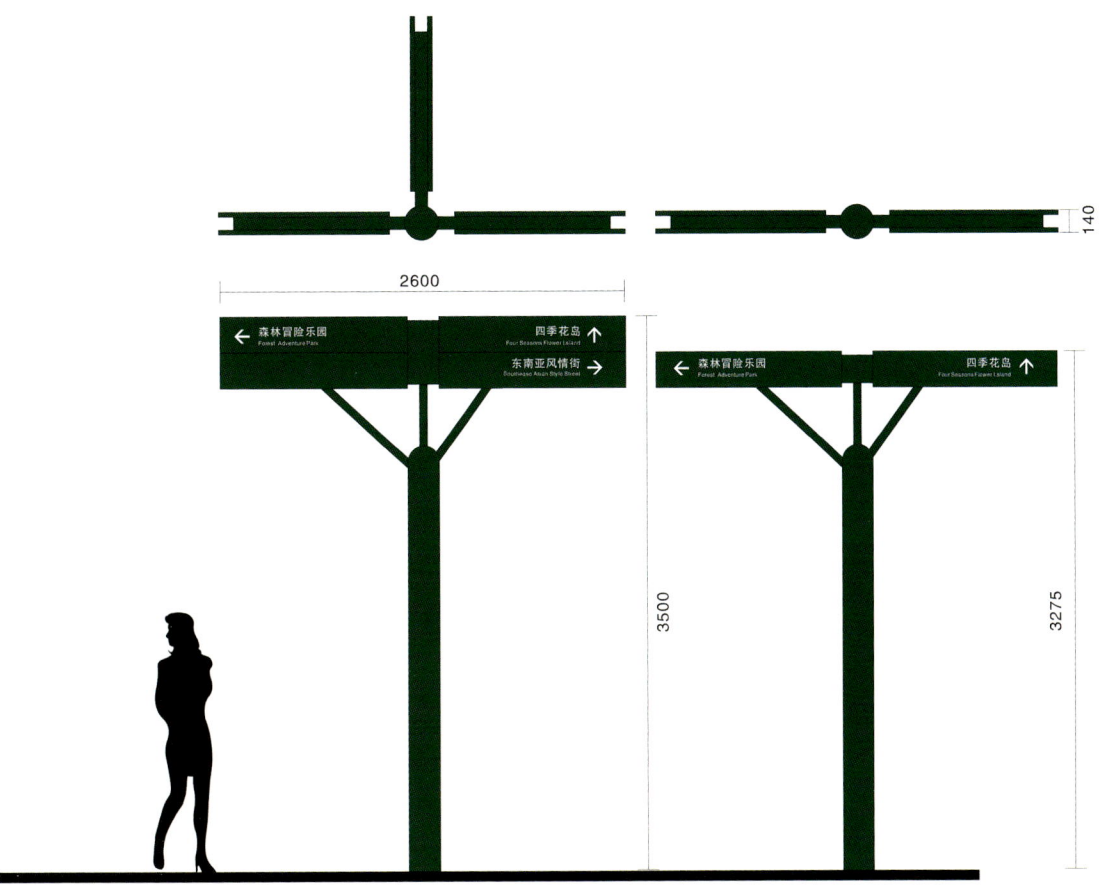

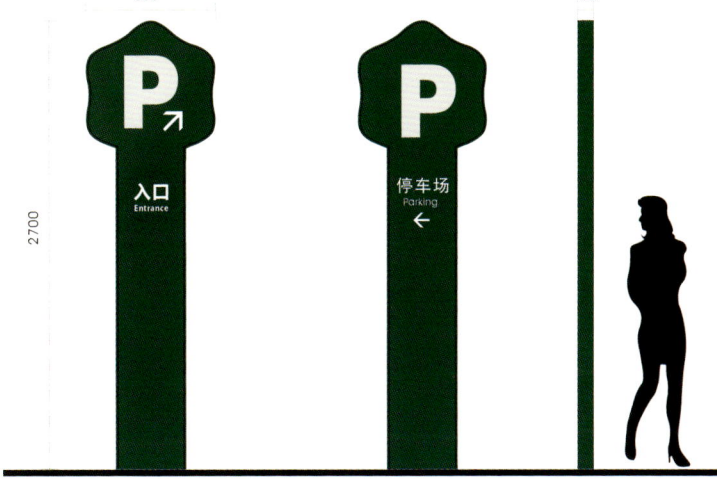

6. 急救设施布置规划

2011西安世界园艺博览会设1个急救调度台、5个急救医疗点、10辆急救车随时候命，为游客提供医疗救助。

6. Emergency Medical Facilities

Expo 2011 Xi'an sets one emergency dispatch center, five emergency treatment spots and ten emergency ambulances available at any time, providing medical treatment for tourists.

图例：
医疗急救服务点 ✚

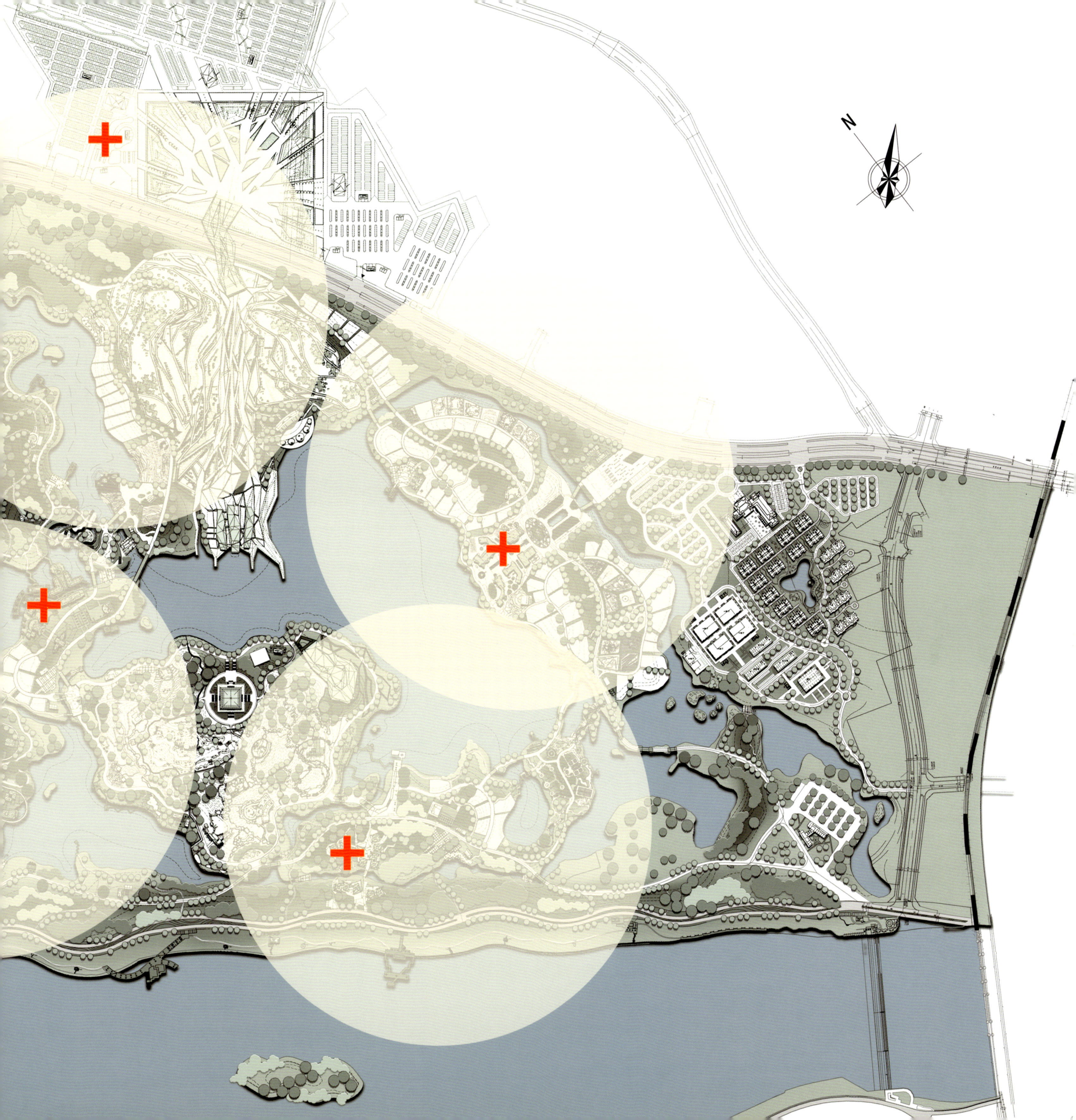

7. 后勤保障中心（世园村）规划

后勤保障中心即世园村，世园村位于园区扩展区域，是各参展国和参展企业的住宿场所，主要功能是为后勤工作人员、演艺人员和安保人员提供展会期间的住宿、餐饮等服务，也可作为后勤物资储存、工作车辆停放场所，包括独栋别墅、双拼别墅、围绕核心景观建设的联排别墅和沿地块东侧的花园洋房片区。毗邻园区的独栋及双拼别墅似两个"岛"的形态伸向园区，双拼别墅设于内侧，或邻岸而居或依岛心绿地而设，户户拥有宽大的空中花园，将室外景观——敛入室内。联排别墅区围绕中心湖景展开，拥有安静宽阔的中心景观。庭院景观与中心湖景相互依托有机联系，出则为开阔的公共交往空间，享受公共交流带来的奔放与豁然；入则为私家小院，闲憩内心的安逸。花园洋房拥有开阔的视野以迎接窗外的绿洲广袤、湖光滟潋。

小区采用了多重景观，层层递进的原则。中心景观系统是一条显著的景观主轴，起于指挥中心，直抵中央湖景。景观主轴上设置景墙、浮雕、柱廊、涌泉等多道景观序列，营造出尊贵大气的景观感受，尽端的湖景豁然开朗，将宽绰开敞的景观推向顶峰。而多层次的组团绿地、邻里中心创造了公共空间、半公共空间、私密空间等多重空间的自然过渡，增加邻里交往的机会，改善小区人居环境。世园村作为生态公寓成为园区中不可或缺的人居典范，科技与生态同时在这里绽放，为2011西安世界园艺博览会增姿添彩。

7. Logistics Support Center (Expo Village) Planning

The Logistics Support Center, that is, the Expo Village, is located in the extension zone of the Expo Site, mainly providing accommodations for the country/company attendees. It serves mostly for logistics staff, entertainers and security personnel around the Expo, and it is also a place for the storage of materials and parking of working vehicles. It consists of single-family detached homes, semi-detached housings, core landscape centered townhouses and garden villas, all lying in the east of the plot. The single-family detached homes and semi-detached housings both stretch to the Expo Site shaped like two islands. The waterfront semi-detached housings scatter around the green heart of the island; each has a large hanging garden, which frames sceneries outside into the room. The townhouse zone encloses the lake, lending it a tranquil ambience. The courtyard landscape sets off the center lake with each other - walking out, you can enjoy the unrestrained and open space for communication; stepping in, you can drink yourself into the comfort and leisure of the private courtyard. While the vast green space and ripples in the lake are framed inside the garden villas.

Multi-layered landscape is rendered in the parcel. The central landscaping system is a main landscape axis starting from the Management Center and ending at the central lake. A series of landscape elements set on the main axis include feature wall, relief, colonnade, fountain, stimulating an honorable feeling toward the landscape. The lake located at the end unfolds a broad view to an ecstasy of the vastness. Multi-layered green land and neighborhood center enable the natural transition of the public/semi-public/private space to increase the meeting opportunities for the users and make it a more livable place. The ecological Expo Village is an essential model of livable environment where technologies and ecology are interwoven to make the International Horticultural Exposition 2011 Xi'an more charming.

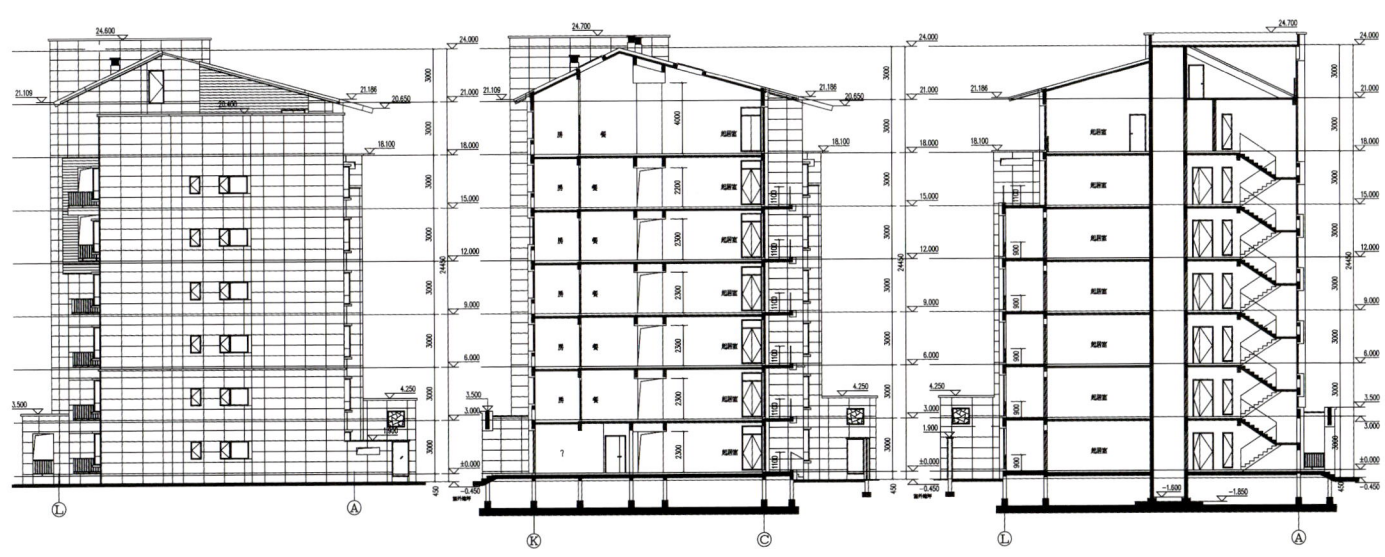

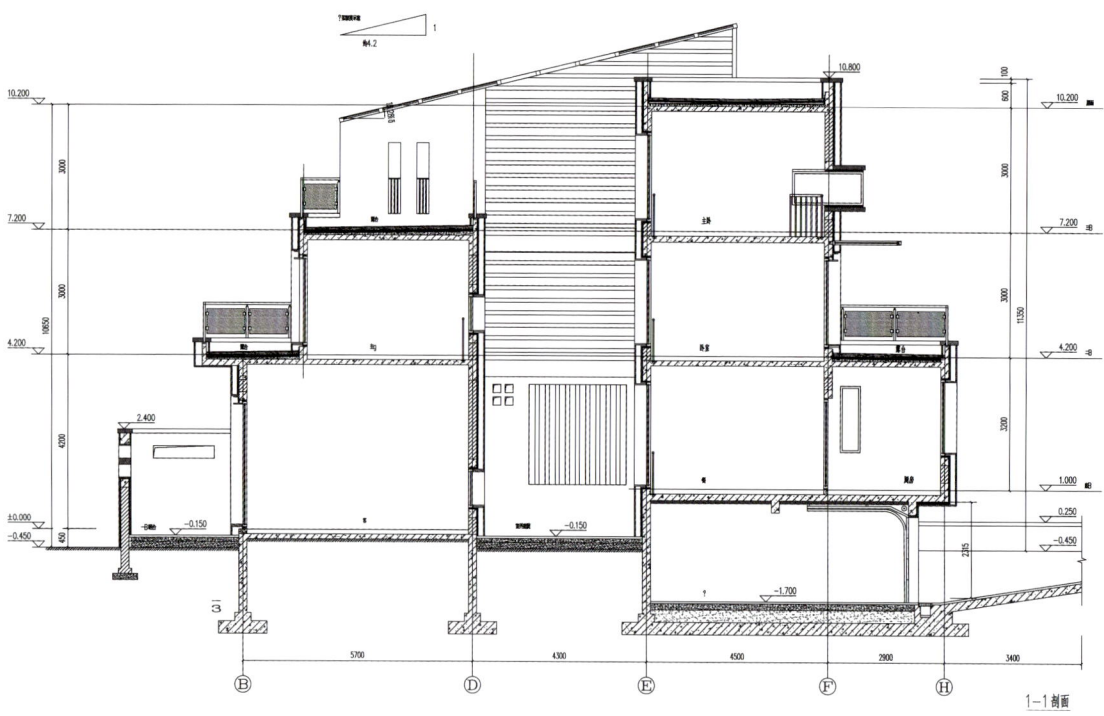

1-1 剖面

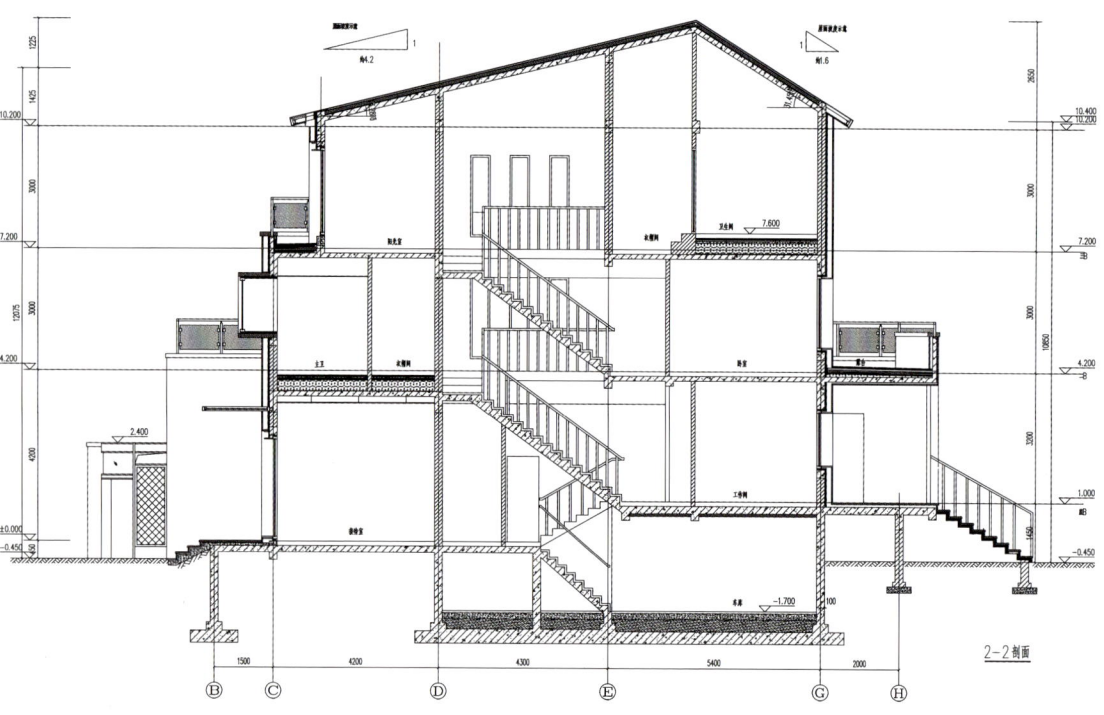

2-2 剖面

Infrastructure Facilities

基础设施规划

1. 管网设备

园区内各种工程管线原则上均沿规划道路敷设，过路管线尽量保持正交。在可能情况下，各种管道在道路两侧的位置遵循下列原则：道路西、北侧为电信、燃气、污水；道路东、南侧为电力、给水、雨水。各种管线竖向交叉时，一般自上而下的顺序为：电信、电力、燃气、给水、雨水、污水。各种管线之间的水平、垂直净距应符合有关规范和标准要求。

2. 通讯、邮政设施规划

在主入口和欧陆风情各设邮政所一个。邮政所除提供信件、明信片、邮票服务外，还可以提供游客自制首日封和明信片服务。在主要的休息服务区设置小型网吧，并提供免费无线网络接入（WIFI）服务。

1. Pipelines Network

In principle, engineering pipelines in the Expo Site should be laid along the planned roads, while pipelines crossing roads should maintain orthogonality. The location of pipelines on both sides of roads should abide by the following principles: telecommunication pipelines, gas pipelines and sewage pipelines should be on the west and north sides of roads, while pipelines for electricity, water supply and rain draining should be on the east and south sides. The minimum buried depth of pipelines under the roads for vehicles should be no less than 0.8 m. When the pipelines cross one another vertically, the order from above to below should be telecommunication pipeline – electricity pipeline – gas pipeline – water supply pipeline – rain draining pipeline – sewage pipeline. The horizontal and vertical net distance between pipelines should meet related standards.

2. Communication & Post Facilities

Two post offices are built at the main entrance and the European Avenue respetively, providing letter, postcard and stamp services as well as self-made first-day covers and postcard services for tourists. Internet Bars and WIFI services are provided in major rest areas.

图例：
邮政服务店
无线网络覆盖区域

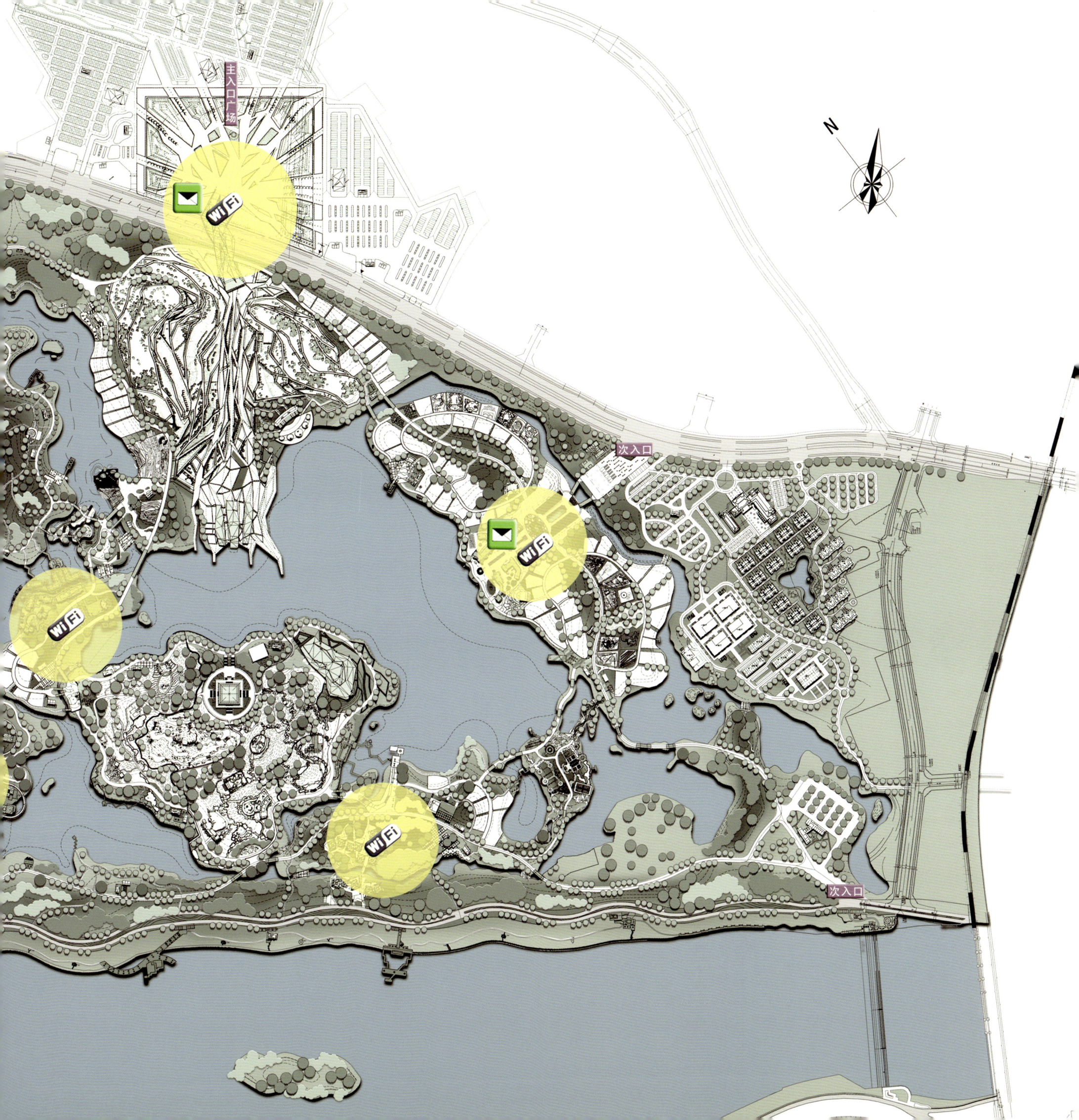

3. 环卫设施及环境卫生系统规划布置规划

一般地段垃圾箱按照20米半径布置，游人集中地段按照10米半径布置。全园共设垃圾收集站5个，均位于一级路旁。垃圾中转站位于园区东南入口，便于垃圾的对外运输。考虑到全园每天产生的垃圾量巨大，每个垃圾收集点都设有垃圾压缩设备。

3. Environmental Health System

Garbage cans will be distributed based on the radius of 20 meters in general sections and 10 meters in tourists-concentrating sections. There are five garbage collection points in the Expo Site, which are all located by the sides of arterial roads. The garbage transfer station is located at the southeast entrance of the Expo Site for the convenience of external transport. Considering the large quantities of garbage produced every day, each garbage collection point is equipped with garbage compression equipment.

图例：
垃圾收集点　垃
垃圾中转站　中转

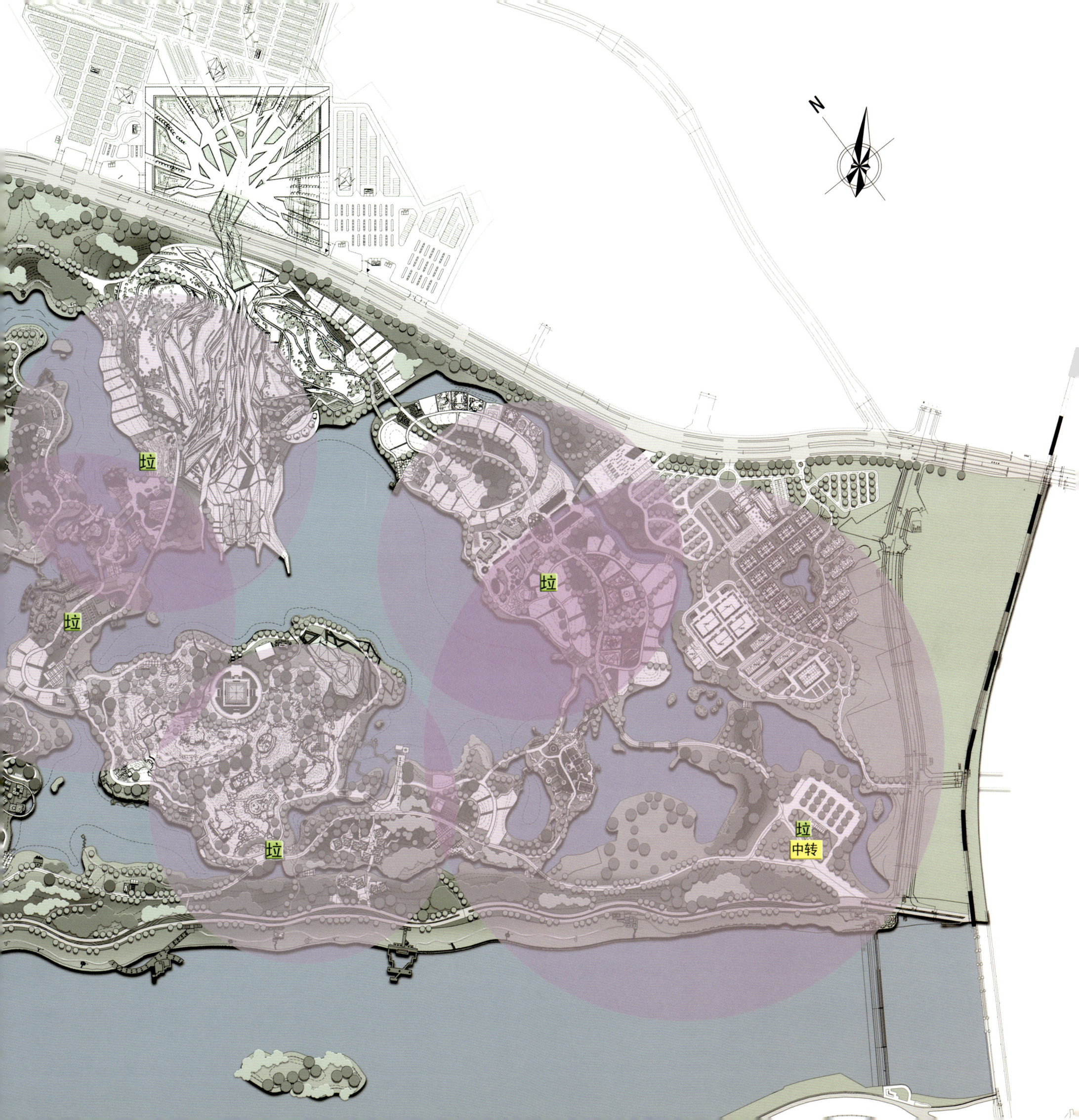

Safety Insurance & Fire Control

安全保障与消防规划

1. 安全保卫体系规划

运行保障系统：即保证世园会正常开园的支持保障系统，包括给排水、电力、通讯、交通等功能保障。
安全防控系统：包括安全情报系统、安全检查和监控系统。

2. 消防规划

世园会全园建筑应以耐火材料为主，严格遵守消防管理，按照国家建筑消防涉及规范，建立消防分区，设立消防栓、自动喷淋设施，配备相应标准的灭火器材。
全园划分为3个消防分区，每个消防分区有专门的消防安全员负责消防器材的日常检查、维护，对本区域的消防安全进行督查，发现问题后亲自监督整改。发现火情，应及时上报，按照消防预案进行扑救。
建立信息化消防报警系统，由安保指挥中心统一协调行动。

1. Safety Insurance

Operation insurance system: the supporting system guarantees the normal opening of the Exposition, including water supply and draining, electricity, communication and traffic, etc.
Security control system: safety intelligence system, safety inspection and monitoring system.

2. Fire-control System

Refractory materials should be used for architecture in the entire Expo Site. Fire prevention zones, fire hydrants, automatic spraying facilities and firefighting equipment should be prepared according to national standards about the firefighting of buildings. The whole Expo Site is divided into three fire prevention zones. Fire safety officers are assigned for each zone to be responsible for daily check and maintenance of firefighting equipment, the supervision of security against fire. They should supervise and solve problems in person, report to leaders immediately in case of fire and put out the fire according to fire control plan. Informationized firefighting and alarm system should be established and the operation should be conducted uniformly under the safety control center.

图例：
指挥中心
消防控制区

3. 安全通道及消防疏散空间规划

游览主环路连接了全园的主要景点，路面宽7米，连接主要的对外出口，适合作为安全疏散通道。紧急情况发生时，游人可根据自己所处的位置，由主环路通往就近的紧急出口。

同时，消防救援车辆沿5个出口进入主环路，可快速到达出事地点，展开救援作业。

3. Emergency Access

The main ring road for sight-seeing connects major scenic spots in the Expo Site. The road is 7 meters wide and connects major exits. Hence, it is suitable to be a channel for safe evacuation. In the state of emergency, tourists can head for nearby emergency exit through the main ring road. Furthermore, fire-fighting and emergency vehicles can reach the main ring road through five exits, thereby arriving in the position of accidents for rescue immediately.

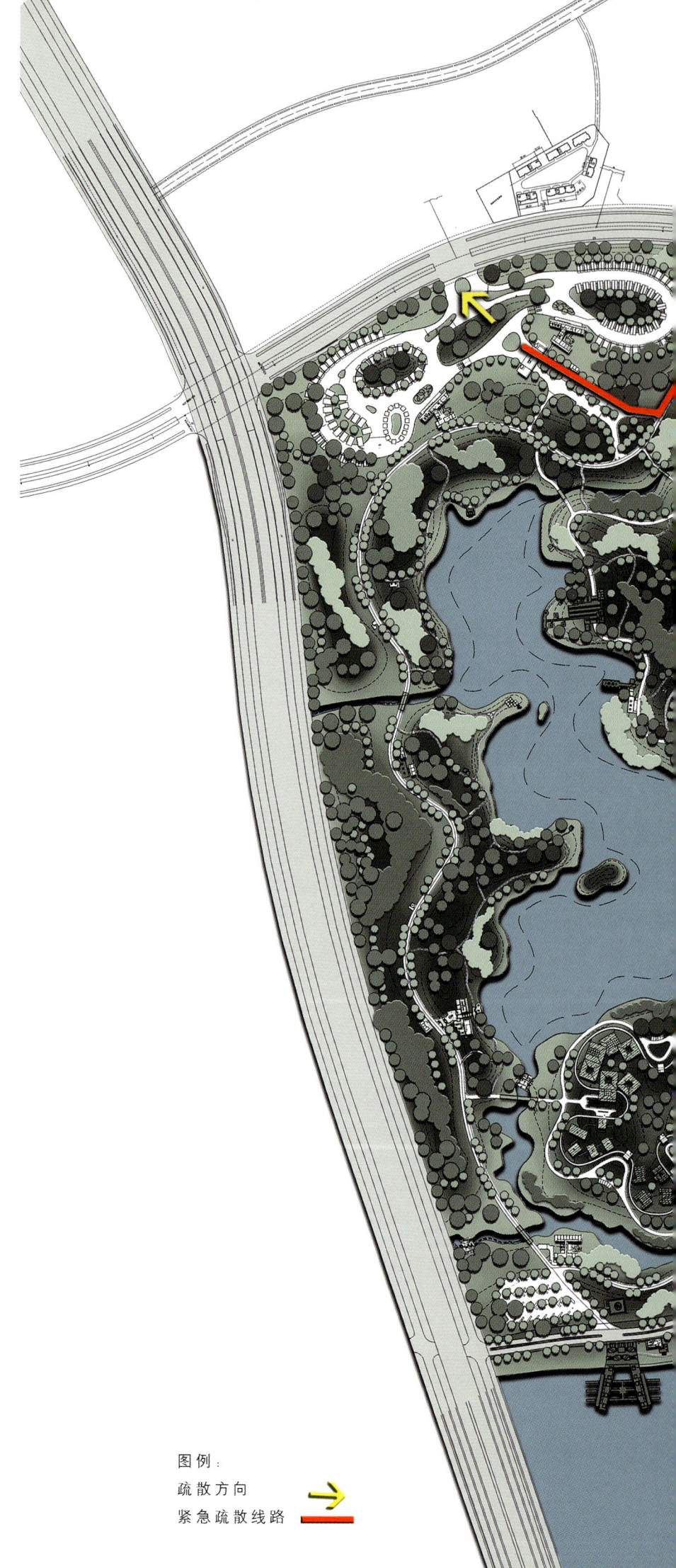

图例：
疏散方向
紧急疏散线路

Command Center Planning

控制指挥中心系统规划

指挥中心设在世园村人行主入口处,并靠近世园会园址次入口,总建筑面积1,500平方米,是保障世园会展会工作高质量运行的枢纽,世园会的日常管理、安全保障、交通疏导等工作均在此进行。主要功能包括:世园会职能机构办公室、新闻会议中心、消防指挥中心、安保指挥中心、交通指挥中心、广播中心、照明控制中心、电力电讯保障中心。

世园会期间作为组织方接待、办公的重要场所,闭幕后指挥中心将成为世园村居民们回家的"门厅",使每个回家的人通过一个温馨而豪华的实体进入自己的社区,为人们创造更多"偶遇"机会,增加邻里关系的和谐。整个指挥中心由办公和配套服务两个实体通过灰空间联系在一起,可分可合。它是世园村景观主轴的中心与起点,统领着小区的景观系统,设有网球场、篮球场和小型高尔夫球场等运动场所,丰富了社区生活,传导健康的生活理念。

The 1,500 m² Command Center is located at the main pedestrian entrance of the Expo Village, adjacent to the Secondary Entrance of the Expo Site. The Command Center is the hub guaranteeing the high-quality operation of the Expo. It is in charge of daily management, safety control and traffic dispersion, etc. for the Expo. It is the functional department of the Expo and the center of conference, fire-fighting, safety control, traffic guidance, broadcasting, lighting control, electrical power and telecommunication guarantee. It is an important reception/office place for the organizers during the Expo period, and a "hallway" for residents of the Expo Village after the Expo ends. For every resident will walk through this luxurious, cozy, solid hallway into its own community, "encounter" is thus created to enhance neighborhood relations. The Command Center consists of office space and service facilities with gray space as a connection node, separating or jointing to some extent. It is the starting point of the main landscape axis of the Expo Village, which plays a leading role in the landscaping system of the community, integrated with sports facilities including tennis court, basketball court and small golf court. It highlights a diversified community life and the concept of healthy life.

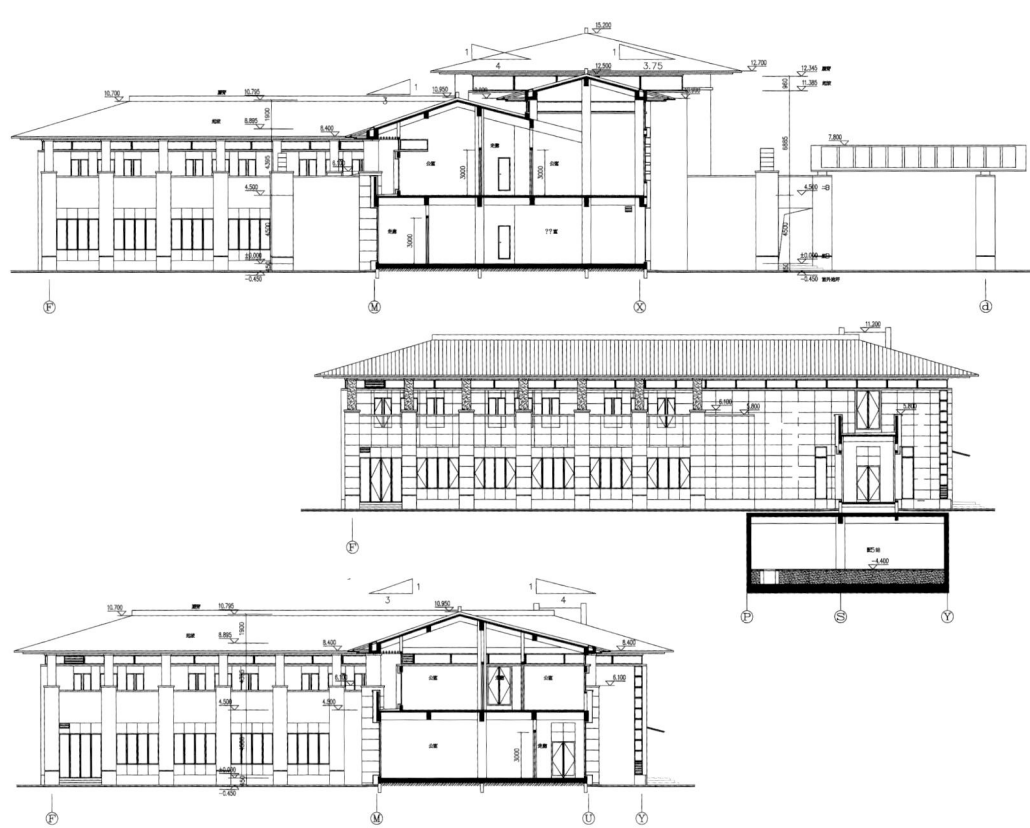

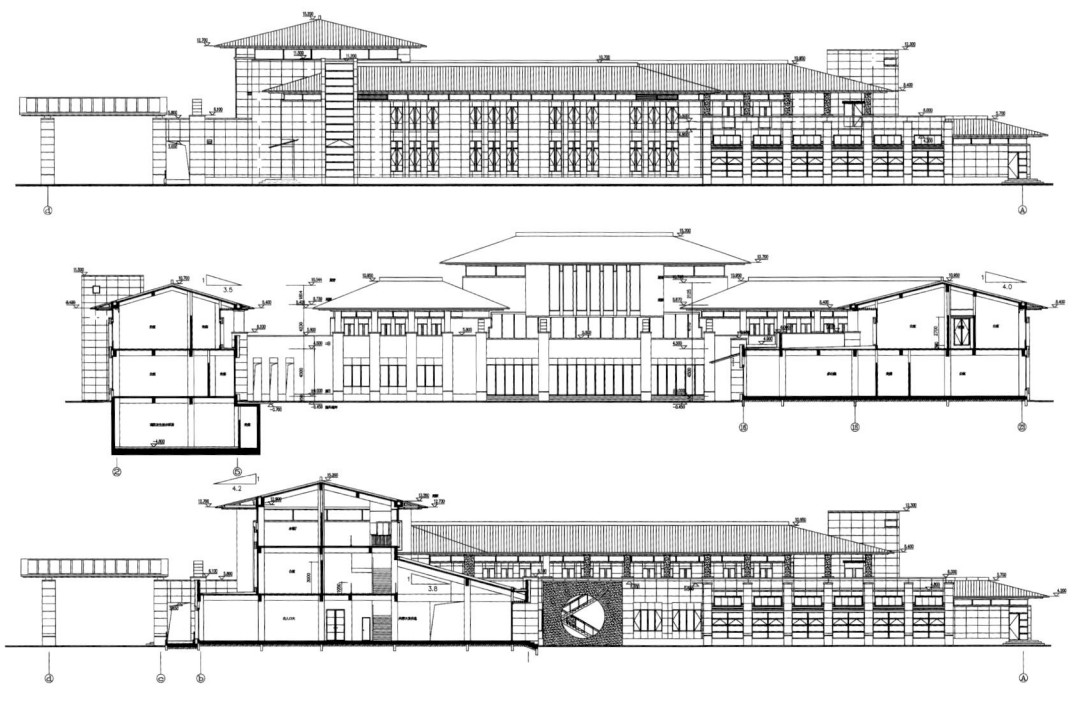

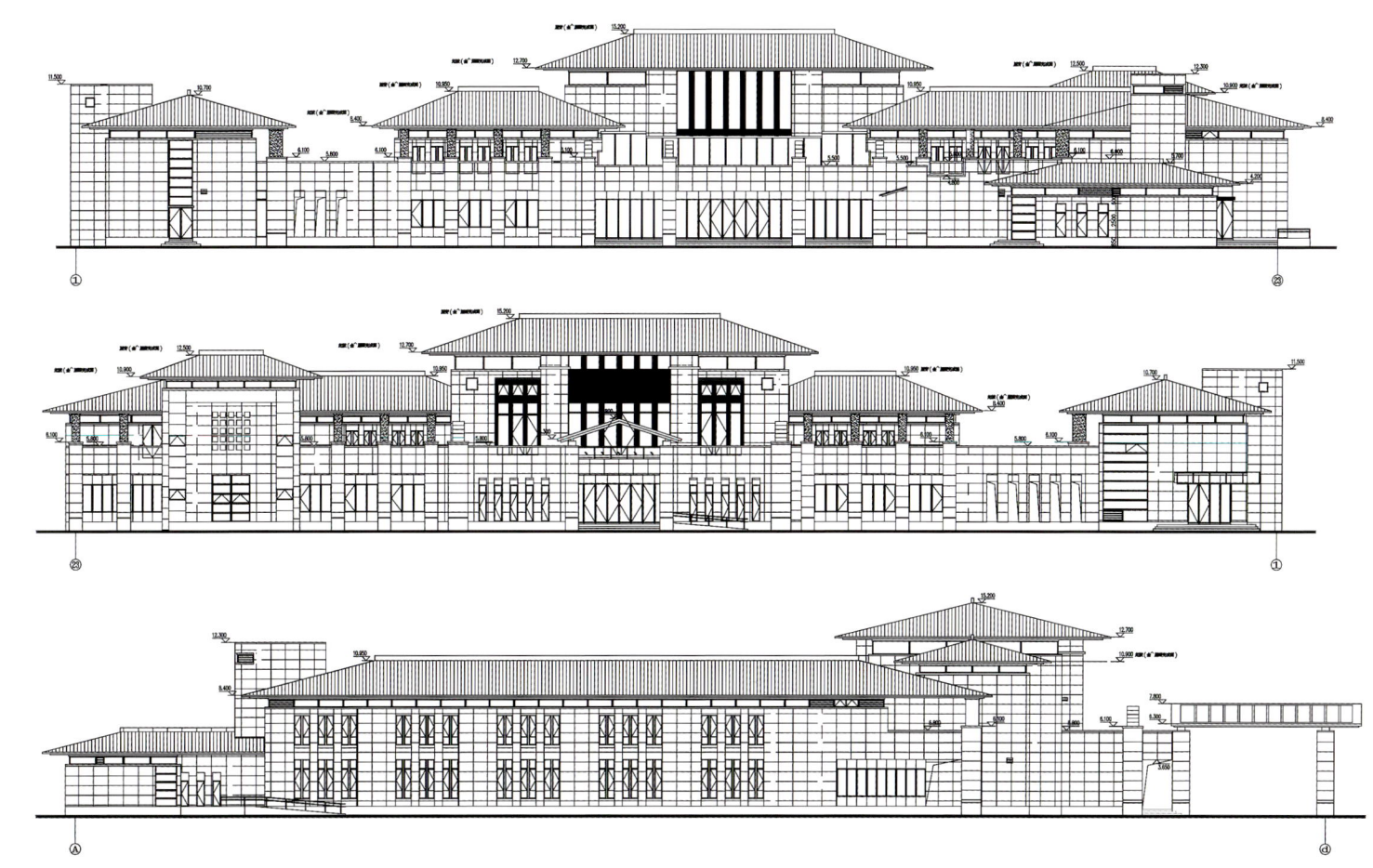

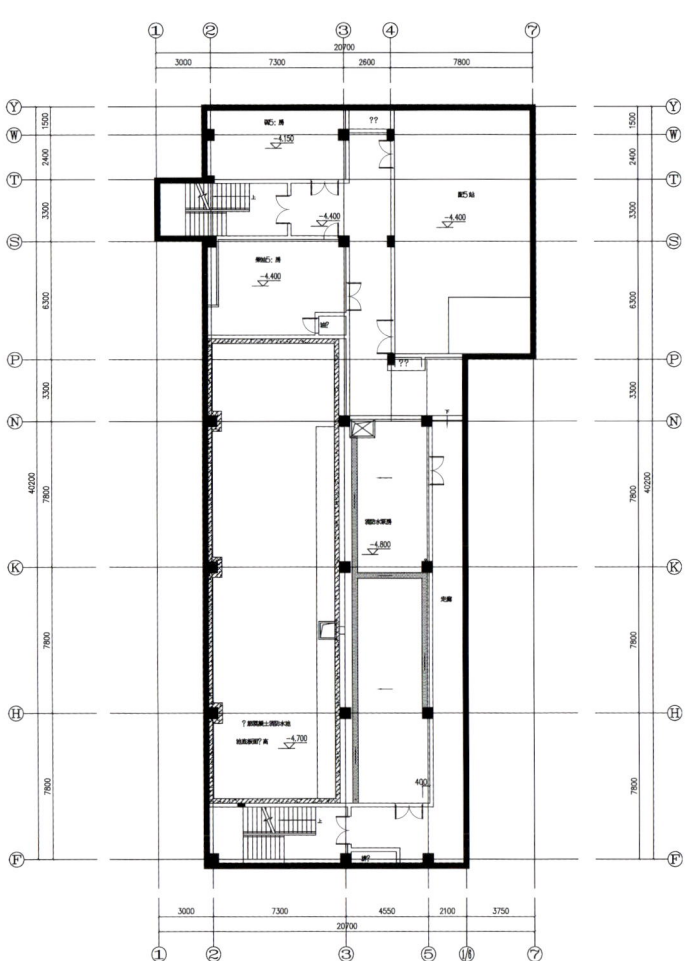

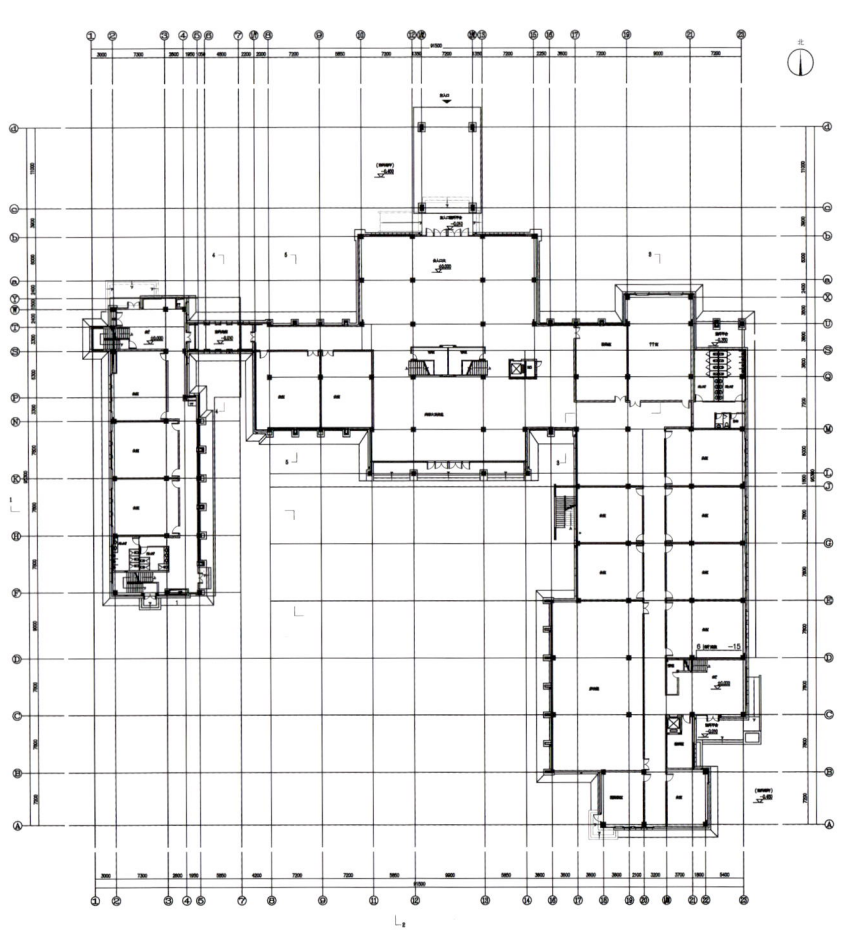
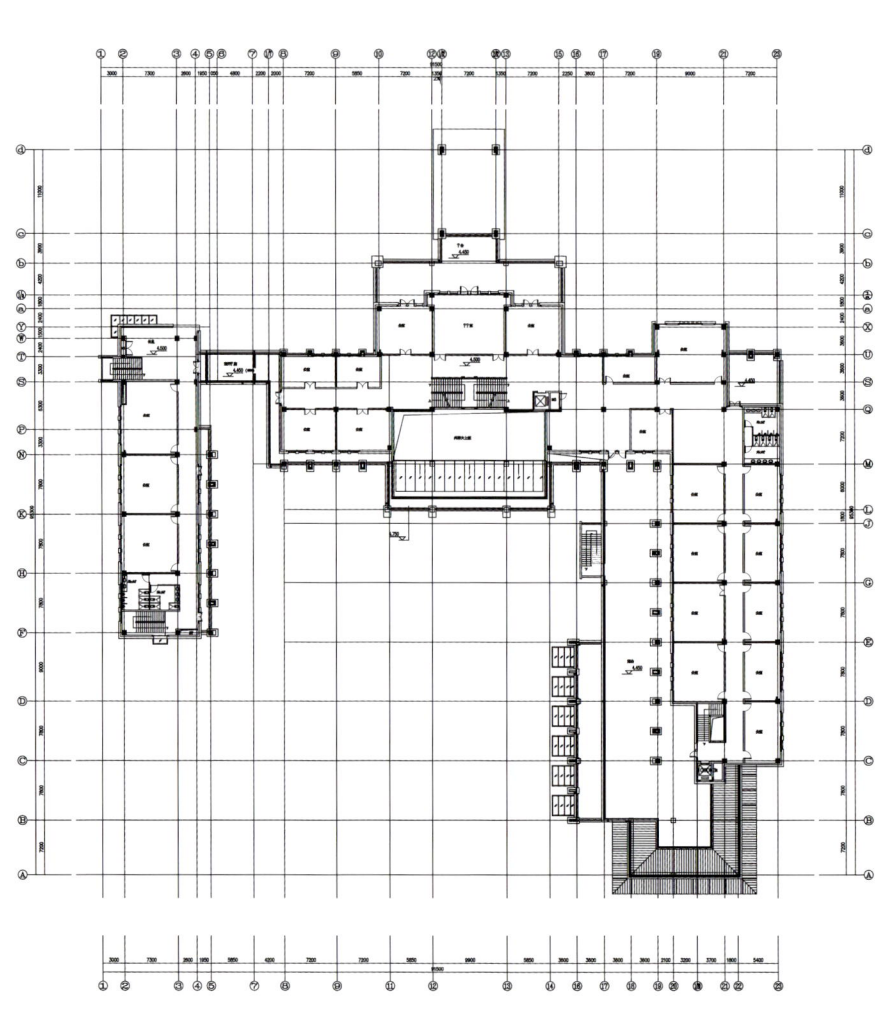

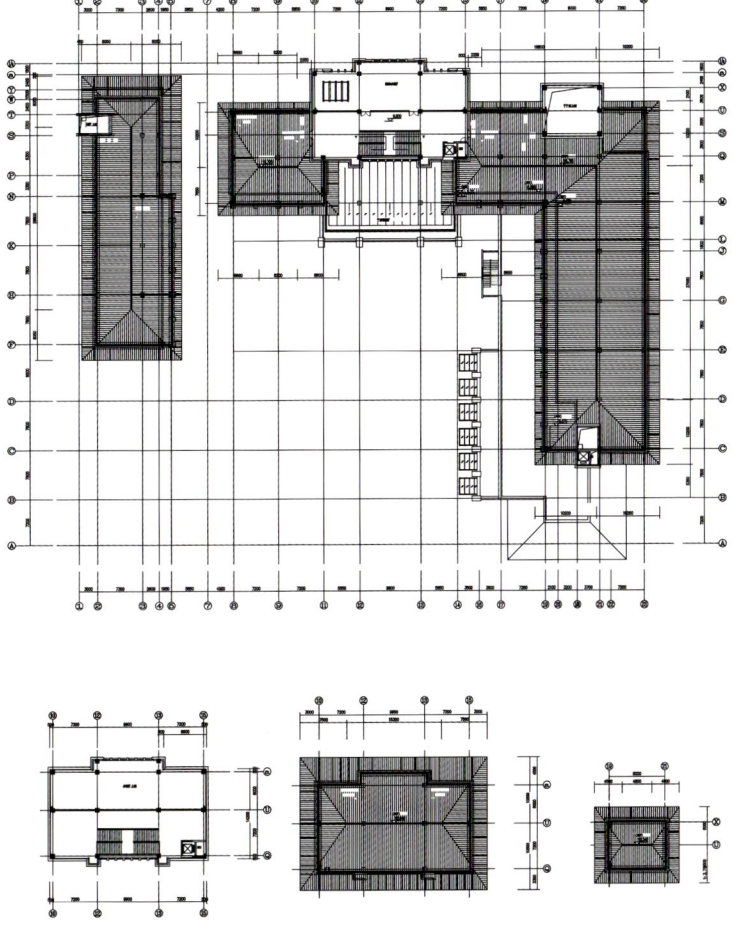
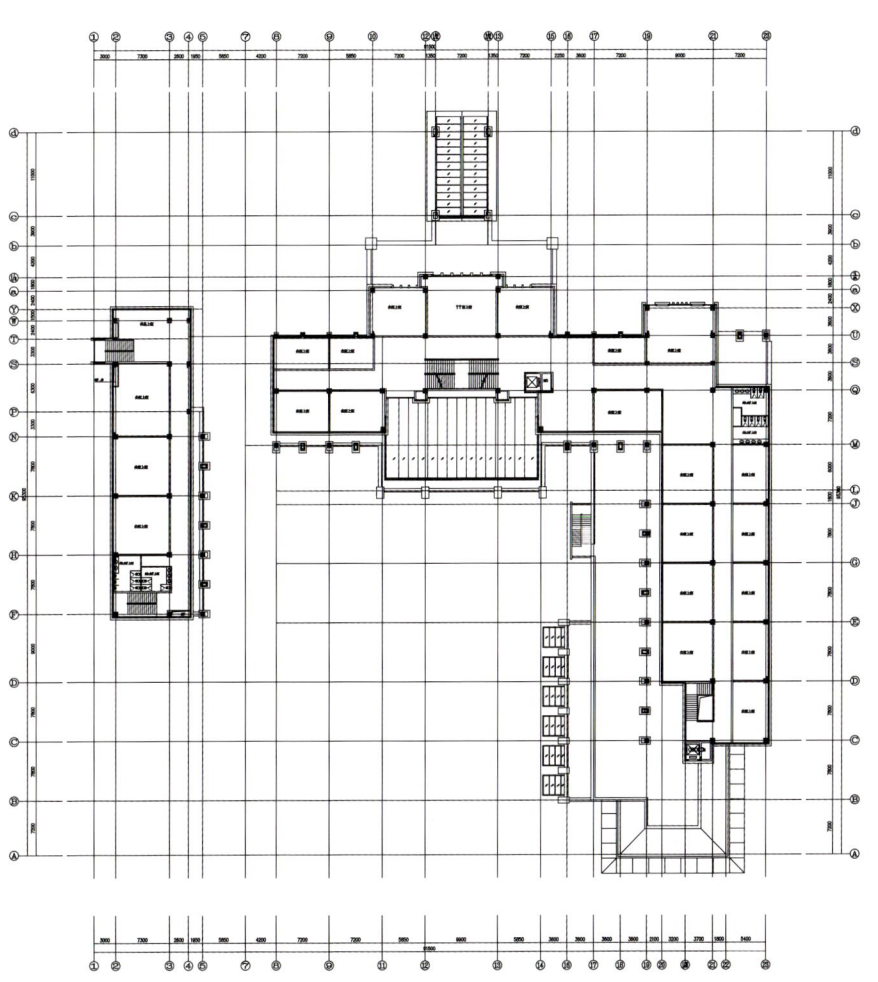

Lighting System

景观照明规划

总体亮度体系概述

根据夜间游览特点将全区分为四级照度体系：主体景区长安塔和广运门为一级照明，椰风水岸、欧陆风情、灞上人家为二级照明；国际园、内地园、专类园、道路及节点广场为三级照明；树林、草坪、滨水空间为四级照明。此外，在制高点建筑上，应增加动态激光照明。

功能性照明设计

• 道路与节点广场照明设计

道路照明的要求首先是安全，提供足够的照度及亮度，保证夜间机动车和行人的安全；其次要高效，采用高效节能的光源及灯具，尽可能防止光污染；第三要美观，充分考虑灯具的造型及光源的光色和显色性。
节点广场是游人休息的地方，照度要温馨宜人，为游人提供方便。

• 滨水地带照明设计

园区主要景点大部分临水，滨水地带的照明首先要保证安全，防止游人因照度不足而意外落水；其次要根据具体地段的景观需要变换照度和光源，做到亮暗有序，冷暖有别。第三要从不同的视角考虑建筑、桥梁、树木、游船等景物倒影变化，体现景观设计的意境。

• 绿地景观照明设计

景观绿地照明是夜景景观形成的总背景和本底，照明设计要求总体统一的前提下微有变化，来突出和渲染主体景物，不可喧宾夺主。

• 广告标识照明设计

夜间的标识是游人游览的必需设施，因此，标识照明要明显、突出，形式上可以采用动态静态相结合，色彩相对鲜艳，使游人容易发现，并方便使用。

The Whole Brightness

According to the characteristics of night tour, the illumination system of the whole site is divided into four levels: first level lighting for Chang'an Tower and Guangyun Entrance; second level for Southeast Asian Street, European Avenue and Romance by the Ba River; third level for International Gardens, Domestic Gardens, Feature Gardens, roads and plaza nodes; fourth level for groves, lawns and waterfront space. Besides, dynamic laser lighting system should be installed on the building at the commanding height.

Functional Lighting

· Road & Plaza Node Lighting

Firstly, the road lighting system should provide enough luminance to ensure the safety of motor vehicles and pedestrian during night. Secondly, it has to apply lights of high efficiency to save energy and prevent light pollution to the greatest extent. Thirdly, the aesthetic design, color and color rendering of lights should be taken into consideration. As the rest places for tourists, plaza nodes have to provide warm and pleasant illuminance to facilitate tourists.

· Waterfront Lighting

Most scenic spots at the Expo Site are adjacent to water. The waterfront lighting system should provide enough illuminance to prevent accidental drowning. Moreover, the illuminance needs to be changed according to landscaping demand in specific sections.
Finally, it has to consider the changes in the inverted images of architecture, bridges, trees and boats from various angles to embody the landscape design.

· Greenbelt Lighting

The lighting of greenbelt is the primary background of the night landscape. On the premise of uniformity in general, slight changes have to be designed for this background lighting system to highlight the main scenery.

· Advertising & Signage System Lighting

Nightscape signs are essential facilities for tourists, so they have to be prominent. Dynamic and static signs can be used. The colors of the signs are designed brightly so that tourists can easily find and make use of.

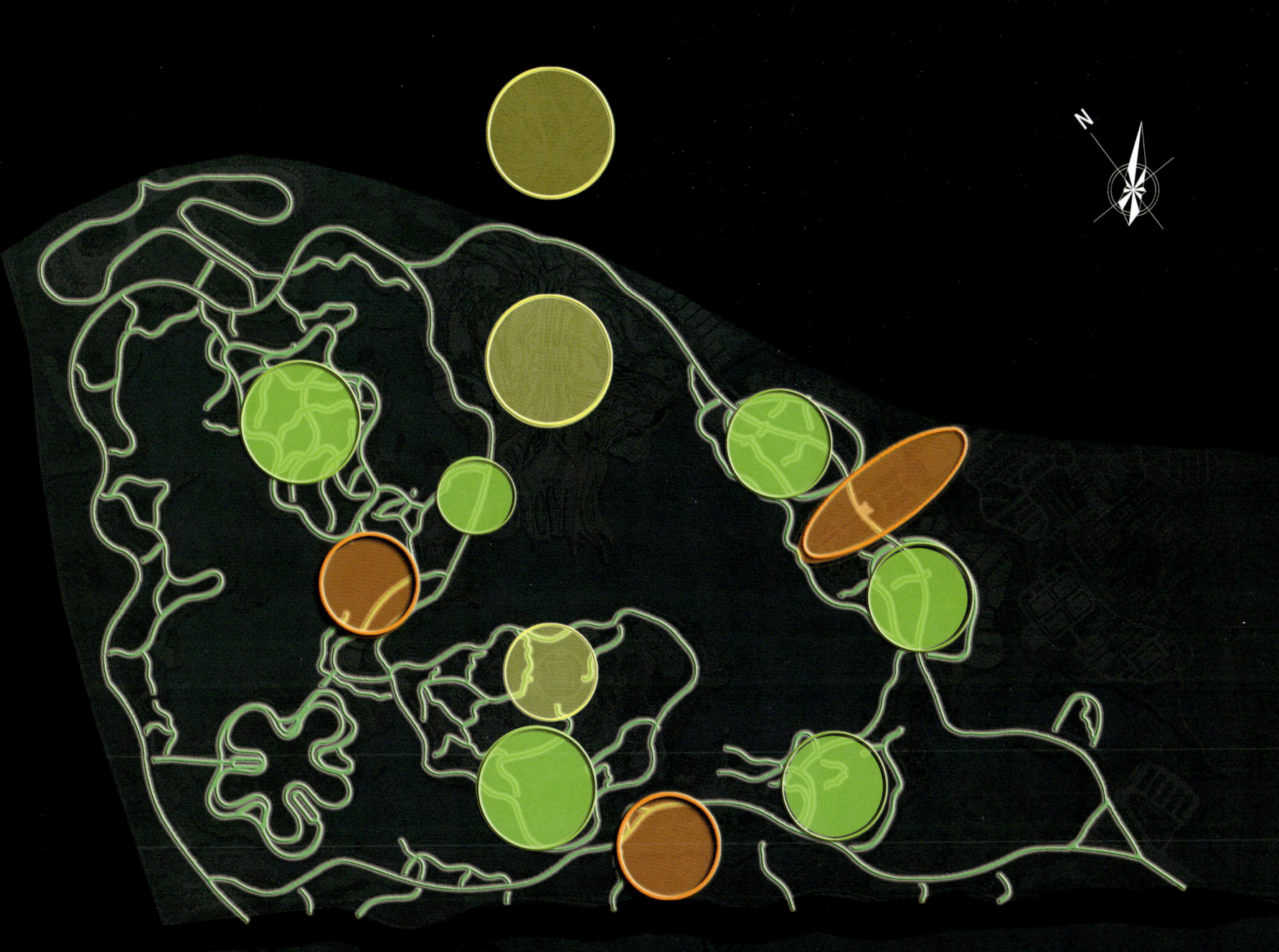

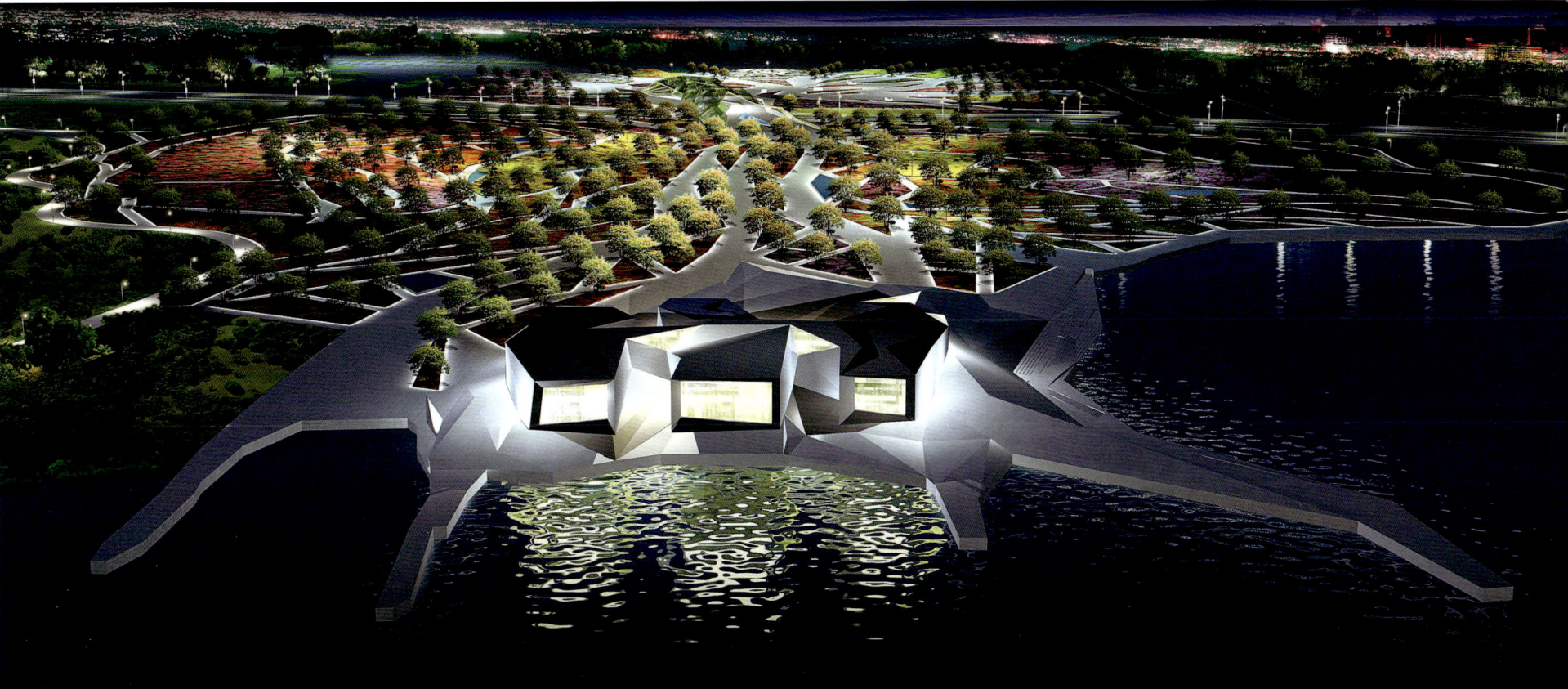

建设中的西安世界园艺博览会
Expo Site under Construction

建设中的西安世界园艺博览会
Expo Site under Construction

Acknowledgement
鸣谢

上海同济城市规划设计研究院丝绸之路研究中心	南加州大学
橙洲（北京）景观规划设计有限公司	加州大学伯克利分校
Plasma Studio	阿根廷托尔夸迪特加大学
中国建筑西北设计研究院有限公司	北京大学
北京市建筑设计研究院（BIAD）	逢甲大学
奥雅纳工程咨询（上海）有限公司北京分公司	建筑联盟学院
清华大学建筑设计研究院	哥伦比亚大学
上海马达思班建筑设计事务所	国家林业局林产工业规划设计院风景园林规划所
麟研部设计事务所	深圳市北林苑景观及建筑规划设计院有限公司
Valleycrest	深圳市陈绍华设计有限公司
陕西省古迹遗址保护工程技术研究中心	北京构易建筑设计有限公司
北京北林地景园林规划设计院有限责任公司	北京市城美绿化设计工程公司
北京多义景观规划设计事务所	碧谱照明设计（上海）有限公司
Miralles Tagliabue EMBT	全景国际照明顾问有限公司
Gross Max Landscape Architects	浙江新中环建筑设计有限公司
TERRAGRAM	西安市自来水公司规划设计院
Topotek 1	西安市水利建筑勘测设计院
Mosbach Paysagistes	西安西京电力建筑设计院
Martha Schwartz Partners	西安市市政设计研究院
West 8 Urban Design & Landscape Architecture	西部建筑抗震勘察设计研究院
SLA	西安市建筑设计研究院
香港大学	陕西通信信息技术有限公司，长安大学工程设计研究院
多伦多大学	沃易森（西安）建筑景观设计有限责任公司
圣若瑟大学	

谨以此书感谢所有为2011西安世界园艺博览会
做出卓越贡献的规划设计人员和单位

Postscript
后记

2008年9月，上海同济城市规划设计研究院正式接受世园会筹备办公室的委托，编制《2011西安世界园艺博览会修建性详细规划》。经慎重考虑，成立了以我院丝绸之路研究中心为骨干力量的规划编制小组。在编制规划期间，我院组织了多次专项学术研究活动，对曾经举办过世园会、园博会的国内外城市进行了数次考察，并专门拜访了曾经亲自参与世园会各项工作的组织者和专家。同时，为了搜集正确的社会信息，专门组织了网上和街头调研，做了大量的分析和研究工作。整个规划编制历时3个月，于12月初完成。2008年12月10日，规划通过了专家评审，2009年1月，通过了西安市规划委员会审查。

2009年5月，世园会进入全面建设阶段。应世园会筹备办特别邀请，为此，我院组织了以丝绸之路研究中心负责人李毅为核心的技术团队，派出规划、建筑、结构、景观、给排水、电气等各专业技术人员30多人，组建世园会技术服务中心，作为世园会工程建设部门的一部分，协调数十个设计和施工单位，与世园会筹备办合署办公开展技术服务。

世园会将于2011年4月28日开园迎客。此书成书之际，世园会的建设收尾工作正在如火如荼地进行之中。世园会规划设计和工程建设工作的如期完成，离不开世园会各级领导的大力支持，离不开各界专家学者的帮助指导，离不开各个设计、施工单位的共同努力。在此，特别感谢张锦秋院士、韩骥教授、吕仁义教授、郭方明先生和刘延江先生，同时要感谢参与世园会规划设计工作的所有设计师及现场服务代表！

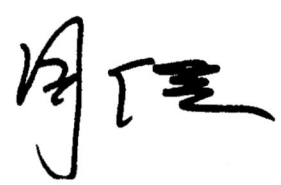

图书在版编目（CIP）数据

2011西安世界园艺博览会规划精粹 /《2011西安世界园艺博览会规划精粹》编委会主编. — 北京：中国建筑工业出版社，2011.4

ISBN 978-7-112-13084-9

Ⅰ.①2… Ⅱ.①2… Ⅲ.①园艺—博览会—景观规划—西安市—2011 Ⅳ.①S68-282.411

中国版本图书馆CIP数据核字（2011）第052983号

策划编辑：石 莹 林佳艺
责任编辑：常 燕 李伟光

2011西安世界园艺博览会规划精粹
*
中国建筑工业出版社出版、发行（北京西郊百万庄）
各地新华书店、建筑书店经销
深圳雅昌彩色印刷有限公司印制
*
开本：635×965毫米 1/12 印张：28.5 字数：240千字
2011年4月第一版 2011年4月第一次印刷
定价：398.00元
ISBN 978-7-112-13084-9
（20505）

版权所有 翻印必究
如有印装质量问题，可寄本社退换
（邮政编码 100037）